Multi-Sensor Systems and Data Fusion in Remote Sensing II

Multi-Sensor Systems and Data Fusion in Remote Sensing II

Guest Editors

Piotr Kaniewski
Stefano Mattoccia
Fabio Tosi

MDPI

Basel • Beijing • Wuhan • Barcelona • Belgrade • Novi Sad • Cluj • Manchester

Guest Editors

Piotr Kaniewski
Faculty of Electronics
Military University of Technology
Warsaw
Poland

Stefano Mattoccia
Department of Computer Science and Engineering (DISI)
University of Bologna
Bologna
Italy

Fabio Tosi
Department of Computer Science and Engineering (DISI)
University of Bologna
Bologna
Italy

Editorial Office

MDPI AG
Grosspeteranlage 5
4052 Basel, Switzerland

This is a reprint of the Special Issue, published open access by the journal *Remote Sensing* (ISSN 2072-4292), freely accessible at: https://www.mdpi.com/journal/remotesensing/special_issues/F9Q0V7FRQQ.

For citation purposes, cite each article independently as indicated on the article page online and as indicated below:

Lastname, A.A.; Lastname, B.B. Article Title. *Journal Name* **Year**, *Volume Number*, Page Range.

ISBN 978-3-7258-5069-3 (Hbk)
ISBN 978-3-7258-5070-9 (PDF)
https://doi.org/10.3390/books978-3-7258-5070-9

Contents

About the Editors

Piotr Kaniewski

Piotr Kaniewski received his M.Sc. in 1994, Ph.D. in 1998, and was habilitated in 2011. Currently he works as an Associate Professor at the Faculty of Electronics at the Military University of Technology, Warsaw, Poland. His research is focused on navigation systems for special purposes, such as supporting synthetic aperture radars and ground-penetrating radars, distributed estimation algorithms for UAV swarms, simultaneous localization and mapping (SLAM), and navigation systems for GNSS-denied environments. He has authored more than 200 scientific papers and 2 books.

Stefano Mattoccia

Stefano Mattoccia is currently an associate professor at the Department of Computer Science and Engineering of the University of Bologna. His research activity concerns computer vision, mainly focusing on depth perception and related tasks. He is Senior IEEE member.

Fabio Tosi

Fabio Tosi received his PhD degree in Computer Science and Engineering from University of Bologna in 2021. Currently, he is a Junior Assistant Professor (RTDA) at the Department of Computer Science and Engineering (DISI) at the University of Bologna. His research interests include deep learning and depth sensing-related topics.

Preface

The Special Issue you hold in your hands gathers contributions to the MDPI Remote Sensing Special Issue "Multi-Sensor Systems and Data Fusion in Remote Sensing II." Recent technological advances—including the introduction of novel sensors, the development of sophisticated sensor platforms, and breakthroughs in signal and data processing—offer scientists and engineers the opportunity to create more capable, integrated multi-sensor systems. Wider frequency bands, enhanced resolution and data rates, and the widespread deployment of distributed sensors have substantially increased data volumes in contemporary multi-sensor configurations. Simultaneously, user demands regarding coverage area, resolution, accuracy, processing speed, and overall system functionality continue to rise. These trends present fresh challenges for data-fusion algorithms, which must now leverage the latest methods from big-data analytics, statistical estimation, and artificial intelligence. The papers collected here provide new insights into recent developments in multi-sensor systems and data fusion and will be of broad interest to the remote-sensing community.

Piotr Kaniewski, Stefano Mattoccia, and Fabio Tosi
Guest Editors

remote sensing

MDPI

Article

A Fast Sequential Similarity Detection Algorithm for Multi-Source Image Matching

Quan Wu [1,*] and Qida Yu [2]

[1] The School of Artificial Intelligence, Nanjing University of Information Science and Technology, Nanjing 210000, China

[2] The School of Electronic and Information Engineering, Nanjing University of Information Science and Technology, Nanjing 210000, China; 003550@nuist.edu.cn

* Correspondence: wuquan@nuist.edu.cn

Abstract: Robust and efficient multi-source image matching remains a challenging task due to nonlinear radiometric differences between image features. This paper proposes a pixel-level matching framework for multi-source images to overcome this issue. Firstly, a novel descriptor called channel features of phase congruency (CFPC) is first derived at each control point to create a pixelwise feature representation. The proposed CFPC is not only simple to construct but is also highly efficient and somewhat insensitive to noise and intensity changes. Then, a Fast Sequential Similarity Detection Algorithm (F-SSDA) is proposed to further improve the matching efficiency. Comparative experiments are conducted by matching different types of multi-source images (e.g., Visible–SAR; LiDAR–Visible; visible–infrared). The experimental results demonstrate that the proposed method can achieve pixel-level matching accuracy with high computational efficiency.

Keywords: similarity measurement; multi-source image; image matching

1. Introduction

With the widespread application of computer technology in geographic information science, image processing techniques have entered the era of multi-resolution [1], multi-source [2], and multi-spectral images [3]. Since the information contained in images is complementary, integrating the data collected by different sensors is necessary to improve the applicability of imagery for Earth observation. Image matching is the process of transforming a target image into the coordinate system of a reference image of the same scene by determining the pixel-based mapping relationship between them. Therefore, the performance of multi-source image matching is crucial for producing image mosaics [4] and for performing feature fusion [5] and change detection [6]. Especially, we can establish correspondence between multiple sensor images of drones, solving the limitation of sensing systems in scene understanding.

After decades of development, automatic image matching has made remarkable progress, and many methods now enable the direct georeferencing of multi-source data using navigation devices on platforms and physical sensor models. These techniques can achieve a matching accuracy of fewer than five pixels [7] and nearly eliminate all the geometric distortion caused by different scales and rotation angles between multi-source images.

In general, area-based methods extract matching information by finding the pixels in the target image that are most similar to the template image, such as mutual information (MI) [8] and Phase Correlation (PC) [9]. Area -based image matching methods obtain region description by calculating the response intensity between the region of interest and surrounding pixels. This type of method is generally applicable to the registration of images with prominent texture features. However, these matching methods usually experience heavy computational complexity and are easily affected by nonlinear radiometric differences in multi-source images [10]. The best-buddies similarity (BBS) theoretically analyzed the effect of cluttered backgrounds, and a patch-based texture was introduced to enhance robustness and accuracy. The caching scheme was optimized and a batch gradient descent algorithm was used to reduce the computational overhead and further speed up the method [11]. Feature-based methods construct an underlying spatial transformation and establish reliable correspondence between two sets of feature points. Considering the complexity of feature presentation, particularly for the radiation difference between multi-source images, it is simple to produce inaccurate or even incorrect matching. Although many excellent algorithms have recently been proposed in this field, it is still challenging to achieve robust and efficient image matching performance. Therefore, if the response intensity characteristics and texture feature can be fully utilized, the influence of complex environment and radiation difference can be effectively solved in multi-source image matching.

Therefore, the main challenge of multi-source image matching is the handling of radiometric differences between the target and reference images. Pixel-level matching has not been achieved to date for multi-source images due to significant radiometric differences. The first row in Figure 1 shows multi-source images with significant radiometric differences between objects, which make it more challenging to detect the same features in both images. However, despite these significant differences, the main structural features are similar. In the second row of Figure 1, although the reflectance values show substantial differences, the structural features extracted using phase congruency are almost the same, which can provide intrinsic structural information and invariant features for transformation estimation. However, the classical phase congruency method only provides the amplitude characteristics, and extending it to a directional representation in the presence of noise is not sufficient [11].

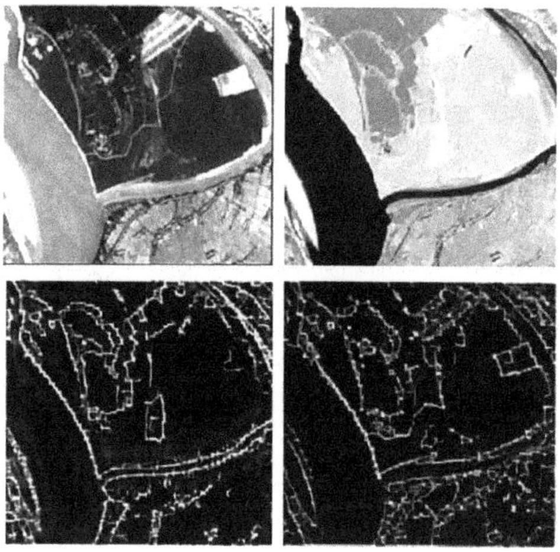

Figure 1. Structural feature extraction for multi-source images.

Phase congruency is an indicator of feature significance, and it has been demonstrated that the phase congruency model conforms to the characteristics of human visual cognition [12], i.e., the model considers structural features. However, it was found that the response intensity of phase congruency is unstable, especially when dealing with multi-source images.

In this paper, a mixture model that combines the features of oriented gradient and response intensity of phase congruency to achieve robust feature description in multi-source image matching. The proposed descriptor, called channel features of phase congruency, enables efficient image matching by extending the classical phase congruency model and creating a pixel-level feature representation of local features. The CFPC descriptor captures the structural properties of an image and remains unaffected by radiometric variations across multiple sources. To reduce computational costs, the feature representation of descriptors is first converted into the frequency domain, followed by similarity measurement. As a result, the corresponding features can be readily detected in multi-source images. The primary contributions of this study are summarized follows:

(1) An efficient and robust image matching framework is proposed based on phase congruency to achieve pixel-level matching. The proposed framework integrates different types of local features for similarity measurement.
(2) A novel feature descriptor (CFPC) is constructed based on the oriented gradient and response intensity of phase congruency to capture structure information.
(3) The proposed matching method uses pixel-level feature representation to evaluate the similarity between multi-source images, and also, a similarity measurement method (F-SSDA) is established to accelerate image matching. Therefore, the intrinsic geometry information is incorporated into the objective function formulation when computing similarity measurements between multi-source images to improve matching performance.

The remaining sections of this paper are structured as follows: Section 2 reviews related work, while Section 3 details the proposed multi-source image matching method, and Section 4 describes the comparative experiments and results. Finally, conclusions are drawn and recommendations are provided in Section 5.

2. Related Work

2.1. Area-Based Methods

Area-based methods perform similarity matching by searching for the most similar gradient or applying a transformation to spatial features in the target window. A relationship is established between the regions surrounding a spatial feature, or pixel-level matching is conducted between the reference image and target image. The normalized correlation coefficient [13] and gradient mutual information [14] are widely used to evaluate the match. Area-based methods are typically used for matching homologous points in images. However, area-based matching is easily affected by nonlinear radiometric differences in multi-source image matching. To address this issue, Shechtman et al. [15] designed a local self-similarity (LSS) descriptor based on texture and gradient information to minimize the radiometric differences. The LSS algorithm was later extended to multi-source image matching due to its excellent performance. Specifically, Liu et al. [16] developed the FLSS-C and LSS-C algorithms and achieved robust geometric invariance using a distribution-based representation. Subsequently, Sedaghat et al. [17] used the LSS for multi-sensor image matching and achieved good performance. Jenkinson et al. [18] constructed geometric structure information based on self-similarity. However, area-based methods typically calculate feature similarity by traversing local pixels, which undoubtedly greatly increases computational consumption. To improve efficiency, a template matching method has recently been proposed to calculate the number of nearest neighbors [19]. Ye et al. [20] converted the similarity measure into the frequency domain using SSD, greatly accelerating three-dimensional feature description. However, the template image did not consider image texture, which will greatly influence the regional representation. If the image window was in an area with low texture or there were no distinctive features, the matching results would be unsatisfactory.

2.2. Feature-Based Methods

Feature-based methods typically detect structural features in an image, such as corners, lines, or texture, rather than spectral information. The selected feature must be distinctive, robust, simple to detect, and must remain invariant for different radiometric values.

The scale-invariant feature transform (SIFT) [21] is a classical image matching algorithm. It is rotation- and scale-invariant, involving Gaussian pyramid and gradient histogram techniques. Rublee et al. [22] proposed the novel oriented fast and rotated (ORB) directional binary robust independent elementary features (BRIEF) for real-time applications. This algorithm has low time complexity and is scale-invariant. Morel et al. [23] simulated two camera orientation parameters and proposed the affine-invariant matching algorithm (Affine-SIFT), which is scale and rotation-invariant. Significant radiometric differences between input images prevent the SIFT algorithm from being used for multi-source image registration despite its strong affine invariance and benefits for processing images with different viewing angles. Dellinger et al. [24] proposed a novel gradient definition for multi-sensor images matching to enhance robustness against speckle noise. Additionally, Sedaghat et al. [25] devised an adaptive binning histogram strategy for characterizing local features, specifically tailored to address distortion in multi-source images. However, SIFT and its extensions lack the computational speed necessary for multi-source image matching in practical applications. Integral graph methods have been developed to improve the algorithm's efficiency.

Generally, deep learning-based methods for multi-source image matching can yield excellent performance. For instance, Zhu et al. [26] proposed a two-stage generative adversarial network-based multi-modal brain image registration method. Fu et al. [27] introduced the texture adaptive observation approach for image depth estimation. Tang et al. utilized the Lovasz-Softmax loss and pre-trained segmentation model to guide the matching network, emphasizing semantic requirements in high-level vision tasks. Tang et al. [28] proposed a versatile image registration and fusion network with semantic awareness. Addressing scenario dependency, Yang et al. [29] proposed a knowledge transfer-based network for multi-source image matching. Sun et al. [30] suggested that the global receptive field provided by Transformer enables dense matches in low-texture areas, where traditional feature detectors often struggle to generate repeatable interest points. However, these methods are sensitive to background clutter and target rotation. They also exhibit high data dependency and often necessitate a large training dataset to achieve high matching accuracy.

The performance of multi-source image matching based solely on gradient features is typically poor because the gradient information of multi-source data is generally sensitive to nonlinear radiometric differences. It is advantageous to combine textural information with spectral information for feature extraction, like how people analyze images. Therefore, many scholars have proposed matching methods based on visual perception to deal with the significant radiometric differences between multi-source images [31]. Meng et al. [32] introduced a new descriptor for vein recognition called local directional code (LDC) based on rich orientation information. Aguilera et al. [33] proposed the edge-oriented histogram (EOH) to capture edge information surrounding key points. Wu et al. [34] proposed the histogram of point-edge orientation (HPEO), a simple and highly robust descriptor to extract structural features. Image matching based on phase congruency has been widely used for multi-source image registration recently. This method extracts structural information from images to obtain invariant features and has achieved better performance than classical matching methods. These methods assume that structural characteristics are consistent across multi-source images and are unaffected by intensity differences. Ye et al. [35] proposed HOPC to extract shape features from multi-source images. Wu et al. [36] extended the phase congruency feature to represent the orientation and constructed a mixed phase congruency descriptor based on gradient information and orientation. However, how to balance fusion features and computational efficiency is a problem that needs to be solved.

3. Methodology and Material

3.1. Basic Framework of the Proposed Matching Algorithm

The classical phase consistency model is extended to produce a pixel-wise feature representation based on the orientation and intensity of phase consistency, enabling robust mixed feature representation of phase consistency for image matching. To improve matching efficiency, the F-SSDA is applied to solve the sequential similarity objective function.

Finally, the matching accuracy can be determined according to the affine transformation model.

3.2. Channel Features of Phase Consistency

The Fourier series expansion of a given signal can be given as follows:

$$f(x) = \sum_n A_n \cos(\phi_n(x)), \tag{1}$$

where ϕ_n and A_n indicate the phase at position x and the amplitude of the frequency component, respectively. Phase congruency can be defined as follows:

$$PC_1(x) = \max_{\varphi(x)\in[0,2\pi)} \{\frac{\sum\limits_n A_n(x)\cos(\phi_n(x) - \overline{\phi}(x))}{\sum\limits_n A_n(x)}\}, \tag{2}$$

where $\overline{\phi}(x)$ indicates the weighted mean in the local region.

Since phase consistency models cannot detect blurred features, some improved methods that calculate phase congruency using the log-Gabor wavelets at multiple orientations and scales. This function is defined in the frequency domain as follows:

$$g(\omega) = \exp(\frac{-(\log(w/w_0))^2}{2(\log(\sigma_w/w_0))}) \tag{3}$$

where σ_w and w_0 indicate the width parameter and central frequency, respectively. The filter of the log-Gabor wavelet can be obtained by an inverse Fourier transform. The "imaginary" and "real" components of the filter are denoted as the log-Gabor odd-symmetric M_{no}^{o} and M_{no}^{e} even-symmetric wavelets, respectively. The convolution results of the input image can be regarded as the response vector:

$$[e_{no}(x,y), o_{no}(x,y)] = [I(x,y) * M_{no}^{e}, I(x,y) * M_{no}^{o}] \tag{4}$$

where $e_{no}(x,y)$ and $o_{no}(x,y)$ indicate the respective responses of the even-symmetric and odd-symmetric wavelets having o as orientation and n as scale. Moreover, the amplitude and phase are, respectively, given as follows:

$$\mathrm{A}_{no} = \sqrt{e_{no}(x,y)^2 + o_{no}(x,y)^2}, \tag{5}$$

$$\phi_{no} = \mathrm{atan}(e_{no}(x,y), o_{no}(x,y)), \tag{6}$$

Considering the blur and noise of multi-source images, the model is defined as follows [35]:

$$PC_2(x,y) = \frac{\sum\limits_n\sum\limits_o \mathrm{W}_o(x,y) + \mathrm{A}_{no}(x,y)\Delta\Phi_{no}(x,y) - T}{\sum\limits_n\sum\limits_o \mathrm{A}_{no}(x,y) + \varepsilon} \tag{7}$$

where $\mathrm{W}_o(x,y)$ represents the weighting factor for the given frequency spread, T is a parameter for controlling noise, $\mathrm{A}_{no}(x,y)$ denotes the amplitude, and ε is a small constant that avoids division by zero. The symbol means that the closed quantity is equal to itself when the value inside the symbol is positive or zero, and $\Delta\Phi_{no}(x,y)$ is phase deviation and it is given as follows:

$$\begin{aligned} &\mathrm{A}_{no}(x,y)\Delta\Phi_{no}(x,y) \\ &= (e_{no}(x,y).\overline{\phi_e}(x,y) + o_{no}(x,y).\overline{\phi_o}(x,y)) \\ &-|e_{no}(x,y)o_{no}(x,y) - o_{no}(x,y).\overline{\phi_e}(x,y)| \end{aligned} \tag{8}$$

where

$$\overline{\phi_e}(x,y) = \sum_n \sum_o e_{no}(x,y) / ((\sum_n \sum_o e_{no}(x,y))^2 + (\sum_n \sum_o o_{no}(x,y))^2)^{1/2} \tag{9}$$

$$\overline{\phi_o}(x,y) = \sum_n \sum_o o_{no}(x,y) / ((\sum_n \sum_o e_{no}(x,y))^2 + (\sum_n \sum_o o_{no}(x,y))^2)^{1/2} \tag{10}$$

However, the phase congruency models given in Equation (7) can only detect response intensity but not the phase orientation. Therefore, it is not suitable to construct robust feature descriptors using solely the phase congruency response intensity. However, in addition to the response characteristic, oriented gradient information was significant for constructing local feature descriptors.

The log-Gabor odd-symmetric convolutions in multiple directions are employed to compute phase congruency. The resulting convolution can then be projected onto the horizontal and vertical axes, yielding the x-direction (horizontal) and y-direction (vertical) image derivatives.

$$\mathrm{X} = \sum_{\theta} (o_{no}(\theta)\sin(\theta)), \tag{11}$$

$$\mathrm{Y} = \sum_{\theta} (o_{no}(\theta)\cos(\theta)). \tag{12}$$

The orientation and response intensity of the phase congruency, which are multi-directional projections, respectively, can be defined as:

$$\Phi = \arctan(\mathrm{X}, \mathrm{Y}), \tag{13}$$

$$\mathrm{P} = \sqrt{\mathrm{X}^2 + \mathrm{Y}^2}. \tag{14}$$

The conventional approach, which builds features based on phase congruency, is efficient in capturing the complex texture structures of images and exhibits robust matching performance for multi-source image. However, using two-dimensional data to represent image information is still sensitive to noisy images and radiation changes, especially for multi-source images. Therefore, several feature vectors are constructed from the oriented gradient information at each pixel and in the Z-direction to construct a 3D pixel-level descriptor ($\mathrm{D}_{x,y,z}^{\mathrm{mixed}}$). The feature representation is convolved with a Gaussian filter to obtain a robust local description. This 3D Gaussian model uses a 2D Gaussian filter in the horizontal and vertical directions and a $[1, 2, 1]^{\mathrm{T}}$ kernel in the Z-direction. The convolution in the Z-direction smooths the orientated gradients and minimizes geometric distortion caused by local radiometric differences. The feature description is normalized using (15) to further reduce the nonlinear radiometric differences:

$$\mathrm{D}_{x,y,z} = (\mathrm{D}_{x,y,z}^{\mathrm{mixed}} / sqrt\left(\sum_{x,y,z=1}^{w} (\mathrm{D}_{x,y,z}^{\mathrm{mixed}})^2 + \varepsilon_1)\right) / \left(\sum_{x,y,z=1}^{w} (\mathrm{D}_{x,y,z}^{\mathrm{mixed}} / sqrt(\sum_{x,y,z=1}^{w} (\mathrm{D}_{x,y,z}^{\mathrm{mixed}})^2 + \varepsilon_1))\right) \tag{15}$$

where the ε_1 small constants avoid the zero denominator.

3.3. The Frequency Sequential Similarity Detection Algorithm

Multi-feature description has been proven to be effective in improving the matching accuracy and reducing the impact of radiation differences [10]. The constructed descriptor contains fused feature information and is represented in the form of 3D data, so the similarity measurement calculation is very expensive. Therefore, to improve matching efficiency, a new F-SSDA algorithm is proposed in calculating feature similarity. The proposed algorithm is used to evaluate the similarity between a pair of signals. Let $D_1(n)$ and $D_2(n)$ represent the corresponding feature representations of control candidate points calculated by Equation (15), respectively. The F-SSDA between the constructed feature presentation within window i is given as

$$S_i(v) = \sum\nolimits_n [(D_1(n) - \overline{D_1(n)}) - (D_2(n-v) - \overline{D_2(n-v)})]^2 T_i(n), \tag{16}$$

where n denotes the coordinates of 3D feature representation, $T_i(x)$ is the masking function over $D_1(n)$, where $T_i(n) = 1$ within the image window, and 0 otherwise. $\overline{D_1(n)}$ is the average value of feature representation within the template window. $S_i(v)$ denotes the F-SSDA calculation between the 3D feature representations translated by vector v for template window i. Consequently, the match between $D_1(n)$ and $D_2(n)$ can be achieved by minimizing $S_i(v)$. Accordingly, the matching function is given as follows:

$$v_i = \arg \min_{v} \left\{ \sum\nolimits_n \left| \left(D_1(n) - \overline{D_1(n)}\right) - \left(D_2(n-v) - \overline{D_2(n-v)}\right) \right|^2 T_i(n) \right\}, \tag{17}$$

where v_i indicates an offset vector while matching signals $D_1(n)$ and $D_2(n)$.

The standard strategy is to calculate the F-SSDA of the feature description for the region around candidate key points. However, this approach substantially increases the computational cost as the feature description contains 3D information. Then, the spatial feature information is converted into the frequency domain to reduce the time consumption.

Herein, the algebraic transformation of similarity function Equation (16) can be present as follows:

$$\begin{aligned} S_i(v) = {} & \Sigma_n (D_1(n) - \overline{D_1(n)})^2 T_i(n) + \Sigma_n (D_2(n-v) - \overline{D_2(n-v)})^2 T_i(n) \\ & -2\Sigma_n (D_1(n) - \overline{D_1(n)})(D_2(n-v) - \overline{D_2(n-v)}) T_i(n) \end{aligned} \tag{18}$$

Since the first term is a constant, similarity is measured through the minimization of the functions of the remaining two terms. The spatial domain convolution is equivalently represented by the dot product in the frequency domain. Therefore, the convolution operation of the last two terms of Equation (18) can be accelerated using FFT, and the offset vector v_i is given as follows:

$$\begin{aligned} v_i = \arg \min_{v} \{ & F^{-1}[F^*(D_{dif2})F(D_{dif2}T_i)](n) \\ & -2F^{-1}\left[F^*\left(D_{dif2}\right)F\left(D_{dif1}T_i\right)\right](n)\}' \end{aligned} \tag{19}$$

where $D_{dif1} = D_1(n) - \overline{D_1(n)}$, $D_{dif2} = D_2(n) - \overline{D_2(n)}$, and F and F^{-1} indicate the FFTs and its inverse transform, respectively; moreover, F^* is the complex conjugate of F.

Through Equation (19), the computational efficiency can be significantly improved. For instance, given a fixed size of $w \times w$ pixels, the corresponding search window is $m \times m$ pixels. The SSD required $O(m^2w^2)$ operations, while the proposed approach needs $O((m+w)^2 \log(m+w))$ operations. Hence, the computational efficiency can be significantly improved when applying Equation (19).

The curve of similarity and computational efficiency is evaluated, and the correct match ratio (CMR) is calculated. Some classical multi-source image data are used for performance analysis.

3.4. Description of Datasets

Several multi-source image data were selected to evaluate the matching performance. We evaluated the matching of (1) visible–SAR, (2) LiDAR–visible, (3) visible–infrared, and (4) visible–map. Various high-resolution and medium-resolution (30 m) images covering different terrains, including suburban and urban areas, were used. There were no differences in translation, rotation, and scale between image pairs. However, significant radiometric differences were unavoidable due to the different sensor wavelengths. Please refer to Table 1 for detailed data.

(a) Visible–SAR: Visible–SAR data 1 and 3 were acquired over urban areas with tall buildings, resulting in significant radiometric differences between them. Visible–SAR 2 is a medium-resolution image in a suburban area. In addition, significant changes had occurred in this area as the SAR image was taken 14 months after the visible image in Visible–SAR 2, thereby complicating the matching process.
(b) LiDAR–Visible: LiDAR–Visible images is collected in urban areas. Significant noise and nonlinear radiometric differences make it more challenging to match LiDAR image data.
(c) Visible–infrared: Both medium- and high-resolution images were used (Daedalus and Landsat 5 TM). The medium-resolution data were acquired over a suburban area.
(d) Visible–map: These data were collected from Google Earth. The images had been rasterized, and there was local distortion between image pairs due to the relief displacement of buildings. In addition, there were radiometric differences between the map data and visible images. The lack of texture features to construct local descriptors makes it challenging to match an image to a map.

Table 1. Description of test images.

Category	Test	Image Pair	Size and GSD	Date	Characteristics
(a) Visible–SAR	1	Google Earth TerraSAR-X	528 × 524, 3 m 534 × 524, 3 m	11/2007 12/2007	These SAR images were collected in urban areas, and these SAR images contain significant noise.
	2	TM band3 TerraSAR-X	600 × 600, 30 m 600 × 600, 30 m	05/2007 03/2008	There is significant noise in these SAR images. The images have a temporal difference of 12 months.
	3	Google Earth TerraSAR-X	628 × 618, 3 m 628 × 618, 3 m	03/2009 01/2008	These SAR images were collected in urban areas and have a temporal difference of 12 months.
(b) LiDAR–visible	1	LiDAR height Airborne visible	524 × 524, 2.5 m 524 × 524, 2.5 m	06/2012 06/2012	These SAR images were collected in urban areas, and these SAR images contain significant noise.
	2	LiDAR intensity WordView2 visible	600 × 600, 2 m 600 × 600, 2 m	10/2010 10/2011	Temporal difference of 12 months; urban area.
	3	LiDAR intensity WordView2 visible	621 × 617, 2 m 621 × 621, 2 m	10/2010 10/2011	Temporal difference of 12 months; urban area.
(c) Visible–infrared	1	Daedalus visible Daedalus infrared	512 × 512, 0.5 m 512 × 512, 0.5 m	04/2000 04/2000	These images were collected in urban areas with high buildings.
	2	Landsat 5 TM band 1 Landsat 5 TM band 4	1074 × 1080, 30 m 1074 × 1080, 30 m	09/2001 03/2002	These SAR images were collected in urban areas and have a temporal difference of 6 months.
(d) Visible–map	1	Google Earth Google Earth	700 × 700, 0.5 m 700 × 700, 0.5 m	/ /	These images were collected in urban areas, and there are some text labels on these SAR images.
	2	Google Earth Google Earth	621 × 614, 1.5 m 621 × 614, 1.5 m	/ /	These images were collected in urban areas, and there are some text labels on these SAR images.

4. Experiments

4.1. Parameters and Evaluation Criteria

Precision, RMSE (Root Mean Square Error), and computational cost were used to evaluate the proposed algorithm. Matches with errors below a given threshold are considered correct matches. Multiple evenly distributed checkpoints were selected in each image to determine the true value. Generally, checkpoints are determined manually. However, due to various texture features and radiometric differences, particularly in the visible–SAR and the LiDAR–visible test, it is difficult to locate the ground truth manually in multi-source images. HOPCncc was to select 200 control points, which were evenly distributed in the image; 50 points with the lowest residuals were chosen as checkpoints. Once a checkpoint was determined, the projective transformation model was used, and the location error was calculated.

4.2. Tests for Similarity Measurement

The similarity obtained from the proposed CFPC was compared to the HPEO, FHOG, and NCC to illustrate the proposed method's advantages for matching multi-source images. Three groups of typical multi-source images were used to calculate likelihood estimation maps. Figure 2 shows the likelihood estimation maps of these description methods. The maximum likelihood estimate represents the pixels of the target center. As can be seen, all methods can detect the pixel of interest. However, the NCC'S accuracy is insufficient. The estimation map calculated by FHOG has multiple peaks, which results in estimation errors. The HPEO and proposed CFPC produce smooth likelihood estimates and achieve precise localization for the three cases. However, the proposed CFPC has a more concentrated location area. The preliminary test results demonstrate that the proposed CFPC is more robust than other algorithms (HPEO, FHOG, and NCC). A more detailed analysis will be presented next.

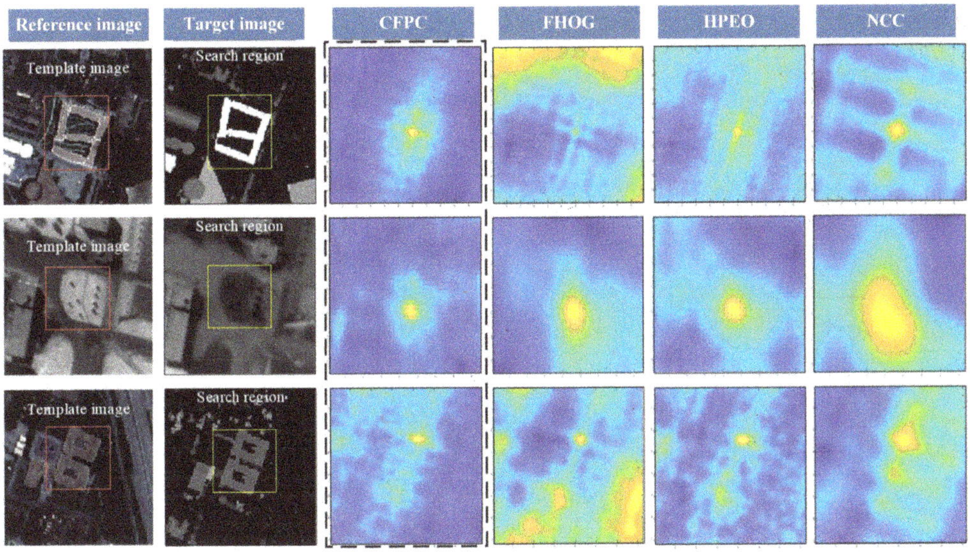

Figure 2. Likelihood estimation obtained from the CFPC and comparison algorithms.

4.3. Precision

To evaluate the matching precision of the proposed algorithm, this section compares the proposed method with some classical methods in terms of CMR. The selected test data correspond to LiDAR–visible 1, visible–infrared 1, 2, and visible–map 1, 2 in Table 1. There are significant radiometric distortions between each pair of images, which will greatly limit the matching performance. In general, artificially synthesized images are more difficult to match than other data as they contain fewer texture features. Different template sizes (ranging from 20 × 20 to 100 × 100 pixels) are also used to analyze the robustness of the similarity measures.

As seen, Figure 3 gives the test results of the proposed algorithm for the five datasets. The first two columns show the target and reference images, respectively, whereas the third column shows the matching results. The points with different colors represent the corresponding pixels of the matching result. As for matching precision, it is reflected in the control points' position. As can be seen, although the radiometric differences between multi-source images are obvious, the corresponding points are in the correct position, and the images are therefore correctly matched. As for the quantization accuracy, it will be shown in Figure 4.

Image matching precision is easily affected by the size of the template window. As its size increases, the matching performance significantly improves. The matching precision curves calculated by different template sizes are illustrated in Figure 5. MI, GMI, and MIND are capable of mitigating nonlinear intensity differences to a certain degree and have been proven suitable for multi-source image matching. While the matching results of NCC remain relatively stable, a primary drawback of the method is its disregard for the structural information of neighboring pixels, leading to a decline in image matching quality. Unlike NCC, CFOG and FHOG enhance the matching performance using the orientated gradient information and are invariant to radiation differences. Moreover, the proposed mixture model, CFPC, which combines both the oriented gradient and the response intensity of phase congruency, has achieved the best performance in all the test results. More specifically, for visible–map experiments with insufficient texture features, the proposed algorithm can also extract weak structural information from images by using the proposed mixed model, as it achieves better performance compared to the other algorithms in multi-source image matching.

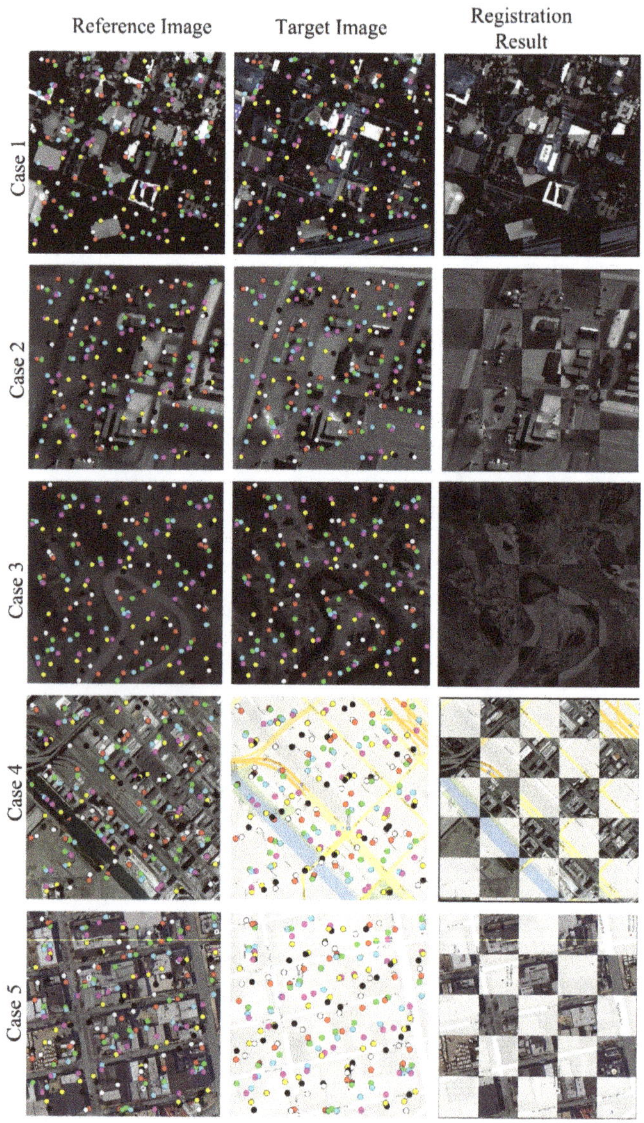

Figure 3. The matching results of the proposed CFPC. (The point pairs with same color in the consistency area of two images represent the matching positions).

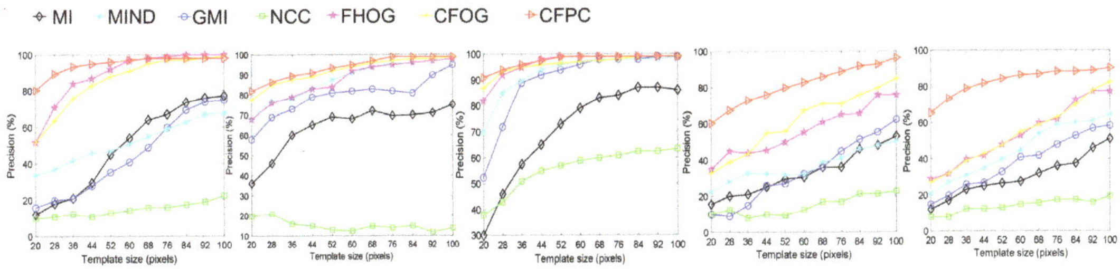

Figure 4. Image matching precision for different template sizes.

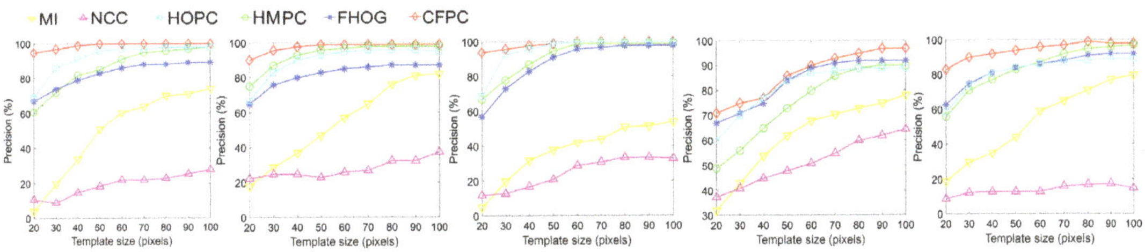

Figure 5. Image matching precision for different template sizes. The HMPC, FHOG, HOPC, and CFPC use similarity measures based on the F-SSDA.

The features of HMPC, FHOG, and HOPC were converted to the frequency domain, and the matching was performed with the proposed F-SSDA technique. Furthermore, some classical methods were chosen for comparison. Figure 6 shows the results for the five datasets (visible–SAR 1, 2, 3 and LiDAR–visible 2, 3). The matching results (i.e., HMPC, FHOG, and HOPC) calculated using the proposed similarity measure are superior to the classical algorithms (NCC and MI) in almost all cases. Obviously, NCC exhibits the worst matching performance, indicating that it is not suitable for multi-source image matching. As for HOPC, it shows better matching precision than NCC and MI because the stable structural features were extracted to evaluate similarity. Although the HMPC converted the similarity measure into the frequency domain, the template image does not consider the image texture, which will greatly influence the regional representation. Contrarily, the proposed CFPC smooths the oriented gradient and response intensity to minimize feature distortions caused by local nonlinear radiometric difference between multi-source images. Added to that, the optimal matching point pair is obtained by calculating the objective function in the frequency domain to ensure high matching performance. The proposed algorithm's matching results are shown in Figure 6, where the matched pixels are shown in different colors and achieve high consistency. From the registration results, the visual matching effect of each module has a high degree of continuity. Regarding the quantitative indicator RMSE, it will be discussed in the next paragraph.

Figure 7 gives the RMSE between the real and measured values. Histograms 1 to 10 illustrate the comparison of test results for all methods on different data, whereas histogram 11 represents the average value calculated from data 1 to 10. Obviously, the proposed CFPC outperforms all other methods (MI, GMI, MIND, FHOG, HOPC, LoFTR, and FLSS), and have the minimum RMSE. This can be attributed to the proposed method being able to extract structural features, which is invariant between multi-source image with radiometric differences. Moreover, the average test result is optimal in terms of RMSE values (histogram 11). The proposed method achieves enhanced robustness due to substantial radiometric variations among the multi-source images and to notable noise in the LiDAR and SAR images, posing challenges in feature matching. However, the proposed method can fully utilize existing information and captures the structural information of the multi-source image; it can effectively establish congruency relationships between corresponding regions.

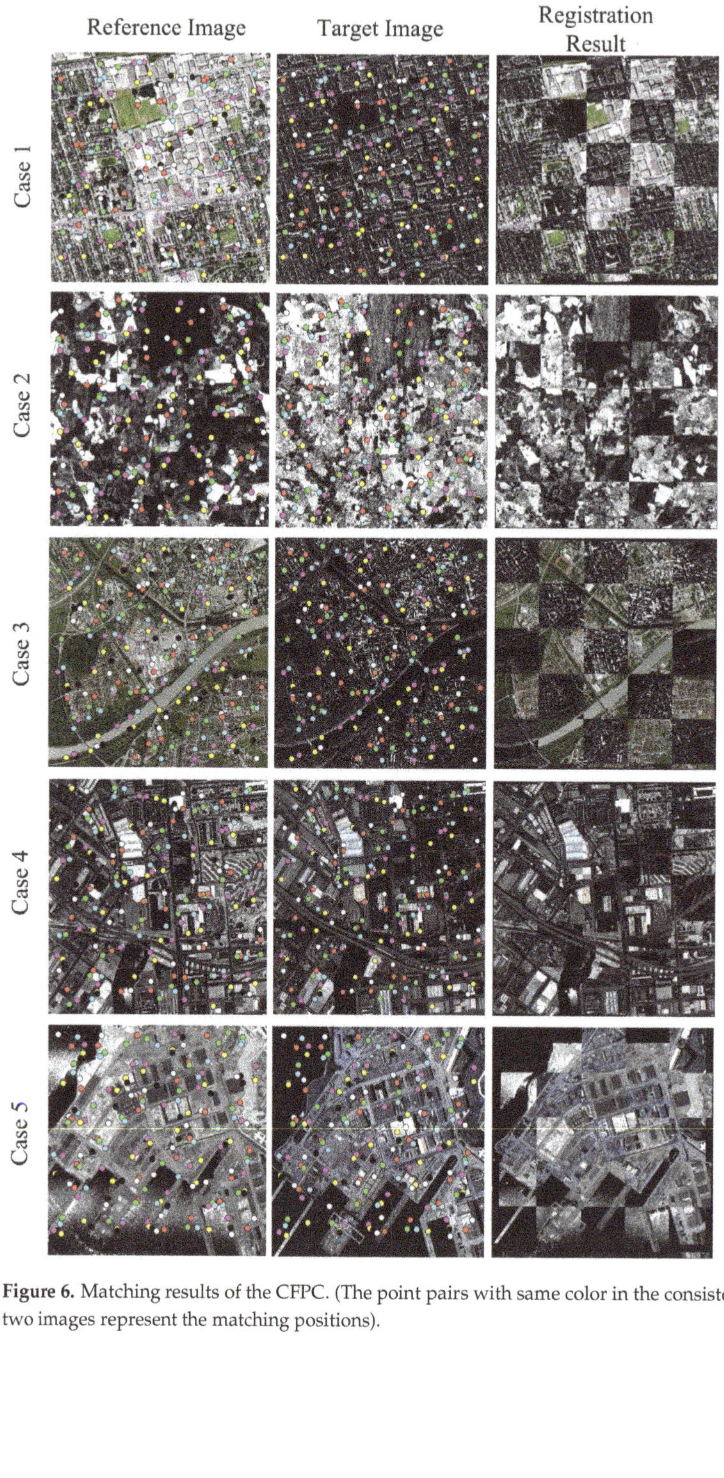

Figure 6. Matching results of the CFPC. (The point pairs with same color in the consistency area of two images represent the matching positions).

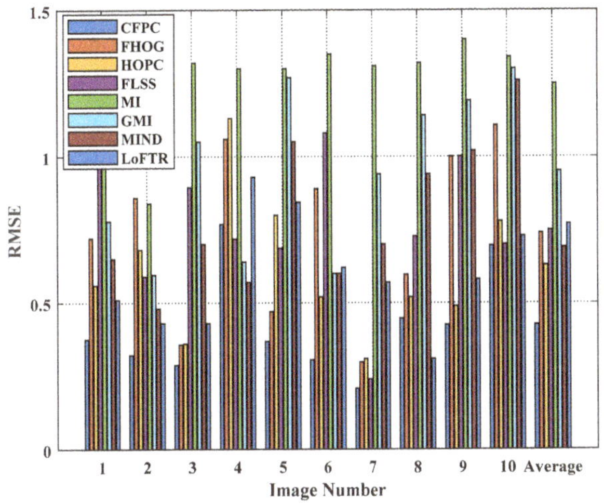

Figure 7. RMSE for the ten test cases.

The proposed method has a potential advantage in terms of computational efficiency. As depicted in Figure 8, GMI and MI stand out as the two most time-consuming techniques among these similarity measures, given that they necessitate the computation of the joint histogram for each matched window pair. However, the CFPC presents lower time consumption than other similarity measure methods (FHOG, FLSS, and HOPC) as the Fast Sequential Similarity Detection Algorithm (F-SSDA) is also developed to accelerate similarity measurement. The proposed matching method uses pixel-wise feature representation to evaluate the similarity between multi-source images. In addition, compared to other algorithms, our method has better performance when dealing with small target matching.

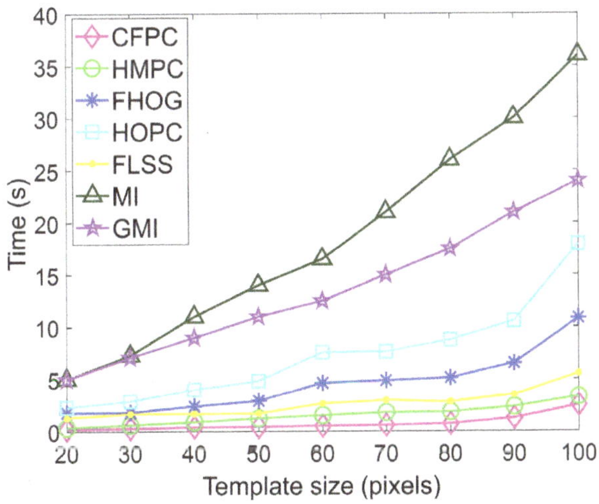

Figure 8. Time consumption for different template images.

In previous studies, SAR images generally had more noise than other images due to imaging mechanisms and device interference, so it was challenging for them to achieve robust matching performance. Then, 40 pairs of SAR–visible images with various Gaussian noises were used in comparative experiments to verify the robustness of the proposed algorithm. As can be seen, Figure 9 shows the matching results of the proposed algorithm. In terms of visual effects, the matching results are very accurate. The detailed statistical results are shown in Figure 10 where MI and GMI achieve robust matching performance under changed noises, but their average precision is lower than EOH and HMPC. For similarity measures, the matching accuracy of proposed CFPC is significantly better than the other methods (MI, GMI, EOH, and HMPC), followed by HPEO and FHOG. Furthermore, compared to HPEO, the proposed CFPC, which constructs robust feature representation by integrating different types of local features, has reduced the influence of radiometric differences and achieved better matching performance in multi-source image matching.

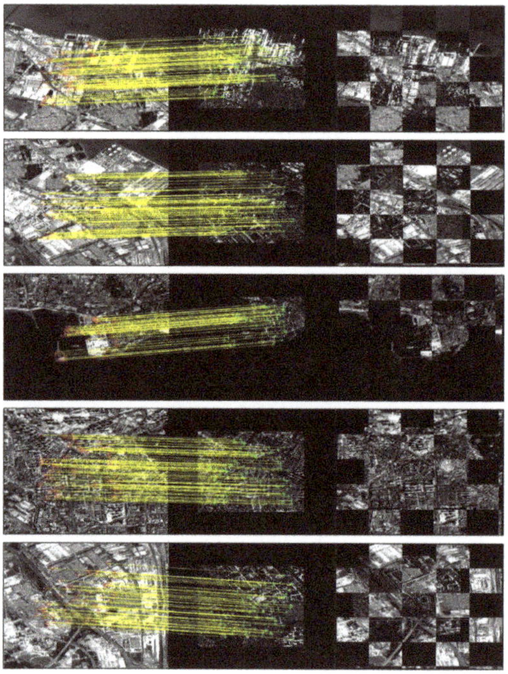

Figure 9. Registration results of the proposed algorithm.

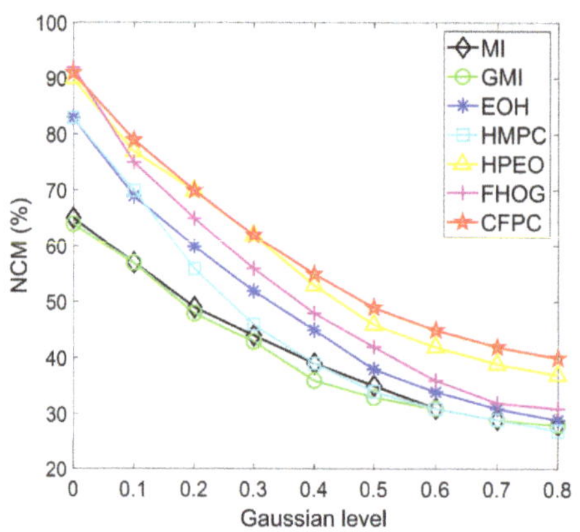

Figure 10. Matching results with increasing noise.

4.4. Ablation Experiment

Finally, Table 2 reports the experimental results of average running time (RT) and precision under different template sizes (TS) and noise levels. The template size varies between 50 and 90 square patches, whereas the noise level from 0.1 to 0.7. The first group of experimental results shows the matching performance of the original algorithm (FHOG, HOPC, and HMPC), while the second group presents the performance of these descriptors using the proposed F-SSDA for similarity measurement. As the size of the template increases, the operation efficiency decreases. Similarly, as noise increases, the accuracy of all test results decreases. Using the F-SSDA algorithm to calculate the similarity of features in FHOG, HOPC, and HMPC, the average matching efficiency was improved by 52.4%, 88.6%, and 11.9%, respectively. Meanwhile, the average matching accuracy was improved by 3.2%, 5.2%, and 1.4%, respectively. In addition, the proposed algorithm achieves the highest computational efficiency and matching accuracy. Therefore, our method can effectively improve matching robustness by constructing a joint feature model, and the matching efficiency obtained through the F-SSDA algorithm is also significantly improved.

Table 2. Ablation experiment.

	Template Size (Pixel) and Running Time (s)			Precision (%) and Noise Level		
Template, Noise	50 × 50	70 × 70	90 × 90	0.1	0.4	0.7
FHOG	4.24	5.14	5.93	86.52	76.96	61.35
HOPC	4.75	7.52	10.21	87.13	69.14	56.35
HMPC	0.43	0.79	1.21	88.19	79.09	64.28
FHOG + F-SSAD	2.73	3.19	4.21	87.50	78.38	65.34
HOPC + F-SSAD	2.61	3.61	5.77	88.12	70.22	63.54
HMPC + F-SSAD	0.41	0.71	1.01	88.81	79.61	66.21
Our method	0.34	0.56	0.88	89.12	79.89	70.14

5. Conclusions

This paper introduced the CFPC descriptor and the similarity measure F-SSDA for multi-source image matching. The CFPC extracts structural information from images using the oriented gradient and response intensity of phase congruency to produce robust local features and handles radiometric differences between multi-source images. The proposed CFPC is not only simple to combine but is also highly efficient and somewhat insensitive to noise and intensity changes. To enhance the matching efficiency, the Fast Sequential Similarity Detection Algorithm is proposed to perform the matching process. Experiments conducted with multi-source remote sensing datasets demonstrated that the proposed method outperformed state-of-the-art methods like MI, GMI, MIND, FHOG, HOPC, and FLSS.

Although the test results are encouraging, there is still some additional work to be done. For instance, there is significant deformation and inconsistent structural features between the reference image and the target image. Therefore, the deformation factor can be taken into account when calculating the similarity of phase congruency to solve the problem of multi-source image deformation.

Author Contributions: Conceptualization, Q.W.; validation, Q.W.; formal analysis, Q.W.; investigation, resources, Q.W.; data curation, Q.W.; writing—original draft preparation, Q.W. and Q.Y.; review and editing, Q.W. and Q.Y.; visualization, Q.W. and Q.Y.; funding acquisition, Q.Y. All authors have read and agreed to the published version of the manuscript.

Funding: This research was funded by the National Natural Science Foundation of China (62073161), the Startup Foundation for Introducing Talent of NUIST (2022r083, 2022r078), and the Natural Science Foundation of the Jiangsu Higher Education Institutions of China (23KJB510012).

Data Availability Statement: Data sharing not applicable.

Acknowledgments: The authors would like to thank Fenghe Lv from the Nanjing University for providing the test environment and open source code.

Conflicts of Interest: The authors declare no conflicts of interest.

References

1. Li, R.; Gao, X.; Shi, F. Scale Effect of Land Cover Classification from Multi-Resolution Satellite Remote Sensing Data. *Sensors* **2023**, *23*, 6136. [CrossRef]
2. Wang, C.; Xu, L.; Xu, R. Triple Robustness Augmentation Local Features for multi-source image registration. *ISPRS J. Photogramm. Remote Sens.* **2023**, *199*, 1–14. [CrossRef]
3. Li, X.; Lyu, X.; Tong, Y. An object-based river extraction method via optimized transductive support vector machine for multi-spectral remote-sensing images. *IEEE Access* **2019**, *7*, 46165–46175. [CrossRef]
4. Lati, A.; Belhocine, M.; Achour, N. Fuzzy correlation based algorithm for UAV image mosaic construction. *Multimed. Tools Appl.* **2024**, *1*, 3285–3311. [CrossRef]
5. Hu, J.; Hu, P.; Wang, Z. Spatial dynamic selection network for remote-sensing image fusion. *IEEE Geosci. Remote Sens. Lett.* **2021**, *19*, 8013205. [CrossRef]
6. Zhang, C.; Wang, L.; Cheng, S. SwinSUNet: Pure transformer network for remote sensing image change detection. *IEEE Trans. Geosci. Remote Sens.* **2022**, *60*, 5224713. [CrossRef]
7. Pallotta, L.; Giunta, G.; Clemente, C. Subpixel SAR image registration through parabolic interpolation of the 2-D cross correlation. *IEEE Trans. Geosci. Remote Sens.* **2020**, *58*, 4132–4144. [CrossRef]
8. Cole-Rhodes, A.; Johnson, L.; LeMoigne, J. Multiresolution registration of remote sensing imagery by optimization of mutual information using a stochastic gradient. *IEEE Trans. Image Process.* **2003**, *12*, 1495–1511. [CrossRef]
9. Ye, Y.; Shan, J.; Bruzzone, L. Robust registration of multimodal remote sensing images based on structural similarity. *IEEE Trans. Geosci. Remote Sens.* **2017**, *55*, 2941–2958. [CrossRef]
10. Wu, Q.; Zhu, S. Multispectral Image Matching Method Based on Histogram of Maximum Gradient and Edge Orientation. *IEEE Geosci. Remote Sens. Lett.* **2021**, *19*, 1–5. [CrossRef]
11. He, H.; Chen, Z.; Liu, H.; Guo, Y.; Li, J. Practical Tracking Method based on Best Buddies Similarity. *Cyborg Bionic Syst.* **2023**, *4*, 50. [CrossRef]
12. Ma, W.; Wu, Y.; Liu, S. Remote sensing image registration based on phase congruency feature detection and spatial constraint matching. *IEEE Access* **2018**, *6*, 77554–77567. [CrossRef]
13. Ma, J.; Chan, C.; Canters, F. Fully automatic subpixel image registration of multiangle CHRIS/Proba data. *IEEE Trans. Geosci. Remote Sens.* **2010**, *48*, 2829–2839.
14. Pluim, P.; Maintz, A. Image registration by maximization of combined mutual information and gradient information. In Proceedings of the Medical Image Computing and Computer-Assisted Intervention–MICCAI 2000: Third International Conference, Pittsburgh, PA, USA, 11–14 October 2000.
15. Shechtman, E.; Irani, M.; Xu, R. Matching local self-similarities across images and videos. In Proceedings of the 2007 IEEE Conference on Computer Vision and Pattern Recognition, Minneapolis, MN, USA, 17–22 June 2007.
16. Liu, J.; Zeng, G.; Fan, J. Fast local self-similarity for describing interest regions. *Pattern Recognit. Lett.* **2012**, *33*, 1224–1235. [CrossRef]
17. Sedaghat, A.; Xu, L.; Xu, R. Distinctive order based self-similarity descriptor for multi-sensor remote sensing image matching. *ISPRS J. Photogramm. Remote Sens.* **2023**, *108*, 62–71. [CrossRef]
18. Jenkinson, P.; Bhushan, M. MIND: Modality independent neighbourhood descriptor for multi-modal deformable registration. *Med. Image Anal.* **2012**, *16*, 1423–1435.
19. Wu, Q.; Xu, G.; Cheng, Y. Robust and efficient multi-source image matching method based on best-buddies similarity measure. *Infrared Phys. Technol.* **2019**, *101*, 88–95. [CrossRef]
20. Ye, Y.; Bruzzone, L.; Shan, J. Fast and robust matching for multimodal remote sensing image registration. *IEEE Trans. Geosci. Remote Sens.* **2019**, *57*, 9059–9070. [CrossRef]
21. Lowe, D. Distinctive image features from scale-invariant keypoints. *Int. J. Comput. Vis.* **2004**, *60*, 91–110. [CrossRef]
22. Rublee, E.; Rabaud, V.; Konolige, K. ORB: An efficient alternative to SIFT or SURFs. *IEEE Comput. Soc. Conf. Comput. Vis. Pattern Recognit.* **2011**, *199*, 2564–2571.
23. Morel, M.; Yu, G. ASIFT: A new framework for fully affine invariant image comparison. *SIAM J. Imaging Sci.* **2009**, *2*, 438–469. [CrossRef]
24. Dellinger, F.; Delon, J.; Gousseau, Y. SAR-SIFT: A SIFT-like algorithm for SAR images. *IEEE Trans. Geosci. Remote Sens.* **2014**, *53*, 453–466. [CrossRef]
25. Sedaghat, A.; Ebadi, H. Remote sensing image matching based on adaptive binning SIFT descriptor. *IEEE Trans. Geosci. Remote Sens.* **2015**, *53*, 5283–5293. [CrossRef]
26. Zhu, X.; Huang, Z.; Ding, M. Non-rigid multi-modal brain image registration based on two-stage generative adversarial nets. *Neurocomputing* **2022**, *505*, 44–57. [CrossRef]

27. Fu, C.; Yuan, H.; Xu, H.; Zhang, H.; Shen, L. TMSO-Net: Texture adaptive multi-scale observation for light field image depth estimation. *J. Vis. Commun. Image Represent.* **2023**, *90*, 103731. [CrossRef]
28. Tang, L.; Deng, Y.; Ma, Y. SuperFusion: A versatile image registration and fusion network with semantic awareness. *IEEE/CAA J. Autom. Sin.* **2022**, *9*, 2121–2137. [CrossRef]
29. Yang, T.; Bai, X.; Cui, X. DAU-Net: An unsupervised 3D brain MRI registration model with dual-attention mechanism. *Int. J. Imaging Syst. Technol.* **2023**, *33*, 217–229. [CrossRef]
30. Sun, J.; Shen, Z.; Wang, Y.; Bao, H.; Zhou, X. LoFTR: Detector-free local feature matching with transformers. In Proceedings of the IEEE/CVF Conference on Computer Vision and Pattern Recognition, Virtual Conference, 19–25 June 2021.
31. Wu, Q.; Xu, G.; Cheng, Y. A deformed contour segment matching for multi-source images. *Pattern Recognit.* **2021**, *117*, 107968. [CrossRef]
32. Meng, X.; Yang, G.; Yin, Y. Finger vein recognition based on local directional code. *Sensors* **2012**, *12*, 14937–14952. [CrossRef] [PubMed]
33. Aguilera, C.; Barrera, F. Multispectral image feature points. *Sensors* **2012**, *12*, 12661–12672. [CrossRef]
34. Wu, Q.; Xu, G.; Cheng, Y. Histogram of maximal point-edge orientation for multi-source image matching. *Int. J. Remote Sens.* **2020**, *41*, 5166–5185. [CrossRef]
35. Ye, Y.; Shen, L. Hopc:A novel similarity metric based on geometric structural properties for multi-modal remote sensing image matching. *ISPRS Ann. Photogramm.* **2016**, *3*, 9–16.
36. Wu, Q.; Li, H.; Zhu, S. Nonlinear intensity measurement for multi-source images based on structural similarity. *Measurement* **2021**, *179*, 109474. [CrossRef]

remote sensing

MDPI

Article

BCMFIFuse: A Bilateral Cross-Modal Feature Interaction-Based Network for Infrared and Visible Image Fusion

Xueyan Gao and Shiguang Liu *

College of Intelligence and Computing, Tianjin University, Tianjin 300350, China; gxy@tju.edu.cn
* Correspondence: lsg@tju.edu.cn

Abstract: The main purpose of infrared and visible image fusion is to produce a fusion image that incorporates less redundant information while incorporating more complementary information, thereby facilitating subsequent high-level visual tasks. However, obtaining complementary information from different modalities of images is a challenge. Existing fusion methods often consider only relevance and neglect the complementarity of different modalities' features, leading to the loss of some cross-modal complementary information. To enhance complementary information, it is believed that more comprehensive cross-modal interactions should be provided. Therefore, a fusion network for infrared and visible fusion is proposed, which is based on bilateral cross-feature interaction, termed BCMFIFuse. To obtain features in images of different modalities, we devise a two-stream network. During the feature extraction, a cross-modal feature correction block (CMFC) is introduced, which calibrates the current modality features by leveraging feature correlations from different modalities in both spatial and channel dimensions. Then, a feature fusion block (FFB) is employed to effectively integrate cross-modal information. The FFB aims to explore and integrate the most discriminative features from the infrared and visible image, enabling long-range contextual interactions to enhance global cross-modal features. In addition, to extract more comprehensive multi-scale features, we develop a hybrid pyramid dilated convolution block (HPDCB). Comprehensive experiments on different datasets reveal that our method performs excellently in qualitative, quantitative, and object detection evaluations.

Keywords: image fusion; bilateral cross-modal feature interaction; hybrid pyramid dilated convolution; infrared image; visible image

1. Introduction

Image fusion technology can generate high-quality fusion images containing rich detail information by integrating two or more source images [1,2]. This technology covers various image types, including medical image fusion (MIF), multi-exposure image fusion (MEIF), and infrared–visible image fusion (IVIF). The main goal is to provide a clear and comprehensive scene representation to enhance scene understanding. Among them, IVIF is the most widespread and challenging since it requires the effective extraction and combination of cross-modal features from different sensors. The core goal of IVIF is to retain the abundant texture details from the visible image and the salient target from the infrared image [3]. By fusing these two, it can avoid problems caused by the low resolution and noise of infrared images, and simultaneously overcome the limitations of visible images in harsh working environments (such as rain, snow, fog, low illumination, etc.). High-quality fusion images are very helpful for downstream high-level visual tasks, including remote sensing [4], object detection [5], image segmentation [6], and autonomous driving [7].

Figure 1 demonstrates the fusion and detection results in a typical smoke scenario. In such conditions, humans and other important targets are often obscured by smoke. Nonetheless, owing to the unique imaging principles inherent to infrared images, they can penetrate smoke and distinctly reveal these targets. As illustrated in Figure 1, the fusion image significantly enhances the visibility of targets in smoky conditions, thereby improving detection accuracy. For target detection, visible images fail to detect critical targets obscured by smoke. The FGAN [8] misses critical target detections, while the DSF [9] results in false detections. SuperF [10] and the proposed method both successfully detected these targets, with our method having higher detection accuracy. This indicates that the proposed method efficiently integrates useful information from infrared and visible images, enhancing subsequent target detection accuracy. IVIF has tremendous potential for applications in remote sensing, especially in monitoring complex environments. For example, in forest fire monitoring, IVIF can help identify fire sources and hotspot areas, clearly showing the location and range of the fire, which is very helpful for decision-makers in formulating more effective rescue plans.

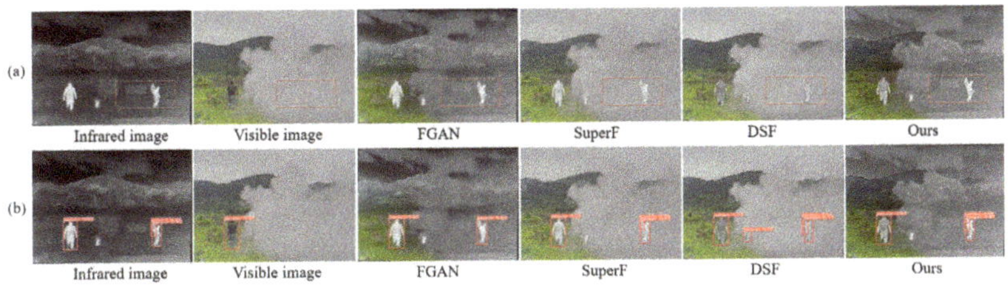

Figure 1. An example of fusion and object detection: (**a**) fusion results; (**b**) detection results. The source images are shown in the first two columns; the fusion and detection results of FGAN, SuperF, DSF, and our method are in the last four columns.

Existing IVIF methods are primarily divided into traditional and deep learning-based methods. Traditional fusion methods encompass methods based on saliency [11], subspace [12], sparse representation (SR) [13], multi-scale transform (MST) [14], and hybrid [15]. Although the above methods obtain satisfactory results, they still have some obvious limitations: (1) These methods depend on manually designed fusion rules that may not comprehensively retain the effective features in the source images, especially as source image complexity increases, necessitating more intricate fusion rules. (2) These methods usually apply uniform feature extraction for images of different modalities, neglecting the differences between modalities, which could cause the loss of some unique modality features in the source images. However, methods based on deep learning exhibit powerful feature extraction capabilities, overcoming inherent deficiencies in traditional methods and garnering attention from numerous scholars. Researchers have developed various models based on deep learning for IVIF, which are generally classified into methods based on the generative adversarial network (GAN) [5,8,16,17], autoencoder (AE) [18–21], convolutional neural network (CNN) [22–25], and transformer [26,27].

While the methods mentioned above have achieved satisfactory fusion performance, several issues remain unresolved. Firstly, most of these methods extract features using a single network or two parallel networks, without considering the differences in source image features, thus ignoring the interaction of features from different modalities and losing some important cross-modal information. It possibly affects the fusion performance to a certain extent. For example, DenseFuse [18] and FusionDN [28] use a single network for feature extraction without accounting for the differences between different modalities' images, which may lead to the loss of some modality-specific information. Secondly, balancing the discrepancies between the fusion image and the source images is difficult because IVIF lacks ground truth. Some researchers have introduced GANs to address this issue, such as FusionGAN [8], DDcGAN [16], and GANMcC [29]. Although GAN-based methods have achieved acceptable fusion results, it is challenging for GAN to effectively utilize the unique information in multi-modal images, and GAN-based methods are difficult to achieve training balance. Finally, they overlook the extraction of diverse features in the source images and may extract some redundant information or miss some important information, thus affecting the fusion performance. For instance, SDNet [30] and PMGI [31] extract features using a single convolution kernel, which has limited receptive fields and may ignore some important features. Considering the extraction of multi-scale features, NestFuse [21] and RFNNest [32] introduce a nested network, which fuses features from different layers using a fusion strategy, but different layers still use a single convolution kernel. However, these methods do not sufficiently account for the inherent characteristics of different modalities, making it challenging to learn diverse feature representations. How to learn diverse feature representation and effectively use the beneficial characteristics of bilateral modality to enhance fusion performance is still a challenge.

To tackle the aforementioned problem, we introduce BCMFIFuse, a network that utilizes bilateral cross-modal feature interaction for IVIF. Firstly, to adequately extract features from infrared and visible images, we construct a two-stream feature extraction network. Given that infrared and visible images pertain to different modalities, we design CMFC to better extract complementary features from these two modalities. CMFC calibrates the current modality features by combining features from different modalities in the channel dimension and spatial dimension. Then, we use FFB to effectively integrate the calibrated features. The FFB is constructed using a cross-attention mechanism to facilitate long-range context information interaction, thereby enhancing global bilateral modality features. Finally, to ensure feature continuity and reduce information loss during transmission, skip connections are utilized between the encoder and decoder. Moreover, to prevent feature loss from using a single convolution kernel, multi-scale feature extraction is essential. Dilated convolution is highly effective in capturing multi-scale features and exploring contextual information. However, regular dilated convolutions detect input features within a square window, which limits their flexibility in capturing diverse features. Simply using a large square dilated convolution window is not an effective solution as it tends to extract redundant features. To tackle this issue, we introduce an HPDCB that integrates rectangular dilated convolutions to collect more diverse and specific contextual information. To capture long-range relationships of isolated areas, we employ a combination of long and narrow kernel shapes in the design of HPDCB. We first employ a long kernel shape with a variable dilation rate along one spatial dimension; then, a narrow kernel shape is employed in the other spatial dimension. The key contributions of the proposed method are outlined below:

(a) A bilateral cross-modal feature interaction-based method for IVIF is suggested. The goal of this method is to provide comprehensive cross-modal interactions and fully leverage the complementary potential of cross-modal features. In addition, we use a hybrid pyramid dilated convolution block (HPDCB) to extract multi-scale features, effectively collect various contextual information, and learn diverse feature representations.
(b) A cross-modal feature correction block (CMFC) is introduced. The module combines the features from different modalities in spatial and channel dimensions to calibrate the current modality features. This enables the two feature extraction branches to better focus on complementary information from both modalities, thereby mitigating uncertainties and noise effects from different modalities and achieving better feature extraction and interaction.
(c) A feature fusion block (FFB) is developed. This module effectively integrates cross-modal features and merges the calibrated features from the CMFC into a single feature for subsequent image reconstruction. This module considers interaction fusion at both the channel and spatial dimensions, which is crucial for the generalization of cross-modal feature combinations.

The remainder of this paper is organized as follows. Section 2 reviews the related work in the field of image fusion. Section 3 describes the proposed method's framework in detail. Section 4 presents the relevant experiments and results. Finally, Section 5 concludes the main content of this paper and outlines directions for future improvements.

2. Related Works

This section outlines a thorough overview of current IVIF methods, encompassing methods based on traditional and deep learning.

2.1. Traditional Image Fusion Methods

Traditional methods vary based on feature extraction, fusion strategy, and feature reconstruction methods, and are classified as methods based on saliency, subspace, SR, MST, and hybrid. The methods based on SR and MST are the most commonly employed.

Methods based on MST (e.g., discrete wavelet transform (DWT) [33] and Laplace pyramid transform (LAP) [34]) primarily decompose the source images into sub-images of varying scales and orientations. These sub-images are then merged following specific fusion rules, and an inverse transformation is employed to generate fusion images. These methods can preserve the source images with their multi-scale features but may lose some detail information, leading to distortion or edge blur of the fusion image.

Methods based on sparse representation mainly rely on the learning of overcomplete dictionaries and the decomposition algorithm of sparse coefficients. Initially, an overcomplete dictionary is obtained through learning, followed by sparse coding of the input images. The obtained sparse coefficients are then fused using various fusion rules, and the image is finally reconstructed with the dictionary and fusion coefficients. These methods are able to preserve the details and structure of source images, whereas they have high computational complexity. Furthermore, the choice of fusion rules and dictionaries is of utmost importance, as they will affect the fusion results.

2.2. Deep Learning-Based Image Fusion Methods

Generally speaking, methods based on deep learning are mainly divided into four types: fusion methods based on AE, GAN, CNN, and transformer.

2.2.1. AE-Based Fusion Methods

These methods leverage encoder and decoder networks for feature extraction and image reconstruction; then, they apply artificially designed fusion rules for fusing features. Densefuse [18] incorporates dense blocks during the encoding process to effectively extract and utilize features, and reconstruct the fusion image with a decoder. Subsequently, Li et al. [32] introduced a residual architecture-based residual fusion network (RFN) to enhance the performance of image fusion. AUIF [35] is based on the principle of algorithm expansion and decomposes the source images into high- and low-frequency information. In addition, to further enhance the fusion performance, NestFuse [21] integrates an attention mechanism into the model. SEDRFuse [36] developed a symmetric network framework with residual blocks and introduced a feature fusion rule based on attention. Res2Fusion [37] incorporates dense Res2Net into the encoder and develops a dual non-local attention-based fusion strategy. Despite the significant fusion performance achieved by these methods, the necessity for manual formulation of fusion rules greatly limits the fusion performance improvement.

2.2.2. CNN-Based Fusion Methods

To tackle the issues of AE-based fusion methods, a number of CNN-based end-to-end fusion techniques were introduced, yielding impressive fusion results. For instance, the PIAFusion [25] network leverages illumination perception for fusing infrared and visible images. STDFusionNet [24] extracts the background regions from visible images and the target regions from infrared images using semantic segmentation, while network optimization is guided by a new loss function. Notably, the key innovation in these approaches is in the design of the loss function. Long et al. [38] introduced RXDNFuse, which has a relatively innovative network structure and uses an aggregated residual dense network to effectively extract and fuse features. To address various image fusion tasks within a single framework, PMGI [31] is proposed, which emphasizes gradient and intensity ratio preservation, enabling multiple image fusion tasks. Additionally, Zhang et al. [30] introduced SDNet, treating the fusion problem as a task of extracting and reconstructing gradient and intensity information. Xu et al. [28,39] contributed to the field with their general image fusion frameworks, U2Fusion and FusionDN. However, despite the noteworthy results these fusion methods have made, they commonly neglect the modality differences inherent between visible and infrared images. They utilize the same network structure for extracting features from different modalities, which limits their ability to distinguish inherent feature differences between modalities, subsequently restricting their fusion performance.

2.2.3. GAN-Based Fusion Methods

IVIF lacks ground truth; to address this challenge, researchers have introduced GAN. FusionGAN [8] is a new beginning in the field of fusion, marking the official application of GAN in this field. Subsequently, numerous GAN-based fusion methods have emerged. For example, ResNetFusion [40] tackles the problems of target edge blurring and texture detail loss in FusionGAN. This method incorporates target edge enhancement loss and detail loss functions to sharpen target edges and enhance detail information. Furthermore, proposed by Ma et al. [16], DDcGAN contains a generator and dual discriminators, which can train the generator more comprehensively and prevent information loss that could occur with a single discriminator. GANMcC [29] introduces a multi-class constraint, transforming the fusion problem into a simultaneous estimation of multiple distributions and achieving excellent fusion results. TarDAL [5] is a dual adversarial learning network using target perception and applies to image fusion and downstream object detection tasks. GAN-FM, proposed by Zhang et al. [41], involves a full-scale skip-connected generator to leverage multi-scale information during the fusion process. Additionally, many researchers introduced attention mechanisms in their frameworks. For example, AttentionFGAN [42] incorporates multi-scale attention into the generator and the discriminator, significantly enhancing fusion performance. TC-GAN [43] applies squeeze and excitation modules in the generator to better retain important texture details in the fusion images. However, achieving training balance in GAN-based methods still poses a challenge.

2.2.4. Transformer-Based Fusion Methods

Models of this kind employ the multi-head self-attention mechanism, exhibiting remarkable performance in capturing global information and overcoming the limited receptive field issue of CNNs. Thanks to its exceptional ability to explore the global context, the transformer model has excelled in computer vision tasks, giving rise to various fusion methods based on transformer [26,27,44]. For instance, SwinFusion [26] introduces a fusion framework that combines the convolutional neural network and the transformer, enabling the utilization of both local and global information. YDTR [44] captures local important information and crucial contextual detail information by utilizing a dynamic transformer, preserving texture details and salient targets in the source images. However, transformer-based methods, while delivering outstanding performance, also entail significant computational costs.

3. Method

In this section, we describe the framework of the BCMFIFuse, which is a method for IVIF based on bilateral cross-modal feature interaction. First, an outline of the overall architecture of BCMFIFuse is presented. Next, we introduce network structures of hybrid pyramid dilated convolution block (HPDCB), cross-modal feature correction block (CMFC), and feature fusion block (FFB). Finally, we delve into the multiple constraint loss functions employed in the BCMFIFuse network model.

3.1. Overall Architecture

IVIF aims to maintain essential target information in infrared images while simultaneously retaining the abundant texture details in visible images. Extracting and leveraging complementary features from the visible and infrared images remains challenging on account of differences in sensor acquisition and imaging mechanisms. Existing fusion methods suffer from limitations in feature representation, potentially leading to information loss and negatively impacting the fusion results. Therefore, achieving comprehensive cross-modal interaction is crucial for fully leveraging the complementary features from the infrared and visible images. In this study, we propose BCMFIFuse, an IVIF network based on bilateral cross-modal feature interaction. Figure 2 demonstrates the overall framework of BCMFIFuse, which is constructed with a two-stream design for feature extraction from infrared and visible images. The architecture of BCMFIFuse primarily comprises four key components: HPDCB, CMFC, FFB, and FR. We use the HPDCB module to extract comprehensive multi-scale feature information, then employ the CMFC to calibrate the extracted features, and subsequently use the FFB to fuse the calibrated features. Additionally, to ensure feature continuity and reduce information loss during transmission, we implement residual connections between the encoder and decoder. The FR is primarily employed to ensure that the reconstructed visible and infrared images contain complete information.

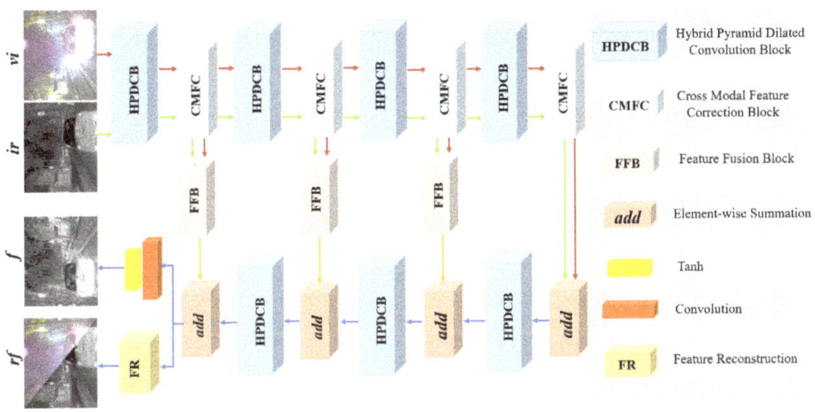

Figure 2. The overall network architecture of our method. vi, ir, and f refer to visible, infrared, and fusion images, respectively. rf stands for reconstructed visible and infrared images.

3.2. Hybrid Pyramid Dilated Convolution Block (HPDCB)

The size and shape of the convolution kernel can affect the features extracted. As the scale changes of the target objects in the dataset may not always be regular, to extract a more comprehensive and richer image feature representation, we need to capture features at different scales. Conventional dilated convolutions mainly detect input features of square windows, which may limit their flexibility in capturing features. Relying solely on large square dilated convolution windows is insufficient to fully tackle this issue and might even lead to the extraction of redundant features. To tackle this issue, we design a module named HPDCB, which aims to obtain more specific and diverse contextual information, as shown in Figure 3. HPDCB is capable of capturing the long-range relationships between isolated areas. During the design phase of HPDCB, rectangular dilated convolution is incorporated into the dilated convolution framework and adopts a method combining long kernel and narrow kernel shapes. First, a long kernel shape with a variable dilation rate along one spatial dimension is deployed; then, a narrow kernel shape is deployed in the other spatial dimension. The process mentioned above can be formulated as follows:

$$\begin{aligned} h_1 &= relu(conv_{3\times1}^{d=1}(relu(conv_{1\times3}^{d=1}(F_a)))) \\ h_2 &= relu(conv_{3\times1}^{d=2}(relu(conv_{1\times3}^{d=2}(F_a)))) \\ h_3 &= relu(conv_{3\times1}^{d=3}(relu(conv_{1\times3}^{d=3}(F_a)))) \\ h_4 &= relu(conv_{3\times3}(F_a)) \\ h &= h_1 + h_2 + h_3 + h_4 \end{aligned} , \quad (1)$$

where F_a denotes the input feature, d denotes dilation rate, $relu$ stands for ReLU activation function, $conv$ stands for convolution, h is the output feature, and 3×1 and 3×1 are the kernel sizes.

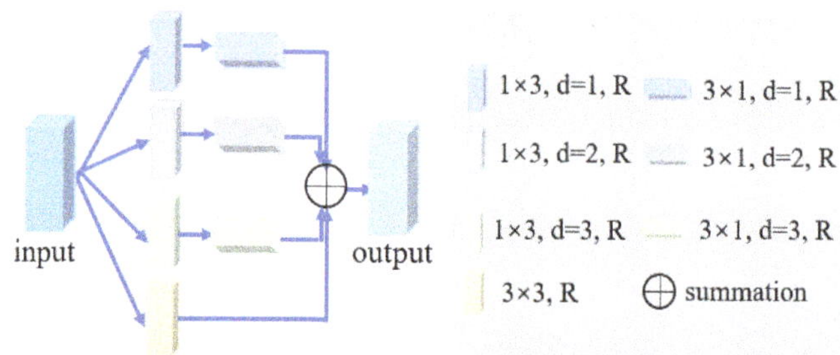

Figure 3. The hybrid pyramid dilated convolution block (HPDCB). d denotes dilation rate. R stands for ReLU activation function.

3.3. Cross-Modal Feature Correction Block (CMFC)

In our work, we construct two parallel branches for extracting features from the visible and infrared images, respectively. As is widely recognized, multimodal images often contain a large amount of noise from different modalities. Nevertheless, infrared and visible images offer complementary information, with the potential for their features to calibrate the noise information pertaining to one another. Therefore, we design CMFC to correct features from different modalities. Figure 4 depicts the structure of CMFC. The calibrated features are then fed into the subsequent phases for further enhancement and refinement of feature extraction. Assuming that the inputs of CMFC are f_i^{ir} and f_i^{vi}, the workflow can be expressed as follows:

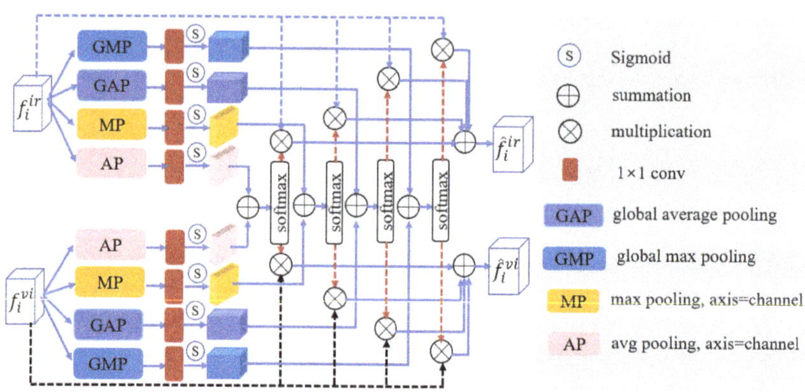

Figure 4. The cross-modal feature correction block.

$$\begin{aligned} M_{ir1}^{i} &= sig(conv(GMP(f_i^{ir}))) \\ M_{ir2}^{i} &= sig(conv(GAP(f_i^{ir}))) \\ M_{ir3}^{i} &= sig(conv(AP(f_i^{ir}))) \\ M_{ir4}^{i} &= sig(conv(MP(f_i^{ir}))) \end{aligned} , \quad (2)$$

$$\begin{array}{l} M^i_{vi1} = sig(conv(GMP(f^{vi}_i))) \\ M^i_{vi2} = sig(conv(GAP(f^{vi}_i))) \\ M^i_{vi3} = sig(conv(AP(f^{vi}_i))) \\ M^i_{vi4} = sig(conv(MP(f^{vi}_i))) \end{array}, \tag{3}$$

$$\begin{array}{l} M^i_{a1} = softmax(M^i_{ir1} + M^i_{vi1}) \otimes f^a_i \\ M^i_{a2} = softmax(M^i_{ir2} + M^i_{vi2}) \otimes f^a_i \\ M^i_{a3} = softmax(M^i_{ir3} + M^i_{vi3}) \otimes f^a_i \\ M^i_{a4} = softmax(M^i_{ir4} + M^i_{vi4}) \otimes f^a_i \end{array}, \tag{4}$$

$$\widehat{f^a_i} = M^i_{a1} + M^i_{a2} + M^i_{a3} + M^i_{a4}, \tag{5}$$

where *conv* stands for convolution, *sig* denotes the sigmoid function, and $\otimes$ represents element-wise multiplication. M^i_{aj} $(a = ir, vi; i, j = 1,2,3,4)$ signifies the result of the j-th branch at the i-th layer. $\widehat{f^s_i}$ represents the calibrated features.

3.4. Feature Fusion Block (FFB)

To extract and integrate the most discriminative features in the infrared and visible images, facilitating long-range contextual interaction and enhancing global cross-modal features, we design a feature fusion block (FFB). This module integrates two features calibrated by CMFC into a feature map, which is then added to the output of different levels of the decoder to transform into the ultimate output feature. The framework of FFB is depicted in Figure 5.

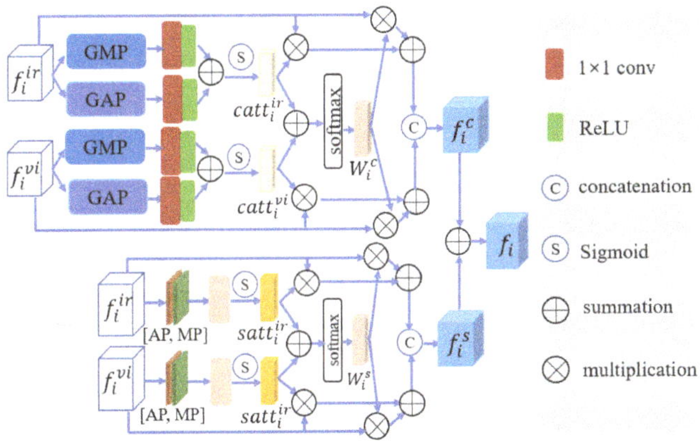

Figure 5. The feature fusion block. *GMP* and *GAP* indicate global average pooling and global max pooling. *AP* and *MP* stand for average pooling and max pooling.

Assuming the i-th HPDCB of the visible and infrared branches generate f^{vi}_i and f^{ir}_i, respectively, we first adopt GAP and GMP to obtain global features. Subsequently, after a series of processing, the channel attention weights $catt^x_i, (x = ir, vi)$ generated by the infrared and visible branches are obtained. These attention weights are then multiplied with the corresponding input features $f^x_i, (i = 1,2,3)$, aiding the model in suppressing unimportant scene features and emphasizing crucial ones. Following this, the attention weights of these two branches are added; then, the softmax function is employed to acquire the cross-channel attention weight W^c_i. The multiplication of the input features and weight W^c_i of the visible and infrared images yields the cross-channel attention output for both modalities. Additionally, f^{ir}_i and f^{vi}_i undergo a spatial attention module (composed of MP, AP, convolution, and sigmoid function), producing the spatial attention output $satt^x_i$ for the visible and infrared images. Subsequent operations are similar to the above process (see Figure 5). The final cross-modal fusion feature f_i is acquired by adding the acquired f^c_i and f^s_i. The process mentioned above can be formulated as follows:

$$\begin{array}{l} gap_i^x = relu(conv1 \times 1(GAP(f_i^x))) \\ gmp_i^x = relu(conv1 \times 1(GMP(f_i^x))) \\ catt_i^x = sig(gap_i^x + gmp_i^x) \\ \widehat{F}_{ci}^x = catt_i^x \otimes f_i^x \\ W_i^c = softmax(catt_i^{ir} + catt_i^{vi}) \\ F_{ci}^x = W_i^c \otimes f_i^x + \widehat{F_{ci}^x} \\ f_i^c = conv1 \times 1(concat(F_{ci}^{ir}, F_{ci}^{vi})) \end{array}, \tag{6}$$

$$\begin{array}{l} satt_i^x = sig(conv7 \times 7(concat(AP(f_i^x), MP(f_i^x)))) \\ \widehat{F}_{si}^x = satt_i^x \otimes f_i^x \\ W_i^s = softmax(satt_i^{ir} + satt_i^{vi}) \\ F_{si}^x = W_i^s \otimes f_i^x + \widehat{F_{si}^x} \\ f_i^s = conv1 \times 1(concat(F_{si}^{ir}, F_{si}^{vi})) \end{array}, \tag{7}$$

$$f_i = (f_i^c + f_i^s) \times 0.5, \tag{8}$$

where $relu$ is the ReLU activation function and $softmax$ denotes the softmax activation function.

3.5. Feature Reconstruction Module (FR)

To ensure that the reconstructed visible and infrared images incorporate complete information, we designed the FR module. This module aims to constrain the final reconstruction features to contain more comprehensive information by reconstructing the visible and infrared images. Figure 6 depicts the architecture of the FR. Assuming X is the input feature, the workflow of FR can be expressed as follows:

$$\begin{array}{l} R_1 = B(conv(relu(B(conv(relu(B(conv(X)))))))) \\ R_2 = B(conv(relu(B(conv(GAP(X)))))) \\ R_3 = sig(R_1 \otimes R_2) \otimes X \\ R = tanh(conv(B(R_3))) \end{array}, \tag{9}$$

where B denotes batch normalization, $conv$ stands for convolution with kernel size 3×3, $relu$ stands for ReLU activation function, GAP represent global average pooling, and $tanh$ is Tanh activation function.

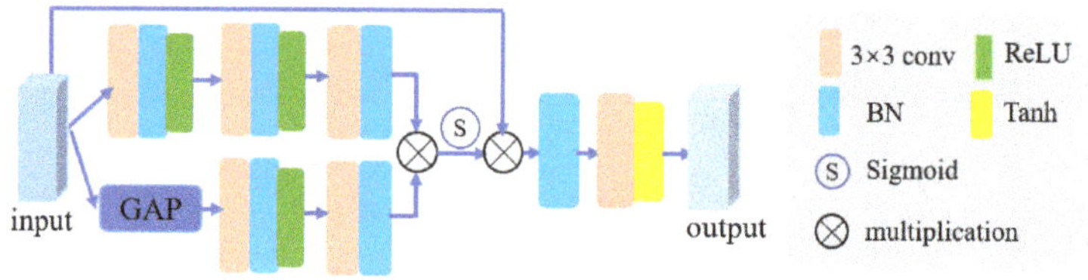

Figure 6. The feature reconstruction module (FR). BN denotes Batch Normalization. GAP represent global average pooling.

3.6. Loss Function

Designing the loss function is pivotal for model training, as it not only guides the optimization direction of the model but also influences the proportion of various information from the source images preserved in the fusion images. The loss function of the proposed method is formed by five terms: intensity loss (L_{int}), detail loss (L_{detail}), structural similarity loss (L_{ssim}), triplet loss (L_{tri}), and reconstruction loss (L_{re}). We express the total loss L_{total} as follows:

$$L_{total} = \alpha L_{\text{int}} + \beta L_{\text{detail}} + L_{ssim} + L_{\text{tri}} + \gamma L_{re}, \tag{10}$$

where the coefficients α, β, and γ are employed to harmonize various loss functions.

To retain important targets in source images, we introduce a saliency-related intensity loss, defined as follows:

$$L_{\text{int}} = MSE(x_f, (W_{ir} \otimes x_{ir} + W_{vi} \otimes x_{vi})), \tag{11}$$

where MSE is the mean squared error; x_f, x_{vi}, and x_{ir} stand for the fusion, visible, and infrared image; W_{iv} and W_{vi} denote weighted maps, $W_{vi} = S_{vi}/(S_{vi} - S_{ir})$ and $W_{ir} = 1 - W_{vi}$; and S denotes the saliency matrix, which is computed by [45].

To maintain abundant texture details in the fusion images, we incorporate a detail loss, defined as follows:

$$L_{\text{detail}} = \left\| \left|\nabla x_f\right| - \max(|\nabla x_{ir}|, |\nabla x_{vi}|) \right\|_1, \tag{12}$$

where $\|\cdot\|_1$ and ∇ denote the l_1-norm and Sobel gradient operator; $\max(\cdot)$ and $|\cdot|$ stand for the element-wise maximum values operation and the absolute operation symbol.

In order to achieve an ideal image that possesses rich texture information, prominent target information, and also retains the overall structure of source images, a modified structural similarity loss [46] L_{ssim} is introduced into L_{total}, defined as follows:

$$L_{\text{ssim}} = \begin{cases} 1 - SSIM\left(x_{vi}, x_f\right) \\ \text{if } \sigma^2(x_{vi}) > \sigma^2(x_{ir}) \\ 1 - SSIM\left(x_{ir}, x_f\right) \\ \text{if } \sigma^2(x_{ir}) >= \sigma^2(x_{vi}) \end{cases}, \tag{13}$$

where σ^2 denotes variance. The use of SSIM provides a metric for quantifying the similarity between two images, where a higher value corresponds to a stronger similarity between the images.

Furthermore, to assist the network in learning more discriminative feature representations and enhance the fusion image quality, we introduce the triplet loss L_{tri}.

$$L_{\text{tri}} = \frac{1}{n}\sum_{i=1}^{3} max(d(f_i, f_i^{pos}) - d(f_i, f_i^{neg}) + b, 0), \tag{14}$$

where the value of n is 3, representing the count of convolutional blocks involved; $d(\cdot)$ denotes the Euclidean distance; $\sum$ stands for summation operator; pos and neg are positive and negative samples; and b represents a parameter and is set to 1.0.

We design an FR module to ensure that the reconstructed source images contain more comprehensive information, thereby optimizing the final fusion result. Accordingly, a reconstruction loss function is introduced, which is defined as follows:

$$\begin{aligned} L_{re} = &\||\nabla x_{ir}| - |\nabla \hat{x}_{v_i}|\|_1 + \||\nabla x_{vi}| - |\nabla \hat{x}_{vi}|\|_1 \\ &+ MSE(x_{ir}, \hat{x}_{ir}) + MSE(x_{vi} - \hat{x}_{vi}), \end{aligned} \tag{15}$$

where $\hat{x}_{ir}$ stands for reconstructed infrared image and $\hat{x}_{vi}$ denotes reconstructed visible image.

4. Experiments and Results

We evaluate the fusion performance of our method by conducting various comparative experiments in this section. Firstly, comprehensive details are presented regarding the experimental datasets, training details, comparison methods, and metrics for evaluation. Then, the exceptional performance of the proposed method is demonstrated through conducting quantitative and qualitative comparisons on different public datasets. Subsequently, we execute ablation studies on each key module to validate their essentiality. Finally, we conduct fusion efficiency experiments. Furthermore, we extend the experiments to object detection, showing that our method enhances the performance of downstream high-level visual tasks.

4.1. Datasets and Training Details

(1) Datasets. *M^3FD:* The M^3FD dataset [5] contains high-resolution visible and infrared image pairs of various object types and scenes. The image pairs cover four typical types in different seasons, daytime, overcast, and night. In our study, we choose 2720 image pairs from this dataset to form our training set. The diversity of M^3FD provides convenience for exploring image fusion algorithms.

TNO: The TNO dataset https://figshare.com/articles/TN_Image_Fusion_Dataset/1008029 (accessed on 30 January 2024), which contains multi-band nighttime images depicting military scenarios, is generally applied in the fusion of visible and infrared. We randomly choose 39 image pairs from it to serve as the test set.

RoadScene: The Roadscene dataset https://github.com/hanna-xu/RoadScene (accessed on 30 January 2024) is composed of 221 aligned pairs of visible and infrared images. These image pairs showcase representative traffic scenes, including pedestrians, traffic signs, vehicles, and roads. The dataset has been formed through careful preprocessing and image registration of some of the most representative scenes from the FILR dataset.

LLVIP: Most of the images in the LLVIP dataset [47] are taken in very dark scenes, making it applicable for low-light vision. This dataset can verify the effectiveness of fusion algorithms under low-light conditions.

(2) Training details. In this work, we utilize 2720 image pairs from the M^3FD dataset for model training. We perform extensive quantitative and qualitative evaluations of our method and comparison methods on the M^3FD, RoadScene, TNO, and LLVIP datasets. To expand the training data, increase data diversity, and mitigate the risk of overfitting [48], we employ random horizontal flipping during the training phase and resize the images to 352 × 352. During the training process, the parameter update employs the Adam optimizer, with a batch size of 16. The learning rate and training epochs are set to 10^{-4} and 100, respectively. The coefficients $\alpha = 50$, $\beta = 50$, and $\gamma = 5$ are determined based on extensive experiments and experience. We employ the PyTorch 1.7 framework to implement our approach and train it on an NVIDIA A100 GPU (NVIDIA, Santa Clara, CA, USA).

4.2. Comparison Methods and Evaluation Metrics

We will assess the fusion images from subjective and objective perspectives to validate the fusion performance of our method. For subjective evaluation, the primary factor is human visual perception. The fusion image's quality is evaluated by the detail richness and target salience, which inherently entails a certain level of subjectivity. On the other hand, objective evaluation methods rely on quantitative metrics to objectively assess image quality.

Comparison methods. We assess the fusion performance of our method by comparing it with twelve state-of-the-art (SOTA) fusion methods (GTF [49], FGAN [8], STDF [24], SwinF [26], TarDAL [5], SeAF [22], SuperF [10], DIDF [19], BDLF [50], SHIP [51], DSF [9], and TCMoA [52]). To ensure fairness, we adhere to the parameter settings from the original publications by the authors in our comparative experiments, without modifying any other configurations.

Evaluation metrics. Given the deficiency of ground truth in IVIF, it is insufficient to rely solely on subjective visual assessments to evaluate the quality of fusion images, particularly when the two fusion images are nearly indistinguishable visually. To make the results more convincing, we select twelve quantitative metrics to comprehensively and objectively evaluate our method, including the sum of the correlation differences (SCD) [53], correlation coefficient (CC) [54], average gradient (AG) [55], visual information fidelity (VIF) [56], mean square error (MSE) [57], peak signal-to-noise ratio (PSNR) [58], edge information (Q_{abf}) [59], multi-scale structural similarity (MSSSIM) [60], structural similarity (SSIM) [61], mutual information (based on wavelet feature (FMI_w), discrete cosine feature (FMI_d), and pixel feature (FMI_p)) [62]. Except for MSE, higher values indicate superior fusion performance.

4.3. Comparison Experiment

We proceed with a comprehensive experiment on the M^3FD dataset to validate the superiority of our method in fusion performance. The following will present both qualitative and quantitative analyses of the results obtained from our method as well as the comparison methods.

Qualitative analysis. Figure 7 shows two groups of qualitative experimental results from the M^3FD dataset. In the first group of images, the visible image fails to provide effective scene details due to contamination. For instance, distinguishing the target person and the car logo inside the red box is challenging in the visible image but clear in the infrared image. The fusion images effectively integrate complementary features from visible and infrared images, but present visual differences across different methods. The results of GTF and FGAN tend towards the infrared image, while STDF, SwinF, TarDAL, SeAF, SuperF, DIDF, and BDLF preserve target information from the source image but lack texture detail. SHIP, DSF, and TCMoA retain significant texture detail but overlook contrast information. Remarkably, the proposed method not only performs well in preserving texture details and salient targets but also has superior visual effects. According to the analysis of the results, our method can effectively utilize infrared image information to complement severely contaminated visible images when they fail to provide scene details. This ability is attributed to CFMR and FFB, which calibrate and supplement the complementary features in source images. Furthermore, our method still demonstrates commendable performance under conditions of low illumination and when targets and backgrounds are similar, as illustrated in the second group in Figure 7. Overall, our method provides more thorough scene data and clearer visuals than other comparison methods.

Figure 7. Qualitative analysis of different methods on the M^3FD dataset. The regions to focus on are marked with red and green rectangles, with magnified views of these regions presented in the lower-left, upper-left, and upper-right corners.

Quantitative analysis. To enhance the credibility of the evaluation results, we conduct a quantitative analysis using twelve evaluation metrics, as illustrated in Figure 8. The testing set comprises 76 image pairs randomly selected from the M^3FD dataset. To make comparison easier, we rank the mean values of the twelve evaluation metrics for different methods to obtain an average ranking, as indicated in Table 1. According to the ranking results, our method excels over others in metrics such as PSNR, MSE, VIF, CC, Q_{abf}, SSIM, MSSSIM, FMI_p, FMI_d, and FMI_w. Regarding SCD, our results outperform all methods except DIDF, and we achieve the third-best in AG. Quantitative analysis demonstrates that our method can ensure the maximum preservation of texture details in visible images while retaining the prominent target in infrared images.

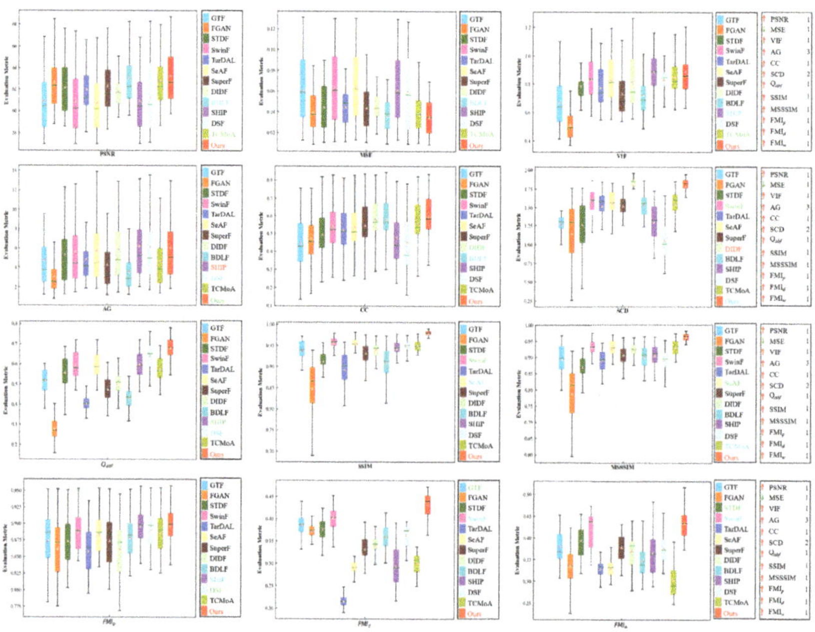

Figure 8. Quantitative analysis of multiple evaluation metrics on the M^3FD dataset. Within each box, the green line indicates the median value, and the orange tangle stands for the mean value. Red serves to highlight the best mean value, blue for the second-best, and green for the third-best. The rankings of the proposed method for each metric are displayed on the right.

Table 1. The ranking results of different metrics on the M^3FD dataset for different methods. Red serves to highlight the best result, blue for the second-best, and green for the third-best.

	PSNR ↑	MSE ↓	VIF ↑	AG ↑	CC ↑	SCD ↑	Q_{abf} ↑	SSIM ↑	MSSSIM ↑	FMI_p ↑	FMI_d ↑	FMI_w ↑
GTF	10	10	12	9	13	10	8	8	9	8	3	4
FGAN	4	4	13	13	11	12	13	13	13	11	5	10
STDF	6	6	8	7	9	11	7	10	12	9	4	3
SwinF	12	12	4	6	6	3	4	3	4	4	2	2
TarDAL	7	8	9	8	8	8	12	12	11	13	13	12
SeAF	13	13	6	5	7	4	5	2	3	6	12	11
SuperF	5	5	10	11	5	7	10	9	6	10	9	5
DIDF	8	7	7	4	3	1	9	7	5	12	8	7
BDLF	2	2	11	12	2	6	11	11	8	7	7	9
SHIP	11	11	2	1	10	9	3	6	7	2	11	8
DSF	9	9	5	2	12	13	2	4	10	3	6	6
TCMoA	3	3	3	10	4	5	6	5	2	5	10	13
Ours	1	1	1	3	1	2	1	1	1	1	1	1

4.4. Generalization Experiment

Generalization assessment plays a vital role in evaluating fusion algorithms. We perform comprehensive experiments on the TNO, RoadScene, and LLVIP datasets to additionally validate the generalization and scalability of our method. The following will present a detailed analysis of the qualitative and quantitative experimental results of different methods on these three datasets.

4.4.1. Results on TNO Dataset

Qualitative analysis. We compare our method against twelve SOTA fusion methods to assess its generalization performance and robustness. Figure 9 presents the fusion results of two representative scenes: one is a smoke scene and the other is a daytime scene with low illumination. To facilitate the observation of the fusion results, we use red and green rectangles to mark some significant targets and texture detail information, respectively, in Figure 9. As observed in Figure 9, all methods can complete the basic fusion task. However, our method exhibits remarkable performance in subjective visual quality and retaining the crucial features in the source image. The fusion images generated by GTF and FGAN tend to infrared images, preserving the significant targets from the infrared images; nevertheless, they fail to integrate the texture detail from the visible images. STDF, SwinF, TarDAL, SeAF, SuperF, DIDF, BDLF, SHIP, and DSF can retain plentiful texture details from the visible images but insufficiently maintain significant targets from the infrared images. TCMoA can maintain the prominent targets from the infrared images and the texture details from the visible images but suffers from low contrast. In comparison, our method has satisfactory visual quality while retaining abundant texture detail from visible images and significant targets from infrared images.

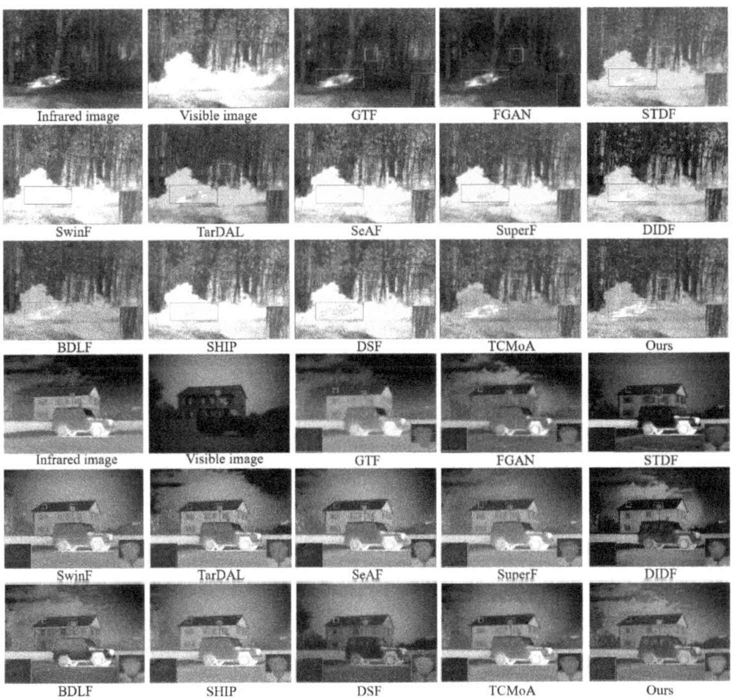

Figure 9. Qualitative analysis of different methods on the TNO dataset. The regions to focus on are marked with red and green rectangles, with magnified views of these regions presented in the lower-right or lower-left corners.

Quantitative analysis. Figure 10 and Table 2 demonstrate the quantitative results of different methods. It is apparent from the results that our method excels over others in terms of PSNR, MSE, CC, SCD, Q_{abf}, SSIM, MSSSIM, FMI_d, and FMI_w, and ranks second only to MPCF on FMI_p. The analysis of these metrics indicates that our fusion results are extremely similar to the source images—with low noise, rich details, and high contrast—and demonstrate excellent visual effects.

Table 2. The ranking results of different metrics on the TNO dataset for different methods. Red serves to highlight the best result, blue for the second-best, and green for the third-best.

	PSNR ↑	MSE ↓	VIF ↑	AG ↑	CC ↑	SCD ↑	Q_{abf} ↑	SSIM ↑	MSSSIM ↑	FMI_p ↑	FMI_d ↑	FMI_w ↑
GTF	10	11	13	12	13	13	12	12	11	9	13	13
FGAN	13	9	12	13	10	12	13	13	13	12	4	6
STDF	8	8	2	4	11	11	7	11	12	10	3	2
SwinF	12	13	5	3	8	5	4	6	7	8	2	3
TarDAL	6	5	6	7	6	6	10	10	8	11	12	10
SeAF	9	10	1	1	4	3	6	7	4	5	10	12
SuperF	4	7	11	11	7	8	8	3	6	7	5	4
DIDF	7	4	8	8	2	2	11	8	3	13	7	5
BDLF	2	2	9	10	3	4	9	9	5	6	6	8
SHIP	11	12	3	2	9	9	3	5	9	1	11	9
DSF	3	3	10	6	12	10	2	2	10	4	8	7
TCMoA	5	6	4	9	5	7	5	4	2	3	9	11
Ours	1	1	7	5	1	1	1	1	1	2	1	1

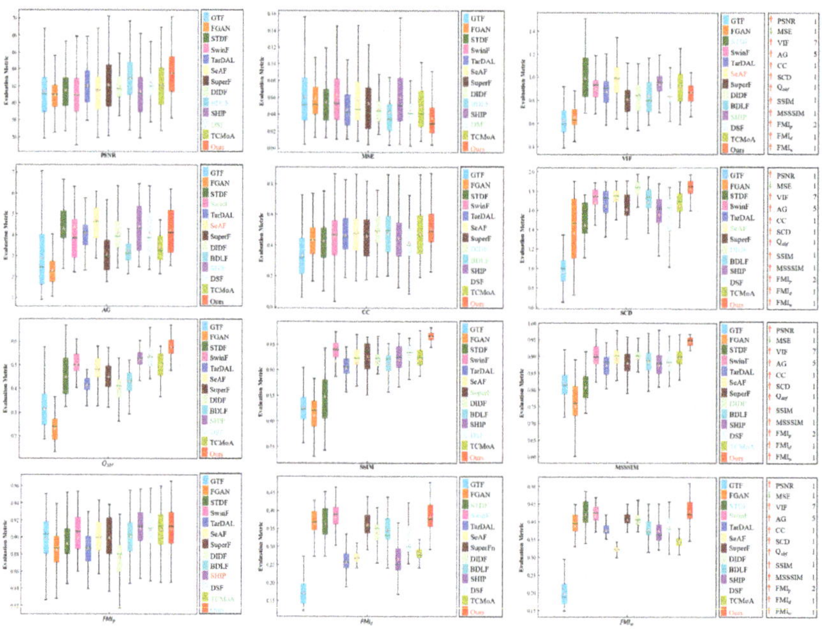

Figure 10. Quantitative analysis of multiple evaluation metrics on the TNO dataset. Within each box, the green line indicates the median value and the orange tangle stands for the mean value. Red serves to highlight the best mean value, blue for the second-best, and green for the third-best. The rankings of the proposed method for each metric are displayed on the right.

4.4.2. Results on RoadScene Dataset

Qualitative analysis. IVIF is instrumental in various traffic applications. Therefore, we choose to test twelve SOTA methods and our method on the RoadScene dataset. Figure 11 displays two sets of the most representative image pairs. The first set is a daytime scene and the second is a night scene with strong light illumination. The details of the utility poles are clearly depicted in our fusion results for the first group of images, while in the second group, even the details of a person concealed in darkness are clearly visible from our result. In comparison, the fusion results from GTF and FGAN are relatively blurry and lose important salient target information, such as the person within the green rectangle of the second set. DIDF, BDLF, and DSF suffer from low contrast, while STDF, SwinF, TarDAL, SeAF, and SuperF lose some texture detail information, marked in red rectangles in Figure 11. While SHIP, TCMoA, and our method retain salient target information and sufficient texture details, SHIP and TCMoA still have some artifacts. Generally, our method maintains essential features from the source images while mitigating interference from useless information, conforming to human visual perception.

Quantitative analysis. Quantitative analysis was conducted on 110 image pairs randomly selected from the RoadScene dataset. Figure 12 and Table 3 present our quantitative results, which reveal that our method excels in multiple evaluation metrics. Specifically, our method excels in MSE, CC, Q_{abf}, SSIM, MSSSIM, and FMI_w, achieving the highest scores, while ranking second in PSNR, SCD, FMI_p, and FMI_d. In general, the proposed method exhibits excellent generalization performance and can still achieve satisfactory fusion results even in environments with small differences between the background and the target and insufficient lighting.

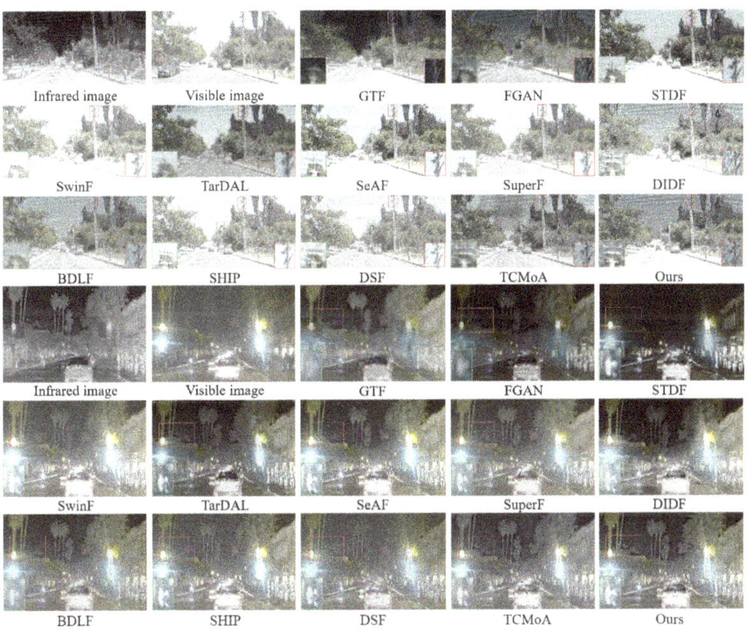

Figure 11. Qualitative analysis of different methods on the RoadScene dataset. The regions to focus on are marked with red and green rectangles, with magnified views of these regions presented in the lower left or lower-right corners.

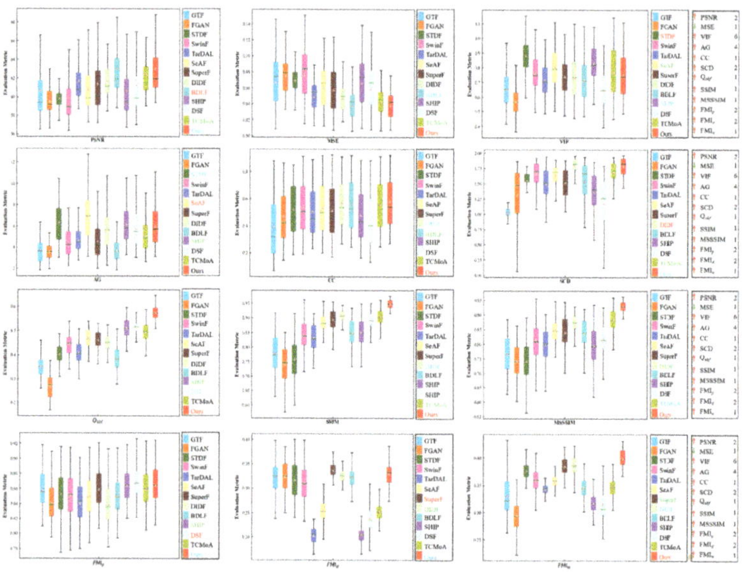

Figure 12. Quantitative analysis of multiple evaluation metrics on the RoadScene dataset. Within each box, the green line indicates the median value and the orange tangle stands for the mean value. Red serves to highlight the best mean value, blue for the second-best, and green for the third-best. The rankings of the proposed method for each metric are displayed on the right.

Table 3. The ranking results of different metrics on the RoadScene dataset for different methods. Red serves to highlight the best result, blue for the second-best, and green for the third-best.

	PSNR ↑	MSE ↓	VIF ↑	AG ↑	CC ↑	SCD ↑	Q_{abf} ↑	SSIM ↑	MSSSIM ↑	FMI_p ↑	FMI_d ↑	FMI_w ↑
GTF	9	10	11	11	13	13	12	11	11	6	4	10
FGAN	13	13	13	13	11	11	13	13	12	11	6	13
STDF	12	11	1	2	7	6	10	12	13	7	7	4
SwinF	11	12	5	10	5	4	8	9	7	9	8	5
TarDAL	4	5	9	8	9	9	9	10	9	12	13	9
SeAF	8	8	3	1	6	5	5	6	4	10	9	6
SuperF	6	6	7	9	8	8	6	4	5	5	1	3
DIDF	5	4	8	6	2	1	7	2	3	13	3	2
BDLF	1	2	10	12	3	7	11	8	6	8	5	8
SHIP	10	9	2	3	10	10	3	7	10	3	12	11
DSF	7	7	12	5	12	12	2	5	8	1	11	12
TCMoA	3	3	4	7	4	3	4	3	2	4	10	7
Ours	2	1	6	4	1	2	1	1	1	2	2	1

4.4.3. Results on LLVIP Dataset

Qualitative analysis. Compared to the fusion of daytime scenes, the low illumination image fusion presents more challenges. When visible images are affected by strong illumination sources and fail to provide sufficient scene information, high-performance fusion methods can effectively utilize infrared images for supplementation. To additionally confirm the superiority of our method, a fusion test on the LLVIP dataset is conducted, with qualitative results depicted in Figure 13. The first group depicts scenes with insufficient illumination at night, while the second group shows nighttime scenes with strong illumination interference. From the displayed results, our method preserves significant targets from the infrared image and the abundant texture details from the visible image while also effectively resisting strong illumination interference to generate high-quality fusion images.

Quantitative analysis. We conduct a quantitative analysis using 96 image pairs randomly selected from the LLVIP dataset. As depicted in Figure 14 and Table 4, the quantitative results demonstrate that our method outperforms others on multiple evaluation metrics. Specifically, it excels in PSNR, MSE, VIF, CC, SCD, Q_{abf}, SSIM, MSSSIM, FMI_p, and FMI_d. On the AG metric, our method is ranked second. On the FMI_w metric, our method ranks fourth. Overall, the proposed method exhibits good generalization ability and robustness, efficiently accomplishing image fusion tasks in various complex environments.

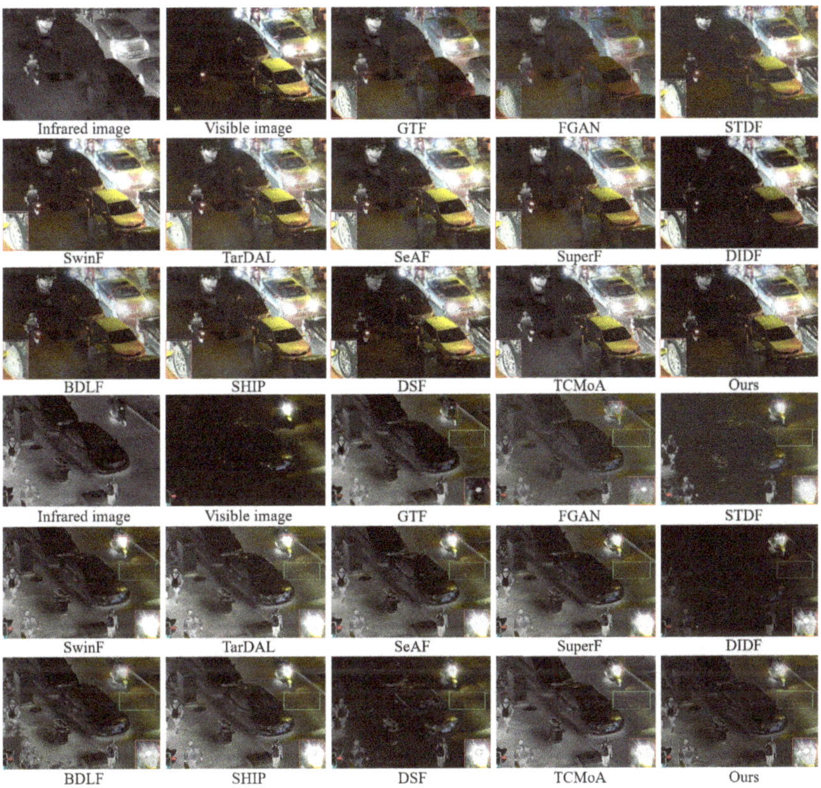

Figure 13. Qualitative analysis of different methods on the LLVIP dataset. The regions to focus on are marked with red and green rectangles, with magnified views of these regions presented in the lower-left or lower-right corners.

Table 4. The ranking results of different metrics on the LLVIP dataset for different methods. Red serves to highlight the best result, blue for the second-best, and green for the third-best.

	PSNR ↑	MSE ↓	VIF ↑	AG ↑	CC ↑	SCD ↑	Q_{abf} ↑	SSIM ↑	MSSSIM ↑	FMI_p ↑	FMI_d ↑	FMI_w ↑
GTF	10	10	8	6	13	11	8	7	9	10	3	1
FGAN	5	5	13	13	10	13	13	13	13	13	11	13
STDF	4	4	11	9	12	12	7	9	12	8	4	7
SwinF	7	7	3	4	3	2	3	2	3	3	2	3
TarDAL	13	13	10	8	9	8	11	11	10	12	13	11
SeAF	11	12	5	5	2	3	5	4	5	5	9	2
SuperF	6	6	9	11	5	7	9	8	7	6	8	10
DIDF	12	11	12	10	6	5	12	12	11	11	10	12
BDLF	2	2	6	12	8	9	10	10	8	9	5	9

Table 4. *Cont.*

	PSNR ↑	MSE ↓	VIF ↑	AG ↑	CC ↑	SCD ↑	Q_{abf} ↑	SSIM ↑	MSSSIM ↑	FMI_p ↑	FMI_d ↑	FMI_w ↑
SHIP	9	9	4	1	7	6	2	3	4	2	7	8
DSF	3	3	7	3	11	10	4	5	6	4	6	5
TCMoA	8	8	2	7	4	4	6	6	2	7	12	6
Ours	1	1	1	2	1	1	1	1	1	1	1	4

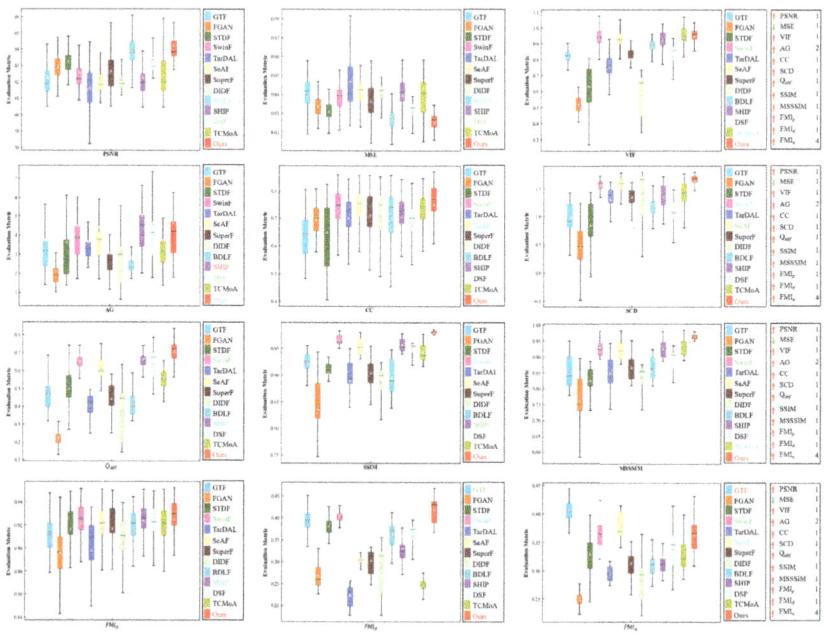

Figure 14. Quantitative analysis of multiple evaluation metrics on the LLVIP dataset. Within each box, the green line indicates the median value and the orange tangle stands for the mean value. Red serves to highlight the best mean value, blue for the second-best, and green for the third-best. The rankings of the proposed method for each metric are displayed on the right.

4.5. Ablation Experiment

We will evaluate the effectiveness of each module in this section. The proposed method comprises three crucial modules, namely, CMFC, FFB, and HPDCB. In our ablation experiments, we verify different combinations of these modules. Additionally, we perform ablation experiments related to FR and l_{tri} loss.

Quantitative analysis. As detailed in Table 5, seven groups of ablation experiments were executed to assess the necessity of our proposed modules. According to the quantitative indicators from Table 5, our method exhibits excellent performance in VIF, AG, CC, SCD, and FMI_w, and ranks second in Q_{abf}, FMI_p, and FMI_d. Overall, every module we proposed contributes to improving the quality of the fusion result.

Qualitative analysis. Figure 15 presents the qualitative results of two groups of ablation experiments. It is evident from the displayed results that the fusion results of M1, M2, M3, and M4 exhibit some blurring, while the contrast of M5 and M6 is also lower compared to our method. Generally speaking, our method excels in maintaining high contrast and complete information while also demonstrating excellent clarity compared to other incomplete combination modules.

Table 5. The quantitative analysis of various modules on the M^3FD dataset. Red serves to highlight the best result, blue for the second-best, and green for the third-best.

Model	M1	M2	M3	M4	M5	M6	M7 (Ours)
CMFC	×	×	✓	✓	✓	✓	✓
FFB	×	✓	×	✓	✓	✓	✓
HPDCB	×	✓	✓	×	✓	✓	✓
FR	×	✓	✓	✓	×	✓	✓
L_{tri}	×	✓	✓	✓	✓	×	✓
PSNR ↑	63.6507	63.4210	63.5573	63.4941	63.3877	63.4001	63.3531
MSE ↓	0.0329	0.0341	0.0334	0.0336	0.0342	0.0342	0.0346
VIF ↑	0.7298	0.8480	0.7603	0.7797	0.8410	0.8491	0.8977
AG ↑	5.9649	6.0204	5.9805	5.9546	5.9800	5.9264	6.0230
CC ↑	0.5946	0.5807	0.5930	0.5984	0.5883	0.5880	0.5995
SCD ↑	1.6584	1.6623	1.6962	1.7565	1.7076	1.7179	1.7835
Q_{abf} ↑	0.6147	0.6857	0.6160	0.6248	0.6766	0.6697	0.6803
SSIM ↑	0.9804	0.9812	0.9800	0.9800	0.9809	0.9801	0.9797
MSSSIM ↑	0.9643	0.9642	0.9674	0.9678	0.9640	0.9638	0.9638
FMI_p ↑	0.8886	0.8988	0.8877	0.8883	0.8973	0.8970	0.8977
FMI_d ↑	0.3684	0.4158	0.3871	0.3804	0.4236	0.4337	0.4317
FMI_w ↑	0.3832	0.4246	0.3920	0.3991	0.4264	0.4301	0.4341

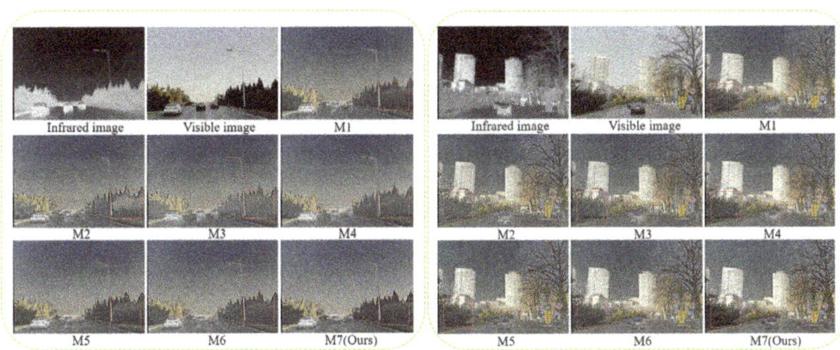

Figure 15. The qualitative results of ablation experiments.

4.6. Efficiency Comparison

The average runtime of various fusion methods on the TNO dataset is presented in Table 6. We conduct all experiments on the same device to ensure fairness. The result from Table 6 reveals that the proposed method surpasses GTF, FGAN, STDF, SwinF, BDLF, TarDAL, SuperF, DIDF, SHIP, and TCMoA in runtime. Overall, the proposed method demonstrates high fusion efficiency while ensuring fusion performance.

Table 6. The average runtime of various fusion methods on the TNO dataset.

Methods	**GTF**	**FGAN**	**STDF**	**SwinF**	**TarDAL**	**SeAF**	**SuperF**
time (s)	2.8868	0.4853	0.3270	2.4847	0.7605	0.0989	0.2218
Methods	DIDF	BDLF	SHIP	DSF	TCMoA	Ours	-
time (s)	1.4027	0.2110	0.4293	0.1614	1.0688	0.2025	-

4.7. Fusion for Object Detection

IVIF lacks ground truth, often necessitating qualitative and quantitative analyses to assess algorithm performance. However, it remains uncertain as to whether fusion images enhance performance in downstream visual tasks (e.g., object detection). To explore this question, experiments were conducted on the M^3FD dataset. To ensure fairness, we employ the YOLOv5 framework to perform object detection on visible, infrared, and fusion images.

Qualitative analysis. To facilitate the observation of differences in object detection performance between various fusion images and source images, two groups of visualized object detection examples are presented (see Figure 16). In the first scenario, YOLOv5 can accurately detect people in the images. However, the results suggest that our fusion results significantly enhance the image clarity and the prominence of the person, leading to higher detection accuracy compared to other fusion images and the source image. In the second scenario, YOLOv5 loses some important target information, such as people, when detecting the source images and the fusion images from other SOTA methods. Our fusion images detect more comprehensive targets, demonstrating that our method effectively integrates useful information from source images, thereby improving the precision of target detection. Our method achieves higher detection precision relative to the twelve SOTA fusion methods.

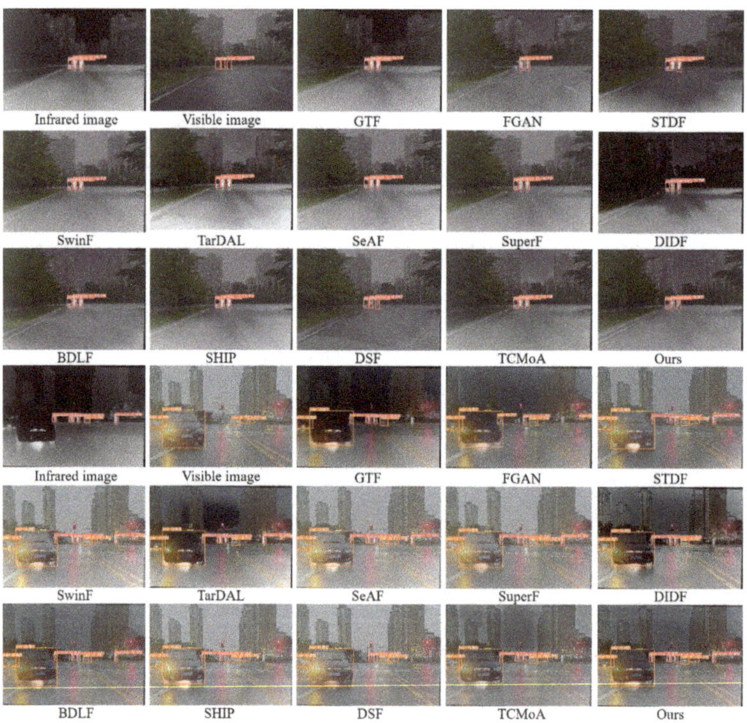

Figure 16. Visualization of object detection results for different images on the M^3FD dataset.

Quantitative analysis. As indicated in Table 7, we calculate the average precision for each fusion method to further assess their detection performance. The results demonstrate that our method obtains the best detection accuracy. This further validates that the proposed method positively impacts practical object detection tasks.

Table 7. The detection precision of different images. Red serves to highlight the best result, blue for the second-best, and green for the third-best. VI and IR represent the visible and infrared images.

	Precision	AP@0.5	mAP@[0.5:0.95]
VI	0.6737	0.5979	0.3916
IR	0.6027	0.5305	0.3003
GTF	0.5351	0.5492	0.3411
FGAN	0.5338	0.5117	0.3148
STDF	0.5760	0.5632	0.3512
SwinF	0.5984	0.5783	0.3716
TarDAL	0.6236	0.5997	0.3820
SeAF	0.5835	0.5772	0.3659
SuperF	0.5699	0.5365	0.3367
DIDF	0.6256	0.5941	0.3774
BDLF	0.6124	0.5751	0.3597
SHIP	0.5619	0.5316	0.3347
DSF	0.6596	0.6421	0.4279
TCMoA	0.6220	0.5890	0.3752
Ours	0.6995	0.6533	0.4341

5. Conclusions

In this study, we propose BCMFIFuse—a network based on bilateral cross-modal feature interaction for IVIF. Firstly, to effectively extract features from the source image, we construct a dual-stream feature extraction network. Next, a CMFC is introduced to calibrate the features of the current modality to better extract complementary features from different modalities. Subsequently, we employ an FFB to effectively integrate the calibrated features. The FFB is built using a cross-attention mechanism, which can realize long-range contextual interaction, thereby enhancing global bilateral modal features. Finally, to ensure the continuity of features and minimize feature loss during transmission, we use shortcut connections between the encoder and decoder. Additionally, for gathering specific and diversified context information and capturing long-range dependencies in isolated areas, we design an HPDCB. Comparison and generalization experiments on multiple datasets indicate that our method has certain advantages in quantitative and qualitative aspects. Furthermore, the evaluation of object detection performance can also reflect the superiority of our method. In upcoming research, we plan to continue optimizing our algorithm to improve fusion efficiency. Moreover, we consider integrating fusion tasks with other high-level visual tasks or modal information (e.g., text information).

Author Contributions: Design and verify experiments, analyze experimental data, and write and revise the paper, X.G.; supervision, S.L. All authors have read and agreed to the published version of the manuscript.

Funding: This study was supported by the National Natural Science Foundation of China under Grant 62072328.

Data Availability Statement: The M^3FD dataset can be download from https://github.com/JinyuanLiu-CV/TarDAL. The TNO dataset can be download from https://figshare.com/articles/dataset/TNO_Image_Fusion_Dataset/1008029. The RoadScene dataset can be download from https://github.com/hanna-xu/RoadScene. The LLVIP dataset can be download from https://bupt-ai-cz.github.io/LLVIP/.

Conflicts of Interest: The authors declare no conflicts of interest.

References

1. Wang, D.; Liu, J.; Liu, R.; Fan, X. An interactively reinforced paradigm for joint infrared-visible image fusion and saliency object detection. *Inf. Fusion* **2023**, 98, 101828. [CrossRef]
2. Rao, Y.; Wu, D.; Han, M.; Wang, T.; Yang, Y.; Lei, T.; Xing, L. AT-GAN: A generative adversarial network with attention and transition for infrared and visible image fusion. *Inf. Fusion* **2023**, 92, 336–349. [CrossRef]
3. Wei, Q.; Liu, Y.; Jiang, X.; Zhang, B.; Su, Q.; Yu, M. DDFNet-A: Attention-Based Dual-Branch Feature Decomposition Fusion Network for Infrared and Visible Image Fusion. *Remote Sens.* **2024**, 16, 1795. [CrossRef]

4. Liu, W.; Yang, J.; Zhao, J.; Guo, F. A Dual-Domain Super-Resolution Image Fusion Method with SIRV and GALCA Model for PolSAR and Panchromatic Images. *IEEE Trans. Geosci. Remote Sens.* **2021**, *60*, 5218814. [CrossRef]
5. Liu, J.; Fan, X.; Huang, Z.; Wu, G.; Liu, R.; Zhong, W.; Luo, Z. Target-aware dual adversarial learning and a multi-scenario multi-modality benchmark to fuse infrared and visible for object detection. In Proceedings of the IEEE/CVF Conference on Computer Vision and Pattern Recognition (CVPR), New Orleans, LA, USA, 18–24 June 2022; pp. 5802–5811.
6. Chen, C.; Wang, C.; Liu, B.; He, C.; Cong, L.; Wan, S. Edge intelligence empowered vehicle detection and image segmentation for autonomous vehicles. *IEEE Trans. Intell. Transp. Syst.* **2023**, *24*, 13023–13034. [CrossRef]
7. Zhang, X.; Wang, L.; Zhang, G.; Lan, T.; Zhang, H.; Zhao, L.; Li, J.; Zhu, L.; Liu, H. RI-Fusion: 3D object detection using enhanced point features with range-image fusion for autonomous driving. *IEEE Trans. Instrum. Meas.* **2022**, *72*, 5004213. [CrossRef]
8. Ma, J.; Yu, W.; Liang, P.; Li, C.; Jiang, J. FusionGAN: A generative adversarial network for infrared and visible image fusion. *Inf. Fusion* **2019**, *48*, 11–26. [CrossRef]
9. Liu, K.; Li, M.; Chen, C.; Rao, C.; Zuo, E.; Wang, Y.; Yan, Z.; Wang, B.; Chen, C.; Lv, X. DSFusion: Infrared and visible image fusion method combining detail and scene information. *Pattern Recogn.* **2024**, *154*, 110633. [CrossRef]
10. Tang, L.; Deng, Y.; Ma, Y.; Huang, J.; Ma, J. SuperFusion: A versatile image registration and fusion network with semantic awareness. *IEEE/CAA J. Autom. Sin.* **2022**, *9*, 2121–2137. [CrossRef]
11. Liu, C.H.; Qi, Y.; Ding, W.R. Infrared and visible image fusion method based on saliency detection in sparse domain. *Infrared Phys. Technol.* **2017**, *83*, 94–102. [CrossRef]
12. Xing, Y.; Zhang, Y.; Yang, S.; Zhang, Y. Hyperspectral and multispectral image fusion via variational tensor subspace decomposition. *IEEE Geosci. Remote Sens. Lett.* **2021**, *19*, 5001805. [CrossRef]
13. Zhang, Q.; Liu, Y.; Blum, R.S.; Han, J.; Tao, D. Sparse representation based multi-sensor image fusion for multi-focus and multi-modality images: A review. *Inf. Fusion* **2018**, *40*, 57–75. [CrossRef]
14. Dogra, A.; Goyal, B.; Agrawal, S. From multi-scale decomposition to non-multi-scale decomposition methods: A comprehensive survey of image fusion techniques and its applications. *IEEE Access* **2017**, *5*, 16040–16067. [CrossRef]
15. Yan, L.; Cao, J.; Rizvi, S.; Zhang, K.; Hao, Q.; Cheng, X. Improving the performance of image fusion based on visual saliency weight map combined with CNN. *IEEE Access* **2020**, *8*, 59976–59986. [CrossRef]
16. Ma, J.; Xu, H.; Jiang, J.; Mei, X.; Zhang, X.P. DDcGAN: A dual-discriminator conditional generative adversarial network for multi-resolution image fusion. *IEEE Trans. Image Process.* **2020**, *29*, 4980–4995. [CrossRef] [PubMed]
17. Chang, L.; Huang, Y.; Li, Q.; Zhang, Y.; Liu, L.; Zhou, Q. DUGAN: Infrared and visible image fusion based on dual fusion paths and a U-type discriminator. *Neurocomputing* **2024**, *578*, 127391. [CrossRef]
18. Li, H.; Wu, X.J. DenseFuse: A fusion approach to infrared and visible images. *IEEE Trans. Image Process.* **2018**, *28*, 2614–2623. [CrossRef] [PubMed]
19. Zhao, Z.; Xu, S.; Zhang, C.; Liu, J.; Li, P.; Zhang, J. DIDFuse: Deep image decomposition for infrared and visible image fusion. In Proceedings of the Twenty-Ninth International Conference on International Joint Conferences on Artificial Intelligence, Yokohama, Japan, 7–15 January 2021; pp. 970–976.
20. Liu, X.; Gao, H.; Miao, Q.; Xi, Y.; Ai, Y.; Gao, D. MFST: Multi-modal feature self-adaptive transformer for infrared and visible image fusion. *Remote Sens.* **2022**, *14*, 3233. [CrossRef]
21. Li, H.; Wu, X.J.; Durrani, T. NestFuse: An infrared and visible image fusion architecture based on nest connection and spatial/channel attention models. *IEEE Trans. Instrum. Meas.* **2020**, *69*, 9645–9656. [CrossRef]
22. Tang, L.; Yuan, J.; Ma, J. Image fusion in the loop of high-level vision tasks: A semantic-aware real-time infrared and visible image fusion network. *Inf. Fusion* **2022**, *82*, 28–42. [CrossRef]
23. Tang, L.; Xiang, X.; Zhang, H.; Gong, M.; Ma, J. DIVFusion: Darkness-free infrared and visible image fusion. *Inf. Fusion* **2023**, *91*, 477–493. [CrossRef]
24. Ma, J.; Tang, L.; Xu, M.; Zhang, H.; Xiao, G. STDFusionNet: An infrared and visible image fusion network based on salient target detection. *IEEE Trans. Instrum. Meas.* **2021**, *70*, 5009513. [CrossRef]
25. Tang, L.; Yuan, J.; Zhang, H.; Jiang, X.; Ma, J. PIAFusion: A progressive infrared and visible image fusion network based on illumination aware. *Inf. Fusion* **2022**, *83*, 79–92. [CrossRef]
26. Ma, J.; Tang, L.; Fan, F.; Huang, J.; Mei, X.; Ma, Y. SwinFusion: Cross-domain long-range learning for general image fusion via swin transformer. *IEEE/CAA J. Autom. Sin.* **2022**, *9*, 1200–1217. [CrossRef]
27. Wang, Z.; Chen, Y.; Shao, W.; Li, H.; Zhang, L. SwinFuse: A residual swin transformer fusion network for infrared and visible images. *IEEE Trans. Instrum. Meas.* **2022**, *71*, 5016412. [CrossRef]
28. Xu, H.; Ma, J.; Le, Z.; Jiang, J.; Guo, X. FusionDN: A unified densely connected network for image fusion. *Proc. AAAI Conf. Artif. Intell.* **2020**, *34*, 12484–12491. [CrossRef]
29. Ma, J.; Zhang, H.; Shao, Z.; Liang, P.; Xu, H. GANMcC: A generative adversarial network with multiclassification constraints for infrared and visible image fusion. *IEEE Trans. Instrum. Meas.* **2020**, *70*, 5005014. [CrossRef]
30. Zhang, H.; Ma, J. SDNet: A versatile squeeze-and-decomposition network for real-time image fusion. *Int. J. Comput. Vis.* **2021**, *129*, 2761–2785. [CrossRef]
31. Zhang, H.; Xu, H.; Xiao, Y.; Guo, X.; Ma, J. Rethinking the image fusion: A fast unified image fusion network based on proportional maintenance of gradient and intensity. *Proc. AAAI Conf. Artif. Intell.* **2020**, *34*, 12797–12804. [CrossRef]

32. Li, H.; Wu, X.J.; Kittler, J. RFN-Nest: An end-to-end residual fusion network for infrared and visible images. *Inf. Fusion* **2021**, *73*, 72–86. [CrossRef]
33. Pu T.; Ni, G. Contrast-based image fusion using the discrete wavelet transform. *Opt. Eng.* **2000**, *39*, 2075–2082. [CrossRef]
34. Zhao, W.; Xu, Z.; Zhao, J. Gradient entropy metric and p-laplace diffusion constraint-based algorithm for noisy multispectral image fusion. *Inf. Fusion* **2016**, *27*, 138–149. [CrossRef]
35. Zhao, Z.; Xu, S.; Zhang, C.; Liu, J.; Zhang, J. Efficient and interpretable infrared and visible image fusion via algorithm unrolling. *arXiv* **2020**, arXiv:2005.05896.
36. Jian, L.; Yang, X.; Liu, Z.; Jeon, G.; Gao, M.; Chisholm, D. SEDRFuse: A symmetric encoder–decoder with residual block network for infrared and visible image fusion. *IEEE Trans. Instrum. Meas.* **2020**, *70*, 5002215. [CrossRef]
37. Wang, Z.; Wu, Y.; Wang, J.; Xu, J.; Shao, W. Res2Fusion: Infrared and visible image fusion based on dense Res2net and double nonlocal attention models. *IEEE Trans. Instrum. Meas.* **2022**, *71*, 5005012. [CrossRef]
38. Long, Y.; Jia, H.; Zhong, Y.; Jiang, Y.; Jia, Y. RXDNFuse: A aggregated residual dense network for infrared and visible image fusion. *Inf. Fusion* **2021**, *69*, 128–141. [CrossRef]
39. Xu, H.; Ma, J.; Jiang, J.; Guo, X.; Ling, H. U2Fusion: A unified unsupervised image fusion network. *IEEE Trans. Pattern Anal. Mach. Intell.* **2020**, *44*, 502–518. [CrossRef] [PubMed]
40. Ma, J.; Liang, P.; Yu, W.; Chen, C.; Guo, X.; Wu, J.; Jiang, J. Infrared and visible image fusion via detail preserving adversarial learning. *Inf. Fusion* **2020**, *54*, 85–98. [CrossRef]
41. Zhang, H.; Yuan, J.; Tian, X.; Ma, J. GAN-FM: Infrared and visible image fusion using GAN with full-scale skip connection and dual Markovian discriminators. *IEEE Trans. Comput. Imaging* **2021**, *7*, 1134–1147. [CrossRef]
42. Li, J.; Huo, H.; Li, C.; Wang, R.; Feng, Q. AttentionFGAN: Infrared and visible image fusion using attention-based generative adversarial networks. *IEEE Trans. Multimed.* **2020**, *23*, 1383–1396. [CrossRef]
43. Yang, Y.; Liu, J.; Huang, S.; Wan, W.; Wen, W.; Guan, J. Infrared and visible image fusion via texture conditional generative adversarial network. *IEEE Trans. Circuits Syst. Video Technol.* **2021**, *31*, 4771–4783. [CrossRef]
44. Tang, W.; He, F.; Liu, Y. YDTR: Infrared and Visible Image Fusion via Y-shape Dynamic Transformer. *IEEE Trans. Multimed.* **2022**, *25*, 5413–5428. [CrossRef]
45. Ghosh, S.; Gavaskar, R.G.; Chaudhury, K.N. Saliency guided image detail enhancement. In Proceedings of the 2019 National Conference on Communications (NCC), Bangalore, India, 20–23 February 2019; pp. 1–6.
46. Rao, D.; Xu, T.; Wu, X.-J. TGFuse: An infrared and visible image fusion approach based on transformer and generative adversarial network. *IEEE Trans. Image Process.* **2023**, 1–12. [CrossRef] [PubMed]
47. Jia, X.; Zhu, C.; Li, M.; Tang, W.; Zhou, W. LLVIP: A visible-infrared paired dataset for low-light vision. In Proceedings of the IEEE/CVF International Conference on Computer Vision, Montreal, QC, Canada, 10–17 October 2021; pp. 3496–3504.
48. Li, L.; Xia, Z.; Han, H.; He, G.; Roli, F.; Feng, X. Infrared and visible image fusion using a shallow CNN and structural similarity constraint. *IET Image Process.* **2020**, *14*, 3562–3571. [CrossRef]
49. Ma, J.; Chen, C.; Li, C.; Huang, J. Infrared and visible image fusion via gradient transfer and total variation minimization. *Inf. Fusion* **2016**, *31*, 100–109. [CrossRef]
50. Liu, Z.; Liu, J.; Wu, G.; Ma, L.; Fan, X.; Liu, R. Bi-level dynamic learning for jointly multi-modality image fusion and beyond. *arXiv* **2023**, arXiv:2305.06720.
51. Zheng, N.; Zhou, M.; Huang, J.; Hou, J.; Li, H.; Xu, Y.; Zhao, F. Probing Synergistic High-Order Interaction in Infrared and Visible Image Fusion. In Proceedings of the IEEE/CVF Conference on Computer Vision and Pattern Recognition, Seattle, WA, USA, 17–21 June 2024; pp. 26384–26395.
52. Zhu, P.; Sun, Y.; Cao, B.; Hu, Q. Task-customized mixture of adapters for general image fusion. In Proceedings of the IEEE/CVF Conference on Computer Vision and Pattern Recognition, Seattle, WA, USA, 17–21 June 2024; pp. 7099–7108.
53. Aslantas, V.; Bendes, E. A new image quality metric for image fusion: The sum of the correlations of differences. *AEU-Int. J. Electron. C* **2015**, *69*, 1890–1896. [CrossRef]
54. Deshmukh, M.; Bhosale, U. Image fusion and image quality assessment of fused images. *Int. J. Image Process. (IJIP)* **2010**, *4*, 484.
55. Zhao, W.; Wang, D.; Lu, H. Multi-focus image fusion with a natural enhancement via a joint multi-level deeply supervised convolutional neural network. *IEEE Trans. Circuits Syst. Video Technol.* **2018**, *29*, 1102–1115. [CrossRef]
56. Zhang, X.; Ye, P.; Xiao, G. VIFB: A visible and infrared image fusion benchmark. In Proceedings of the IEEE/CVF Conference on Computer Vision and Pattern Recognition Workshops, Seattle, WA, USA, 14–19 June 2020; pp. 104–105.
57. Ma, J.; Ma, Y.; Li, C. Infrared and visible image fusion methods and applications: A survey. *Inf. Fusion* **2019**, *45*, 153–178. [CrossRef]
58. Poobathy, D.; Chezian, R.M. Edge detection operators: Peak signal to noise ratio based comparison. *Int. J. Image Graph. Signal Process.* **2014**, *10*, 55–61. [CrossRef]
59. Petrovic, V.; Xydeas, C. Objective image fusion performance characterisation. In Proceedings of the Tenth IEEE International Conference on Computer Vision, Beijing, China, 17–21 October 2005; Volume 1, pp. 1866–1871.
60. Wang, Z.; Simoncelli, E.P.; Bovik, A.C. Multiscale structural similarity for image quality assessment. In Proceedings of the Thrity-Seventh Asilomar Conference on Signals, Systems & Computers, Pacific Grove, CA, USA, 9–12 November 2003; Volume 2, pp. 1398–1402.

61. Wang, Z.; Bovik, A.C.; Sheikh, H.R.; Simoncelli, E.P. Image quality assessment: From error visibility to structural similarity. *IEEE Trans. Image Process.* **2004**, *13*, 600–612. [CrossRef] [PubMed]
62. Haghighat, M.B.A.; Aghagolzadeh, A.; Seyedarabi, H. A non-reference image fusion metric based on mutual information of image features. *Comput. Electr. Eng.* **2011**, *37*, 744–756. [CrossRef]

remote sensing

MDPI

Article

CPROS: A Multimodal Decision-Level Fusion Detection Method Based on Category Probability Sets

Can Li, Zhen Zuo *, Xiaozhong Tong, Honghe Huang, Shudong Yuan and Zhaoyang Dang

College of Intelligence Science and Technology, National University of Defense Technology, Changsha 410073, China; lican22@nudt.edu.cn (C.L.); tongxiaozhong@nudt.edu.cn (X.T.); huanghonghe16@nudt.edu.cn (H.H.); yuanshudong21@nudt.edu.cn (S.Y.); dzy0329@nudt.edu.cn (Z.D.)

* Correspondence: z.zuo@nudt.edu.cn

Abstract: Images acquired by different sensors exhibit different characteristics because of the varied imaging mechanisms of sensors. The fusion of visible and infrared images is valuable for specific image applications. While infrared images provide stronger object features under poor illumination and smoke interference, visible images have rich texture features and color information about the target. This study uses dual optical fusion as an example to explore fusion detection methods at different levels and proposes a multimodal decision-level fusion detection method based on category probability sets (CPROS). YOLOv8—a single-mode detector with good detection performance—was chosen as the benchmark. Next, we innovatively introduced the improved Yager formula and proposed a simple non-learning fusion strategy based on CPROS, which can combine the detection results of multiple modes and effectively improve target confidence. We validated the proposed algorithm using the VEDAI public dataset, which was captured from a drone perspective. The results showed that the mean average precision (mAP) of YOLOv8 using the CPROS method was 8.6% and 16.4% higher than that of the YOLOv8 detection single-mode dataset. The proposed method significantly reduces the missed detection rate (MR) and number of false detections per image (FPPI), and it can be generalized.

Keywords: improved Yager formula; category probability set; fusion strategy; multimodal

1. Introduction

Target detection is a typical computer vision problem, and because of their broad application potential in image processing and pattern recognition, target detectors have been widely used in various safety systems over the past few decades, such as safe autopilot and drone detection [1–3]. To overcome the impact of smoke interference and illumination damage during the detection of ground targets by UAV [4,5], we studied the use of visible and infrared dual-light pods for multimodal target detection as they can provide stronger object features in the case of smoke interference and insufficient illumination [6].

The core problem of multimodal detection is multimodal information fusion, which can be divided into pixel-, feature-, and decision-level fusion according to different multimodal fusion stages [7], as shown in Figure 1. Pixel-level fusion forms a four-channel output by superimposing three-channel RGB and one-channel IR, which then produces the detection result through the detector [8]. Feature-level fusion involves inputting visible and infrared images into the feature extractor and then merging the extracted features into the subsequent detection network [8,9]. Decision-level fusion refers to the fusion of detection results obtained by separately detecting visible and infrared modalities using a fusion decision based on mathematical theorems [10].

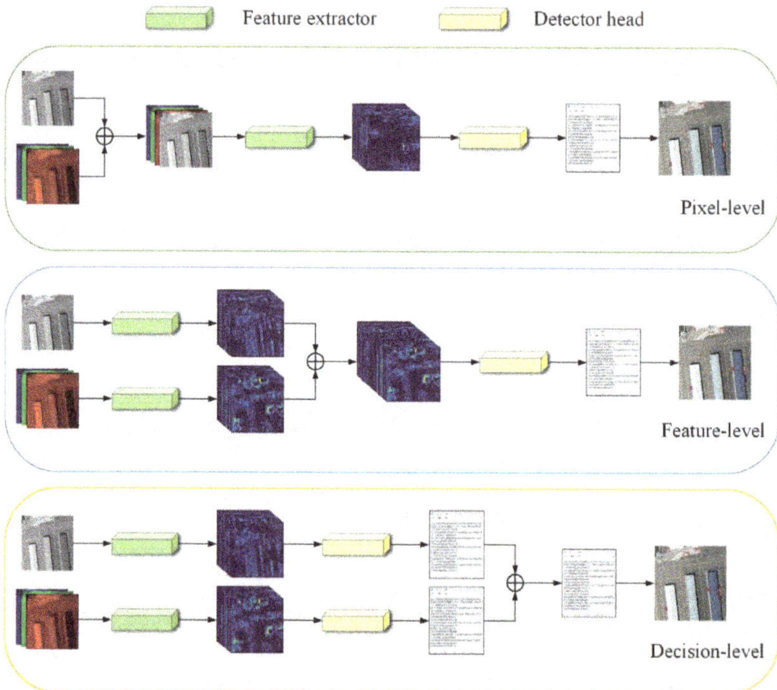

Figure 1. Schematic diagram of fusion in different stages.

In practical applications, pixel- and feature-level fusion often require a large number of image pairs of different modes with high alignment accuracy for training, which results in a significant workload [11,12], and the advantages of non-learning decision-level fusion are self-evident. Figure 2 illustrates the fusion strategy for multimodal detection by fusing the detection results of the single-mode detector. (a) A simple method involves averaging the scores of overlapping tests; however, this results in lower confidence scores for the test results [13]. (b) To avoid score degradation, non-maximum suppression (NMS) can be applied to suppress overlapping detections from different modes, which always obtains the detection with the highest confidence. Although NMS is a simple and effective fusion strategy, it can only avoid a reduction in confidence scores and cannot effectively improve it [14]. (c) A strategy that can effectively enhance confidence scores is the D-S evidence theory, which can improve the multimodal detection and detection accuracy of targets [15,16]. However, D-S can only improve the scores of test results with high confidence and cannot be applied to test results with low confidence. (d) Therefore, a simple fusion strategy—CPROS—is derived in this study; it fuses a set of category probabilities of the test results to improve the scores for tests with strong evidence from multiple modes.

The aim of this study is to obtain detection results under different modes through efficient single-mode detectors trained on different modal datasets and to fuse the detection results of different modes according to fusion strategies designed by mathematical theorems to obtain optimal decision results. Our non-learning CPROS is not only more interpretable than fusion algorithms that need to be learned but also better than previous work. Although this concept is simple, its effectiveness is good.

The main contributions of this study are summarized below. (1) A multimodal decision-level fusion detection method (CPROS) based on category probability sets is proposed. Experimental results on benchmark datasets prove the effectiveness of CPROS. (2) Ablation studies were conducted to verify the superiority of the CPROS by combining different confidence and detection box fusion methods. The experimental results show that CPROS can achieve higher detection accuracy than the three most advanced fusion decisions. (3) The proposed decision-level fusion method, CPROS, is applied to different single-mode detectors, and it is found that the multimodal detector using CPROS has better detection performance than the single-mode detector, proving the generality of CPROS. The remainder of this article is organized as follows. The second section describes the recent progress of research on decision-level fusion. In the third section, the two aspects of probability fusion and box fusion are discussed. In the fourth section, the validity, superiority, and generality of the proposed CPROS method are verified using the VEDAI dataset. Finally, the fifth section presents conclusions and prospects for future work.

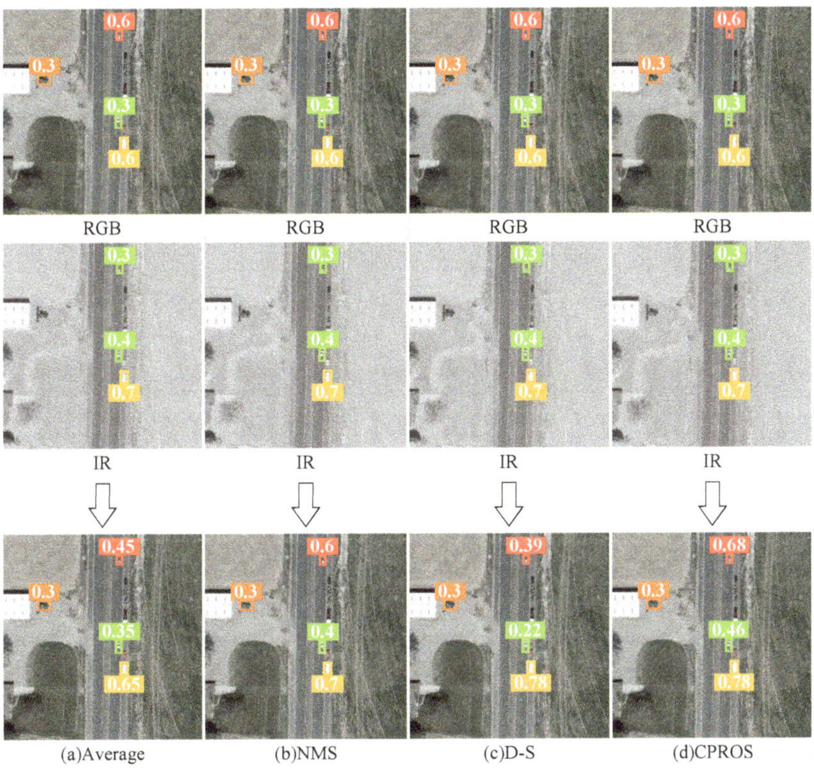

Figure 2. Illustration of various fusion strategies. (**a**) Average; (**b**) NMS; (**c**) D-S; and (**d**) CPROS. A single-mode detector is used to detect visible and infrared images separately, obtaining detection results (category, confidence, and detection box). They are then fused through different fusion strategies to output the final result. Different colors represent various categories.

2. Related Works

2.1. Multimodal Fusion

Currently, research on multimodal fusion mainly focuses on pixel- and feature-level fusion, with relatively little research on decision-level fusion, which requires strict logic. Pixel-level fusion can achieve fine operations on an image in image processing, but the information of each pixel needs to be considered. Therefore, it requires high computational resources and processing power and is easily affected by noise, resulting in unstable results. Additionally, pixel-level fusion uses relatively little spatial information, which may result in a lack of visual coherence and structure in processed images [17]. Feature-level fusion can better capture the abstract information of an image, reduce data dimensions, and improve computational efficiency. Additionally, the semantic information of the images can be better retained, making subsequent image recognition and classification tasks more accurate and reliable [18]. Most importantly, feature-level fusion can better adapt to inputs of different sizes and shapes, has greater generalization ability, and is suitable for a broader range of image-processing tasks [19].

Pixel- and feature-level fusions based on deep learning require a large amount of alignment data for training to achieve good results in practical image-processing tasks [20,21]. Pixel-level fusion involves the consideration of the information of each pixel; therefore, it requires training data that contain rich image details and data samples from various scenes to ensure that the model can accurately capture and fuse pixel-level information. This means that large-scale image datasets are required for training pixel-level fusion models, and it is necessary to ensure the diversity and representativeness of datasets to improve the generalization ability and adaptability of the models. Although feature-level fusion can reduce the data dimension compared with pixel-level fusion, large-scale image data are still needed to train the feature extractor or feature fusion model [22].

In actual situations, the amount of data in different modes is often unequal, sometimes not even of the same order of magnitude, and the need to obtain image pairs of different modes with high alignment accuracy brings more significant challenges to the training of pixel- and feature-level fusion, and the advantages of non-learning decision-level fusion are self-evident.

2.2. Decision-Level Fusion

Decision-level fusion is the fusion of single-mode detection results, which have been the subject of few research studies and are currently difficult to achieve. For multisensor systems, information is diverse and complex; therefore, the basic requirements for decision-level fusion methods are robustness and parallel processing capabilities. Other requirements include the speed, accuracy, and information sampling capabilities of the algorithm. Generally, mathematical methods based on nonlinear systems can be used as decision-level fusion methods if they exhibit fault tolerance, adaptability, associative memory, and parallel processing capabilities [23]. The following mathematical methods are commonly used.

1. Weighted average method [13,24]: The most straightforward and intuitive method is the weighted average method, which weighs the redundant information provided by a group of sensors, and the result is used as the fusion value. This method operates directly utilizing a data source.
2. Multi-Bayesian estimation method [25,26]: Each sensor is regarded as a Bayesian estimator, and the associated probability distribution of each object is synthesized into a joint posterior probability distribution function. By minimizing the likelihood function of the joint distribution function, the final fusion value of the multisensor information is obtained, and a prior model of the fusion information and environment is developed to provide a feature description of the entire environment.
3. D-S evidence reasoning method [27]: This method is an expansion of Bayesian reasoning; its reasoning structure is top-down and divided into three levels. The first level is the target synthesis, and its role is to synthesize the observation results from the independent sensor into a total output result. The second stage is inference, whose function is to obtain the observation results of the sensor, make inferences, and expand the observation results to the target report. The third level is updated, and the sensors are generally subject to random errors. Therefore, a set of successive reports from the same sensor that is sufficiently independent in time is more reliable than any single report. Therefore, before inference and multisensor synthesis, it is necessary to update the sensor observation data.

Additionally, fuzzy set theory [28,29], rough set theory [30], Z-number theory [31], and D-number theory [32] have been proposed. Among the various decision-level fusion methods using mathematical theory proposed at the present stage, a more straightforward decision-level fusion method is used to aggregate the detection of each mode and then weight the average scores of overlapping detection rather than the less inhibited detection, such as non-maximum suppression; however, this operation will inevitably reduce the reported scores compared with NMS. Intuitively, if two patterns agree on candidate detection, the score of one should improve [13,14,24]. For this reason, Chen et al. introduced the D-S evidence theory into fusion decision-making [33]. When detectors of different modes detect the same object, D-S can gracefully deal with the missing information and significantly improve multimodal detection and the detection accuracy of the target. However, D-S can only improve the scores of detection results with high confidence and is not applicable to detection results with low confidence. Moreover, when the categories detected by the detector are inconsistent, no concrete or feasible decision methods are provided.

In summary, all the fusion decision methods mentioned above use only confidence information in the detection results, focus more on the fusion of confidence, and do not provide effective fusion strategies for categories. Additionally, confidence fusion is less effective at low confidence levels. Accordingly, a decision-level fusion method based on a category probability set is proposed in this study, aiming to make full use of the output of a single-mode detector to provide more efficient and interpretable detection results.

3. Fusion Strategies for Multimodal Detection

An overview of the detection pipeline using CPROS is presented in Figure 3. First, the visible and infrared images and their corresponding labels were input into the YOLOv8 detection network for training, and a single-mode detector suitable for different modes with good detection performance was developed. In subsequent practical tasks, the detector can obtain detection results based on the characteristics of the target in various modes, and then we can apply our decision-level fusion method to achieve more accurate detection.

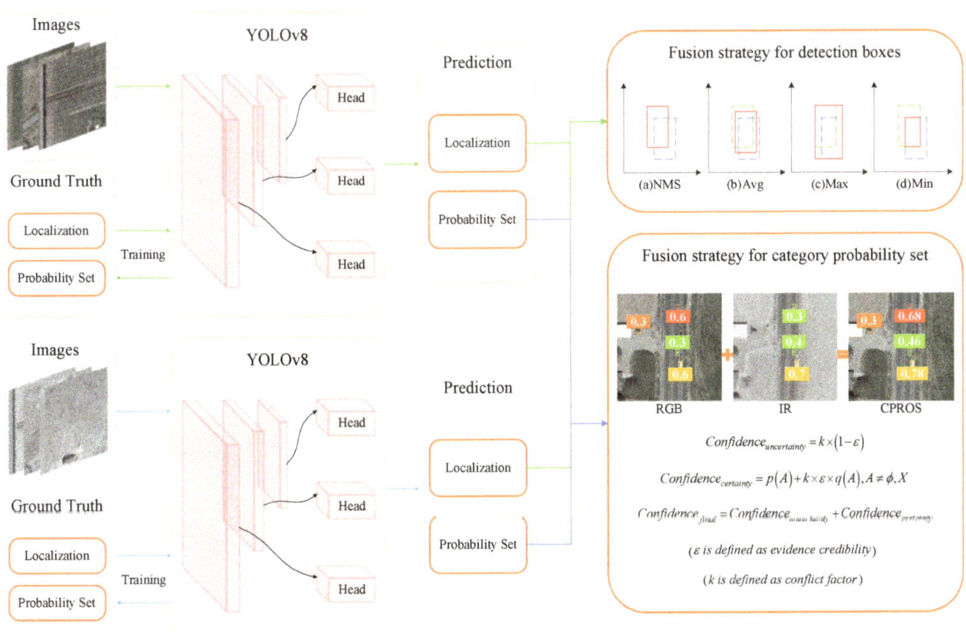

Figure 3. Overall framework of CPROS.

We derive our fusion method, CPROS, from multiple theorems and rules, which combine the advantages of various fusion decisions proposed at this stage and innovatively introduce the improved Yager formula to solve the problem of poor fusion effects under low-confidence conditions [34,35]. Importantly, CPROS can gracefully handle "missing detection" through probabilistic integration. Because the CPROS is a non-learning strategy based on mathematical rules, it does not require multimodal data for training. Therefore, the CPROS is a general-purpose technology used for integrated detectors.

3.1. Fusion Strategy for Detection Boxes

We explored four methods for fusing the detection boxes, as shown in Figure 4. The first method, "NMS", eliminates redundant detection results by comparing the confidence scores of targets in the overlapping region, but it is not applicable to the problem of box size imbalance, which may lead to missing detection of small targets. The second method, "average", is equivalent to simply averaging bounding box coordinates, which can reduce the bias of a single detection box and improve the overall position estimation accuracy, but it ignores the differences between different detection boxes. For detection boxes with low overlap or significant size differences, the averaging method may not be applicable. The third method, "union", can cover the overall position and size of the target more comprehensively, but it will lead to the final detection box being too large, and the box is selected not to belong to the target area. The fourth method, "intersection", can determine the location and size of the target more accurately but results in the final detection box being too small to cover the entire area of the target entirely. Different methods are suitable for different detection scenarios, so we created a plug-and-play module to obtain the most accurate detection results.

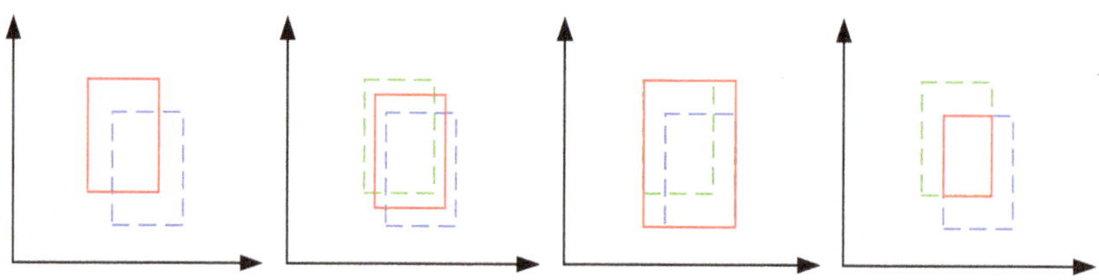

Figure 4. Different fusion strategies for detection boxes. The blue and green boxes represent the detection boxes of other detectors, while the red box represents the final detection box selected for different decisions. (**a**) NMS, (**b**) average, (**c**) union, and (**d**) intersection.

3.2. Fusion Strategy for Category Probability Set (CPROS)

The decision-level fusion method of CPROS proposed by us is mainly aimed at solving four problems: (1) the single-mode detector has serious missed detections, (2) we hope to obtain higher confidence for the detection results with high confidence, (3) the existing fusion methods have poor fusion effect for the detection results with low confidence, and (4) we aim to solve the category conflict detected by the single-mode detector. In the previous discussion, we knew that when using D-S evidence theory to identify targets by multimodal fusion, there may be abnormal recognition results; that is, it is considered that highly conflicting evidence cannot make a reasonable decision. Therefore, we introduced the improved Yager formula and proposed a more effective fusion decision.

Suppose that $m_1, m_2, \cdots, m_t$ is a detector of different modes, and the corresponding evidence set (probability distribution) is $F_1, F_2, \cdots, F_t$, then A_t represents the decision result of A_t (single category) in the evidence set F_t, and $m_t(A_t)$ is the mass function of A_t. To quickly describe the degree of conflict in the detection results of different mode detectors, we defined the conflict factor k as follows:

$$k = 1 - \sum_{A_i \in F_i, \cap_{i=1}^{t} A_i \neq \phi} m_1(A_1) \cdot m_2(A_2) \cdots m_t(A_t). \tag{1}$$

The magnitude of the conflict between the detection results of the modes m_i and m_j is:

$$k_{ij} = 1 - \sum_{A_i \in F_i, A_j \in F_j, A_i \cap A_j \neq \phi} m_i(A_i) \cdot m_j(A_j). \tag{2}$$

Here, ε is defined as evidence credibility and $\varepsilon = e^{-\widehat{k}}$, where $\widehat{k} = \frac{1}{t(t-1)/2} \sum_{i<j\leq t} k_{ij}$, $i, j \leq t$, and t are the number of pieces of evidence. $\widehat{k}$ is the average sum of the t evidence sets, reflecting the degree of conflict between evidence pairs. ε is a decreasing function of $\widehat{k}$, reflecting the credibility of the evidence; that is, the credibility of the evidence decreases as the conflict between the pieces of evidence increases.

When a conflict is detected between categories, we believe that the statement "a% belongs to category A, $(1-a)$% belongs to category B" is unreasonable because there should be a portion of the probability that is wavering, which we call uncertainty probability. Therefore, the statement should be changed to "a% belongs to category A, b% belongs to category B, $(1-a-b)$% is the uncertainty probability, and other decision rules need to be introduced to determine which category this part of the probability belongs to". The uncertainty probability depends on the degree of conflict and credibility of the evidence and is calculated as follows:

$$Confidence_{uncertainty} = k \times (1 - \varepsilon). \tag{3}$$

In addition to calculating the uncertainty probability, we must also calculate the certainty probability, which is the a% and b% in "a% belongs to category A, b% belongs to category B". It depends on two parts. Part of this is the agreement of the evidence of the different modes in category A; we define it as $p(A)$, and the calculation formula is as follows:

$$p(A) = \sum_{A_i \in F_i, \cap_{i=1}^{t} A_i = A} m_1(A_1) \cdot m_2(A_2) \cdots m_t(A_t). \tag{4}$$

Part is the average support of the evidence for different modes for category A; we define it as $q(A)$, and the calculation formula is as follows:

$$q(A) = \frac{1}{t} \sum_{i=1}^{t} m_i(A). \tag{5}$$

Then, the certainty probability is calculated as follows:

$$Confidence_{certainty} = (1-k)\frac{p(A)}{1-k} + k \times \varepsilon \times q(A), A \neq \phi, X. \tag{6}$$

The first item that can be found, $p(A)/(1-k)$, is precisely the D-S evidence theory synthesis formula. Thus, the new composition formula is actually a weighted sum form, where $1-k$ and k are the weighting coefficients. In the first term of the formula, the result of synthesis is similar to that of D-S synthesis. When $k = 0$, the new synthesis formula is equivalent to the D-S synthesis formula. When $k \to 1$, the evidence is highly conflicting, and the resultant result will be determined mainly by A. Therefore, for highly contradictory evidence, the consequent results are primarily determined by the evidence confidence $\varepsilon \times q(A)$ and the mean support of the evidence $q(A)$.

The uncertainty probability is not discarded, and its attributes depend on the size of different categories of certainty probability. The uncertainty probability belongs to the category with the largest certainty probability. Therefore, the final confidence calculation formula is

$$Confidence_{final} = Confidence_{uncertainty} + Max(Confidence_{certainty}). \tag{7}$$

3.3. Fusion

After determining the fusion method, we combined the detection results of the two modalities. We assume that the detection results for the different modalities are $(x_1, y_1, w_1, h_1, a_1, a_2, \cdots a_n)$ and $(x_2, y_2, w_2, h_2, b_1, b_2, \cdots b_n)$. In order to make the probability sum of detection result categories for each detector equal to 1, we introduce the "background" category, as shown in Table 1.

Table 1. Category probability set.

	0	**1**	...	**n**	**Background**
RGB	a_0	a_1	...	a_n	a_{n+1}
IR	b_0	b_1	...	b_n	b_{n+1}

Among them, $\sum_{i=0}^{n+1} a_i = 1$ and $\sum_{i=0}^{n+1} b_i = 1$. In practical situations, the detection probabilities of specific categories are extremely low. To simplify the calculation process, we kept all probabilities to two decimal places. The specific implementation process of the algorithm is shown in Algorithm 1.

Algorithm 1. Fusion Strategies for Multimodal Detection

```
Input: detections from multiple modes. Each detection d = (pro, box, cls, conf) contains
category probability set pro = (p1, p2...pn), box coordinates box = (x, y, w, h), tag cls =
(0/1/.../n) and confidence conf = (x).
Integrate the detection results of the same image corresponding to different modes. Set
D = (d1, d2, ... dn).
Traverse the set and place boxes with IOU greater than the threshold together to form a
detection set at the same position. Set H = {D1, D2, ..., Dn}.
if len(Di) > 1:
    Take the two elements with the highest confidence in Di
    Fusion strategy for detection boxes
    Fusion strategy for category probability set(CPROS)
if len(Di) = 1:
    No need for fusion
return set F of fused detections
```

4. Experiments

In this section, we briefly introduce the evaluation metrics used to measure the algorithm performance, dataset, and experimental settings. The single-mode detector YOLOv8 with good detection performance was selected as the baseline to validate the effectiveness and generalization ability of the proposed CPROS on the public dataset VEDAI, which was captured from a drone perspective [36]. A series of ablation studies were conducted under the same conditions for evaluation.

4.1. Evaluating Indicator

The accuracy of the prediction depends on whether the IoU between the predicted and actual boxes is greater than 0.5. The calculation method for the IoU is shown in Figure 5, which illustrates the intersection union ratio between the predicted and actual boxes. The subsequent calculations of the evaluation metrics were based on this.

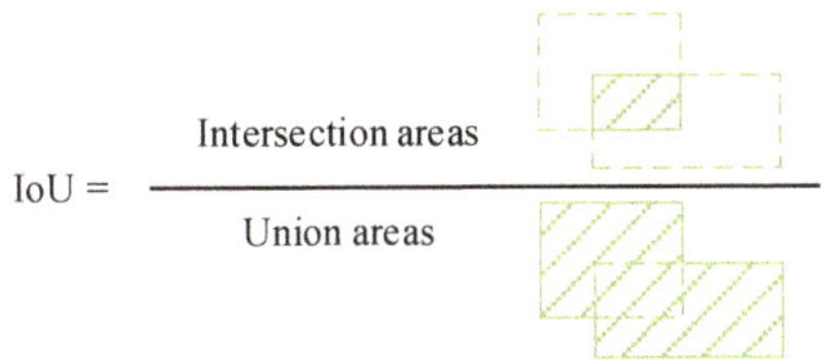

Figure 5. Schematic diagram of intersection and union ratio calculation.

For the proposed decision-level fusion method, we calculated the mAP, missed detection rate, and false detections per image *FPPI* for each image based on the final detection results and label files, which is different from typical evaluation index calculation methods. We used the 11-point interpolation method to calculate the average precision (*AP*) using the following formula:

$$AP = \frac{1}{11} \sum_{r \in \{0,0.1,0.2,\ldots 1.0\}} MAX_{\hat{r}:\hat{r} \geq r} P(\hat{r}) \tag{8}$$

The 11-point interpolation calculation method selects 11 fixed thresholds $\{0, 0.1, 0.2, \cdots, 1.0\}$ because only 11 points are involved in the calculation, $V = 11$, and v is the threshold index. The $MAX_{\hat{r}:\hat{r} \geq r} P(\hat{r})$ is the maximum value in the sample after the sample point corresponding to the vth threshold.

Assuming that the detection results contain *TP* (true positives), *FN* (false negatives), *TN* (true negatives), and *FP* (false positives),

$$MR = \frac{FN}{TP + FN} \tag{9}$$

$$FPPI = \frac{FP}{Num(Images)} \tag{10}$$

4.2. Experiment Settings and Datasets

In this study, we employed VEDAI, a well-known dataset for vehicle detection in aerial imagery, as a tool for benchmarking automatic target recognition algorithms in an unconstrained environment. The VEDAI dataset is a dual-mode image dataset containing 1246 pairs of 512 × 512 pixel visible and infrared remote sensing images, which are divided into 935 pairs of training set images and 311 pairs of validation set images. The VEDAI dataset for a single mode contains six categories of vehicle targets that, in addition to their small size, exhibit various variabilities such as multiple orientations, lighting/shadow changes, specular reflection, and occlusion. The number of targets in each category is shown in Figure 6.

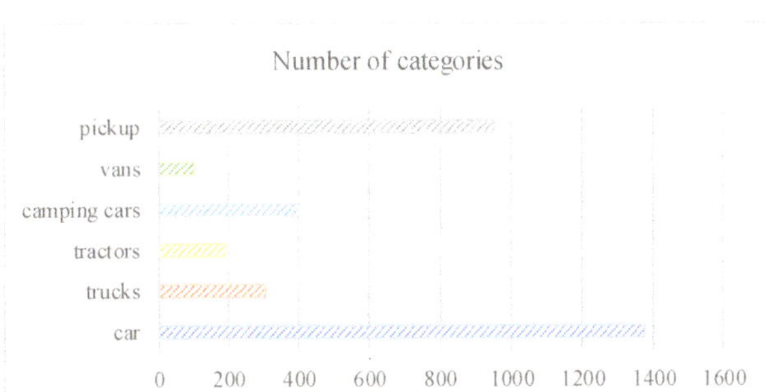

Figure 6. Statistical chart of the number of targets for six types of vehicles.

The system environment of this study was Ubuntu 22.04LTS, and the software environments were CUDA 11.7 and cuDNN 8500. All models were trained on the PyTorch 2.0.1 framework using a single GPU and an NVIDIA GeForce RTX 3070. The number of epochs was set to 300, and the batch size to 4. SGD was used as the optimizer with a learning rate of 0.01.

4.3. Experimental Results

In this section, we present comparative experiments on single-mode detection, pixel-level fusion detection, feature-level fusion detection, and decision-level fusion detection to evaluate the effectiveness of fusion detection at different stages. Meanwhile, the evaluation of the proposed CPROS and several existing fusion strategies was mainly conducted in the form of ablation studies on the VEDAI using YOLOv8, which has good detection performance and was selected as the baseline method. Finally, we applied the proposed CPROS method to different single-mode detectors to verify its generality. In the tables, the best results for each column are highlighted in bold orange, the second best in bold blue, and the third best in bold green. The decision-level fusion detection results obtained using the proposed method are highlighted in gray in the tables.

4.3.1. Comparative Experimental Results

From Tables 2–4, it can be seen that the mAP of YOLOv8 using the CPROS method was 8.6% and 16.4% higher than that of YOLOv8 in detecting single-mode datasets. The missed detection rates were reduced by 3.6% and 8.4%, respectively. The number of false detections per image was reduced by 0.032 and 0.019 (32 and 19 false detections per 1000 frames). From a single-category perspective, our method significantly reduced the missed detection rate, with a maximum reduction of 35.8%. Meanwhile, for most categories, our method can also improve the mAP and reduce the number of false positives per image (the maximum improvement in average detection accuracy can reach 35.1%, and the maximum reduction in false positives per image can reach 0.061). Additionally, by comparing the effects of single-mode detection, pixel-level fusion detection, feature-level fusion detection, and decision-level fusion detection, we found that pixel-level and feature-level fusion detection were only effective for certain specific targets. Overall, the detection effect was not as good as decision-level fusion detection, and the generalization ability was also not as good as decision-level fusion detection.

Table 2. Comparison of experimental results between fusion detection and non-fusion detection in different stages based on the VEDAI dataset (mAP).

AP/mAP	IOU = 0.5						
	Car	Trucks	Tractors	Camping Cars	Vans	Pickup	All
RGB	**62.3%**	37.5%	**35.6%**	**51.4%**	**58.0%**	**64.8%**	**51.6%**
IR	62.0%	36.3%	17.0%	39.4%	45.4%	62.9%	43.8%
SeAFusion	57.0%	**44.1%**	**38.7%**	**55.3%**	44.2%	57.2%	49.4%
MMIF-CDDFuse	**69.9%**	**38.1%**	30.8%	**51.2%**	55.8%	**63.3%**	**51.5%**
RFN-Nest	56.9%	33.3%	19.4%	45.4%	**57.1%**	53.2%	44.2%
YDTR	49.7%	23.4%	15.4%	40.7%	49.2%	52.8%	38.5%
CPROS (ours)	**71.8%**	**47.1%**	**52.1%**	50.4%	**70.0%**	**69.7%**	**60.2%**

However, as shown in Figure 7, when we focus on the category of camping cars, we find that single-mode detectors generate numerous false alarms when detecting visible and infrared images in the VEDAI dataset. On this basis, if we apply our method again, only the false alarms will increase. In this regard, our algorithm is not as effective as some pixel- or feature-level fusion detection algorithms such as SeAFusion and YDTR. We found a high similarity in shape and color between real shopping car targets and house targets that were falsely detected as camping cars in some VEDAI datasets, which was not conducive to distinguishing between the two. Moreover, there are few camping car targets available for training single-mode detectors in the dataset, and the detectors have not learned enough features, which may lead to an insufficient generalization ability of the model. In practical applications, in addition to improving the detection performance of single-mode detectors, solving such problems can also be considered using Kalman filtering to eliminate false alarms, thereby achieving better fusion results.

Table 3. Comparison of experimental results between fusion detection and non-fusion detection in different stages based on the VEDAI dataset (missed detection rate).

MR	IOU = 0.5						
	Car	Trucks	Tractors	Camping Cars	Vans	Pickup	All
RGB	20.2%	43.0%	39.4%	25.0%	36.7%	21.3%	24.8%
IR	23.2%	48.7%	65.6%	41.5%	43.3%	20.9%	29.6%
SeAFusion	27.6%	36.7%	35.3%	28.6%	40.0%	25.9%	29.0%
MMIF-CDDFuse	18.0%	44.2%	50.0%	28.9%	26.7%	18.8%	23.7%
RFN-Nest	27.9%	45.9%	67.7%	25.3%	26.7%	24.9%	30.2%
YDTR	30.6%	52.7%	77.4%	41.8%	33.3%	23.4%	34.1%
CPROS (ours)	17.7%	34.9%	29.8%	19.1%	28.1%	19.7%	21.2%

Table 4. Comparison of experimental results between fusion detection and non-fusion detection in different stages based on the VEDAI dataset (the number of false detections per image).

FPPI	IOU = 0.5						
	Car	Trucks	Tractors	Camping Cars	Vans	Pickup	All
RGB	0.186	0.087	0.055	0.096	0.006	0.167	0.598
IR	0.170	0.080	0.048	0.077	0.013	0.196	0.585
SeAFusion	0.273	0.087	0.051	0.074	0.023	0.170	0.678
MMIF-CDDFuse	0.177	0.077	0.048	0.096	0.029	0.196	0.624
RFN-Nest	0.238	0.103	0.029	0.141	0.032	0.222	0.765
YDTR	0.257	0.132	0.023	0.061	0.026	0.222	0.720
CPROS (ours)	0.164	0.087	0.045	0.129	0.006	0.135	0.566

Figure 8 shows a visualization of the real labels and the results of our algorithm. The yellow line represents the fusion of low-confidence detection results, the green line represents the fusion of high-confidence detection results, the blue line represents the fusion of detection results with missed detections, and the red line represents the fusion of detection results with category conflicts. This corresponds to the four issues mentioned in Section 3.2, which must be addressed. From the graph, it can be observed that our method can significantly improve confidence when the categories detected by the detector are identical. When the categories detected by the detector are different, the proposed method eliminates the impact of erroneous detection and improves confidence. Meanwhile, our method can elegantly handle "missing detection" through probability integration.

Figure 9 shows a visualization of the results of the fusion detection algorithms for each stage. Comparing the visualization of real labels in Figure 8, it can be seen that our algorithm has fewer missed detections and false positives than the other algorithms, and overall has a higher confidence in object detection.

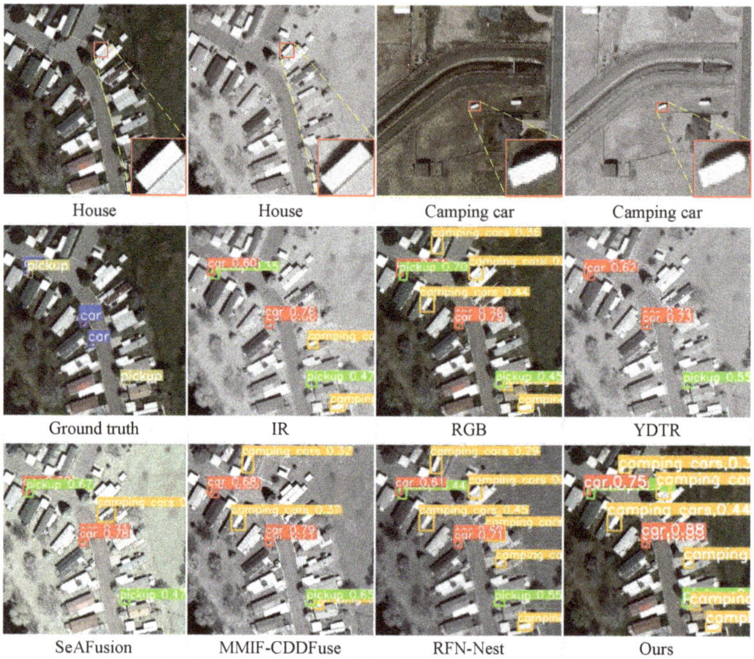

Figure 7. Analysis of experimental results for the category of camping cars.

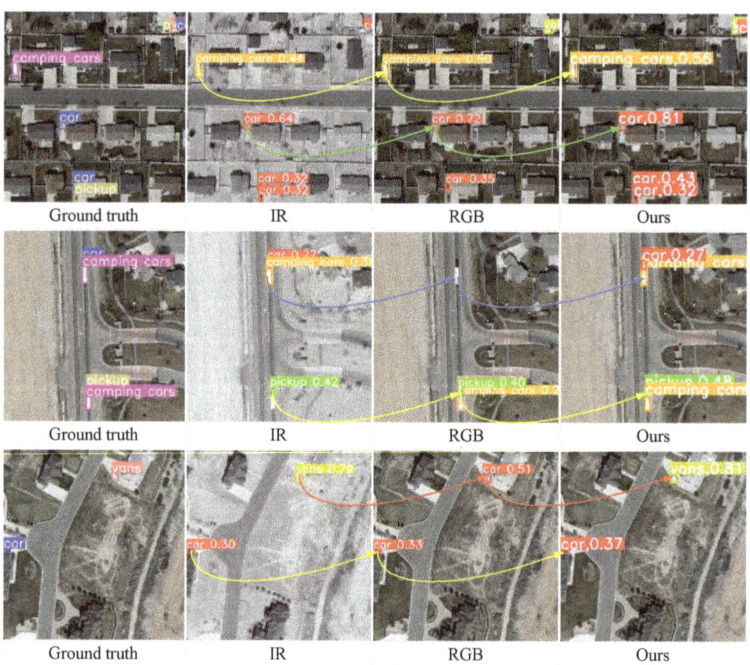

Figure 8. Realistic labels and visualization of results from our algorithm.

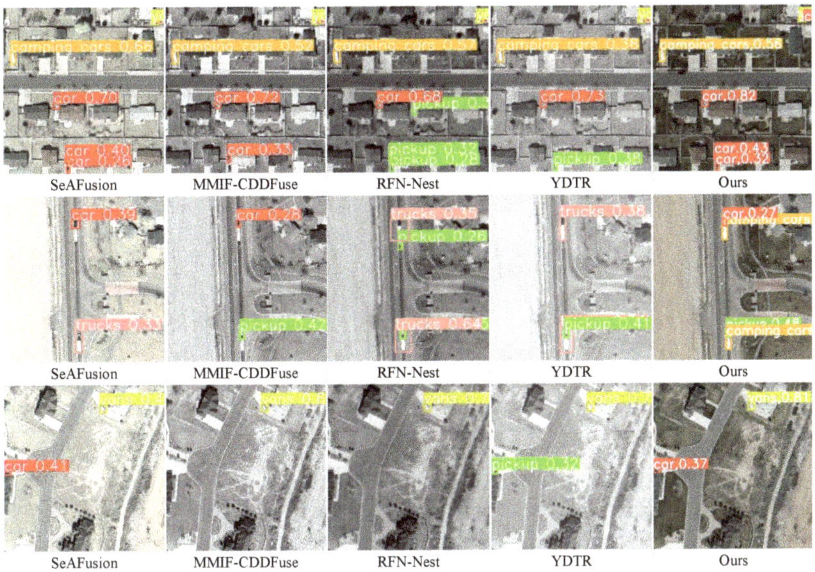

Figure 9. Visualization of the results of fusion detection algorithms in each stage.

4.3.2. Results of Ablation Experiment

To verify the superiority of CPROS, we conducted ablation studies by combining four confidence fusion methods and four detection box fusion methods. From Table 5, it can be seen that when the detection box fusion method is fixed, the mAP of the detection results using the four probability fusion methods of average, max, D-S, and our CPROS improves overall in sequence. Taking the averaging method for detection box fusion as an example, the CPROS method for confidence fusion increased by 1%, 0.8%, and 0.6% compared to the average, maximum, and D-S methods, respectively. However, Tables 6 and 7 show that when the detection box fusion method is fixed, changes in the probability fusion method do not affect the MR and FPPI. The reason behind this is that among the three indicators of mAP, MR, and FPPI, only mAP was affected by changes in confidence. The experimental results show that the proposed CPROS is a reliable decision-level fusion method that improves the detection accuracy of objects compared to the other three methods.

Table 8 presents the comparison results of different single-mode detectors before and after using the CPROS method. The results indicate that the experimental results using the CPROS decision-level fusion method are superior to that of single-mode detectors for all indicators. Specifically, by using CPROS to fuse the results obtained from YOLOv8 for detecting visible images with those obtained from YOLOv8 for detecting infrared images, the mAP was 8.6% and 16.4% higher, the MR was 3.6% and 8.4% lower, and the FPPI was 0.032 (reduced by 32 false positives every 1000 frames) and 0.019 lower, respectively. Using CPROS to fuse the results obtained from YOLOv5 detection of visible images with those obtained from YOLOv5 detection of infrared images, the mAP was 6.8% and 12% higher, the MR was 2.1% and 7.7% lower, and the FPPI was 0.045 and 0.051 lower, respectively. Using CPROS to fuse the results obtained from YOLOv8 in detecting visible images with those obtained from YOLOv5 in detecting infrared images, the mAP was 6.8% and 12.3% higher, the MR was 4.1% and 8.9% lower, and the FPPI was 0.003 and 0.067 lower, respectively. The experimental results show that the proposed CPROS is a reliable decision-level fusion method with plug-and-play capabilities and can be widely used for fusion processing between different mode detectors, significantly improving detection performance.

Table 5. Comparison of mAP between fusion methods with different confidence levels and detection box fusion methods.

Score -Fusion	Box -Fusion	AP/mAP (IOU = 0.5)						
		Car	Trucks	Tractors	Camping Cars	Vans	Pickup	All
Avg	NMS	73.3%	45.7%	43.7%	49.4%	66.9%	70.7%	58.3%
	Avg	71.5%	46.2%	51.7%	49.4%	66.9%	69.2%	59.2%
	Max	61.5%	46.2%	51.7%	49.4%	66.9%	61.2%	56.2%
	Min	73.6%	45.7%	43.7%	49.4%	66.9%	70.1%	58.2%
Max	NMS	73.4%	46.0%	43.9%	50.2%	66.9%	70.9%	58.6%
	Avg	71.5%	46.8%	51.8%	50.2%	66.9%	69.4%	59.4%
	Max	61.5%	46.8%	51.8%	50.2%	66.9%	61.3%	56.4%
	Min	73.6%	46.0%	43.9%	50.2%	66.9%	70.4%	58.5%
D-S	NMS	73.7%	46.4%	44.0%	50.2%	66.9%	71.1%	58.7%
	Avg	71.9%	47.1%	51.8%	50.2%	66.9%	69.7%	59.6%
	Max	61.9%	47.1%	51.8%	50.2%	66.9%	61.6%	56.6%
	Min	73.8%	46.4%	44.0%	50.2%	66.9%	70.6%	58.7%
CPROS (ours)	NMS	73.6%	46.4%	44.2%	50.4%	70.0%	71.0%	59.3%
	Avg	71.8%	47.1%	52.1%	50.4%	70.0%	69.7%	60.2%
	Max	61.9%	47.1%	52.1%	50.4%	70.0%	61.7%	57.2%
	Min	74.0%	46.4%	44.2%	50.4%	70.0%	70.5%	59.3%

Table 6. Comparison of missed detection rates between fusion methods with different confidence levels and detection box fusion methods.

Score -Fusion	Box -Fusion	MR (IOU = 0.5)						
		Car	Trucks	Tractors	Camping Cars	Vans	Pickup	All
Avg	NMS	15.7%	36.1%	31.9%	20.2%	28.1%	18.5%	20.4%
	Avg	17.7%	34.9%	29.8%	19.1%	28.1%	19.7%	21.2%
	Max	20.3%	34.9%	29.8%	20.2%	28.1%	20.2%	22.5%
	Min	15.1%	36.1%	31.9%	20.2%	28.1%	18.9%	20.3%
Max	NMS	15.7%	36.1%	31.9%	20.2%	28.1%	18.5%	20.4%
	Avg	17.7%	34.9%	29.8%	19.1%	28.1%	19.7%	21.2%
	Max	20.3%	34.9%	29.8%	20.2%	28.1%	20.2%	22.5%
	Min	15.1%	36.1%	31.9%	20.2%	28.1%	18.9%	20.3%
D-S	NMS	15.7%	36.1%	31.9%	20.2%	28.1%	18.5%	20.4%
	Avg	17.7%	34.9%	29.8%	19.1%	28.1%	19.7%	21.2%
	Max	20.3%	34.9%	29.8%	20.2%	28.1%	20.2%	22.5%
	Min	15.1%	36.1%	31.9%	20.2%	28.1%	18.9%	20.3%
CPROS (ours)	NMS	15.7%	36.1%	31.9%	20.2%	28.1%	18.5%	20.4%
	Avg	17.7%	34.9%	29.8%	19.1%	28.1%	19.7%	21.2%
	Max	20.3%	34.9%	29.8%	20.2%	28.1%	20.2%	22.5%
	Min	15.1%	36.1%	31.9%	20.2%	28.1%	18.9%	20.3%

Table 7. Comparison of false positives per image using fusion methods with different confidence levels and detection box fusion methods.

Score -Fusion	Box -Fusion	FPPI (IOU = 0.5)						
		Car	Trucks	Tractors	Camping Cars	Vans	Pickup	All
Avg	NMS	0.135	0.090	0.048	0.129	0.006	0.122	0.531
	Avg	0.164	0.087	0.045	0.129	0.006	0.135	0.566
	Max	0.196	0.087	0.045	0.129	0.006	0.141	0.605
	Min	0.125	0.090	0.048	0.129	0.006	0.125	0.524

Table 7. *Cont.*

Score -Fusion	Box -Fusion	FPPI (IOU = 0.5)						
		Car	Trucks	Tractors	Camping Cars	Vans	Pickup	All
Max	NMS	0.135	0.090	0.048	0.129	0.006	0.122	0.531
	Avg	0.164	0.087	0.045	0.129	0.006	0.135	0.566
	Max	0.196	0.087	0.045	0.129	0.006	0.141	0.605
	Min	0.125	0.090	0.048	0.129	0.006	0.125	0.524
D-S	NMS	0.135	0.090	0.048	0.129	0.006	0.122	0.531
	Avg	0.164	0.087	0.045	0.129	0.006	0.135	0.566
	Max	0.196	0.087	0.045	0.129	0.006	0.141	0.605
	Min	0.125	0.090	0.048	0.129	0.006	0.125	0.524
CPROS (ours)	NMS	0.135	0.090	0.048	0.129	0.006	0.122	0.531
	Avg	0.164	0.087	0.045	0.129	0.006	0.135	0.566
	Max	0.196	0.087	0.045	0.129	0.006	0.141	0.605
	Min	0.125	0.090	0.048	0.129	0.006	0.125	0.524

Table 8. Comparison of performance between different single-mode detectors before and after using the CPROS method.

Detector	IOU = 0.5		
	mAP	MR	FPPI
YOLOv8(RGB)	51.6%	24.8%	0.598
YOLOv8(IR)	43.8%	29.6%	0.585
YOLOv8(RGB) YOLOv8(IR)	60.2%	21.2%	0.566
YOLOv5(RGB)	51.3%	24.0%	0.656
YOLOv5(IR)	46.1%	29.6%	0.662
YOLOv5(RGB) YOLOv5(IR)	58.1%	21.9%	0.611
YOLOv8(RGB)	51.6%	24.8%	0.598
YOLOv5(IR)	46.1%	29.6%	0.662
YOLOv8(RGB) YOLOv5(IR)	58.4%	20.7%	0.595

5. Conclusions

This study followed the approach of first detection and then fusion. The performance of a single-mode detector directly affects the fusion performance in multimodal detection. Based on this, we first explored different fusion strategies for multimodal detection in visible and infrared images using highly tuned YOLOv8 trained on large-scale single-mode datasets and proposed a multimodal decision-level fusion detection method based on category probability sets (CPROS). Numerous experimental results show that our proposed decision-level fusion method based on CPROS is significantly better than a single-mode detector without the decision-level fusion method in terms of detection accuracy. Moreover, it gracefully handles the missed detections of specific modes, significantly reducing the MR and FPPI.

Second, to prove the superiority and generality of the proposed decision-level fusion method, we combined different confidence and detection box fusion methods to perform ablation experiments. We also applied the proposed method to different single-mode detectors to compare detection performance before and after decision-level fusion. The results show that the proposed CPROS is significantly superior to previous methods in terms of detection accuracy. Compared to the single-mode detector, the mAP of multimodal detection using the fusion strategy was improved considerably, and the MR and FPPI were significantly reduced.

In the future, our goals are to (1) study object association methods so that the proposed decision-level fusion method can be applied to unaligned multimodal detection datasets; (2) mount the proposed algorithm framework on the UAV edge computing platform and apply it to real-time target detection tasks; (3) research accurate positioning methods to enable the UAV platform to achieve high positioning performance; and (4) use the Kalman filter algorithm to eliminate noise and improve the detection performance of the algorithm.

Author Contributions: Conceptualization, C.L.; methodology, C.L.; software, C.L.; validation, H.H. and S.Y.; formal analysis, X.T. and C.L.; investigation, Z.D.; resources, C.L.; data curation, S.Y. and Z.D.; writing—original draft preparation, C.L.; writing—review and editing, C.L., H.H. and X.T.; visualization, C.L.; supervision, X.T.; project administration, C.L.; funding acquisition, Z.Z. All authors have read and agreed to the published version of the manuscript.

Funding: This work was supported by the National Natural Science Foundation of China under Grant 52101377 and the National Natural Youth Science Foundation of China under Grant No. 62201598.

Data Availability Statement: The original contributions presented in the study are included in the article, further inquiries can be directed to the corresponding author.

Conflicts of Interest: The authors declare no conflicts of interest.

References

1. Xie, Y.; Xu, C.; Rakotosaona, M.-J.; Rim, P.; Tombari, F.; Keutzer, K.; Tomizuka, M.; Zhan, W. SparseFusion: Fusing Multi-Modal Sparse Representations for Multi-Sensor 3D Object Detection. In Proceedings of the IEEE/CVF International Conference on Computer Vision (ICCV), Paris, France, 1–6 October 2023.
2. Khosravi, M.; Arora, R.; Enayati, S.; Pishro-Nik, H. A Search and Detection Autonomous Drone System: From Design to Implementation. *arXiv* **2020**, arXiv:2211.15866. [CrossRef]
3. Geiger, A.; Lenz, P.; Urtasun, R. Are We Ready for Autonomous Driving? The KITTI Vision Benchmark Suite. In *Proceedings of the 2012 IEEE Conference on Computer Vision and Pattern Recognition, Providence, RI, USA, 16–21 June 2012*; IEEE: New York, NY, USA, 2012; pp. 3354–3361.
4. Devaguptapu, C.; Akolekar, N.; Sharma, M.M.; Balasubramanian, V.N. Borrow from Anywhere: Pseudo Multi-Modal Object Detection in Thermal Imagery. In Proceedings of the 2019 IEEE/CVF Conference on Computer Vision and Pattern Recognition Workshops (CVPRW), Long Beach, CA, USA, 16–17 June 2019; pp. 1029–1038.
5. Wu, W.; Chang, H.; Zheng, Y.; Li, Z.; Chen, Z.; Zhang, Z. Contrastive Learning-Based Robust Object Detection under Smoky Conditions. In *Proceedings of the 2022 IEEE/CVF Conference on Computer Vision and Pattern Recognition Workshops (CVPRW), New Orleans, LA, USA, 19–20 June 2022*; IEEE: New York, NY, USA, 2022; pp. 4294–4301.
6. Mustafa, H.T.; Yang, J.; Mustafa, H.; Zareapoor, M. Infrared and Visible Image Fusion Based on Dilated Residual Attention Network. *Optik* **2020**, *224*, 165409. [CrossRef]
7. Zhang, X.; Demiris, Y. Visible and Infrared Image Fusion Using Deep Learning. *IEEE Trans. Pattern Anal. Mach. Intell.* **2023**, *45*, 10535–10554. [CrossRef]
8. Wagner, J.; Fischer, V.; Herman, M.; Behnke, S. Multispectral Pedestrian Detection Using Deep Fusion Convolutional Neural Networks. In Proceedings of the ESANN 2016: 24th European Symposium on Artificial Neural Networks, Computational Intelligence and Machine Learning, Bruges, Belgium, 27–29 April 2016.
9. Liu, J.; Zhang, S.; Wang, S.; Metaxas, D.N. Multispectral Deep Neural Networks for Pedestrian Detection. *arXiv* **2016**, arXiv:1611.02644.
10. Li, Q.; Zhang, C.; Hu, Q.; Fu, H.; Zhu, P. Confidence-Aware Fusion Using Dempster-Shafer Theory for Multispectral Pedestrian Detection. *IEEE Trans. Multimed.* **2023**, *25*, 3420–3431. [CrossRef]
11. Zhang, X.X.; Lu, X.Y.; Peng, L. A Complementary and Precise Vehicle Detection Approach in RGB-T Images via Semi-Supervised Transfer Learning and Decision-Level Fusion. *Int. J. Remote Sens.* **2022**, *43*, 196–214. [CrossRef]
12. Tziafas, G.; Kasaei, H. Early or Late Fusion Matters: Efficient RGB-D Fusion in Vision Transformers for 3D Object Recognition. In Proceedings of the 2023 IEEE/RSJ International Conference on Intelligent Robots and Systems (IROS), Detroit, MI, USA, 1–5 October 2023.
13. Guan, D.; Cao, Y.; Liang, J.; Cao, Y.; Yang, M.Y. Fusion of Multispectral Data Through Illumination-Aware Deep Neural Networks for Pedestrian Detection. *Inf. Fusion* **2019**, *50*, 148–157. [CrossRef]
14. Zhang, C.; Ma, W.; Xiao, J.; Zhang, H.; Shao, J.; Zhuang, Y.; Chen, L. VL-NMS: Breaking Proposal Bottlenecks in Two-Stage Visual-Language Matching. *ACM Trans. Multimed. Comput. Commun. Appl.* **2023**, *19*, 166. [CrossRef]
15. Xiao, F. A New Divergence Measure of Belief Function in D–S Evidence Theory. *Inf. Sci.* **2019**, *514*, 462–483. [CrossRef]
16. Sentz, K.; Ferson, S. *Combination of Evidence in Dempster-Shafer Theory*; U.S. Department of Energy: Oak Ridge, TN, USA, 2002; SAND2002-0835; p. 800792.
17. Li, S.; Kang, X.; Fang, L.; Hu, J.; Yin, H. Pixel-Level Image Fusion: A Survey of the State of the Art. *Inf. Fusion* **2017**, *33*, 100–112. [CrossRef]
18. Zhang, Z.; Jin, L.; Li, S.; Xia, J.; Wang, J.; Li, Z.; Zhu, Z.; Yang, W.; Zhang, P.; Zhao, J.; et al. Modality Meets Long-Term Tracker: A Siamese Dual Fusion Framework for Tracking UAV. In *Proceedings of the 2023 IEEE International Conference on Image Processing (ICIP), Kuala Lumpur, Malaysia, 8–11 October 2023*; IEEE: New York, NY, USA, 2023; pp. 1975–1979.

19. Wu, Y.; Guan, X.; Zhao, B.; Ni, L.; Huang, M. Vehicle Detection Based on Adaptive Multimodal Feature Fusion and Cross-Modal Vehicle Index Using RGB-T Images. *IEEE J. Sel. Top. Appl. Earth Obs. Remote Sens.* **2023**, *16*, 8166–8177. [CrossRef]
20. Yang, K.; Xiang, W.; Chen, Z.; Zhang, J.; Liu, Y. A Review on Infrared and Visible Image Fusion Algorithms Based on Neural Networks. *J. Vis. Commun. Image Represent.* **2024**, *101*, 104179. [CrossRef]
21. Liu, J.; Fan, X.; Jiang, J.; Liu, R.; Luo, Z. Learning a Deep Multi-Scale Feature Ensemble and an Edge-Attention Guidance for Image Fusion. *IEEE Trans. Circuits Syst. Video Technol.* **2022**, *32*, 105–119. [CrossRef]
22. Feng, D.; Haase-Schutz, C.; Rosenbaum, L.; Hertlein, H.; Glaser, C.; Timm, F.; Wiesbeck, W.; Dietmayer, K. Deep Multi-Modal Object Detection and Semantic Segmentation for Autonomous Driving: Datasets, Methods, and Challenges. *IEEE Trans. Intell. Transp. Syst.* **2021**, *22*, 1341–1360. [CrossRef]
23. Chen, X.; Zhang, H.; Zhang, S.; Feng, J.; Xia, H.; Rao, P.; Ai, J. A Space Infrared Dim Target Recognition Algorithm Based on Improved DS Theory and Multi-Dimensional Feature Decision Level Fusion Ensemble Classifier. *Remote Sens.* **2024**, *16*, 510. [CrossRef]
24. Solovyev, R.; Wang, W.; Gabruseva, T. Weighted Boxes Fusion: Ensembling Boxes from Different Object Detection Models. *Image Vis. Comput.* **2021**, *107*, 104117. [CrossRef]
25. Yang, F.-J. An Implementation of Naive Bayes Classifier. In *Proceedings of the 2018 International Conference on Computational Science and Computational Intelligence (CSCI), Las Vegas, NV, USA, 12–14 December 2018*; IEEE: New York, NY, USA, 2018; pp. 301–306.
26. Zhou, H.; Dong, C.; Wu, R.; Xu, X.; Guo, Z. Feature Fusion Based on Bayesian Decision Theory for Radar Deception Jamming Recognition. *IEEE Access* **2021**, *9*, 16296–16304. [CrossRef]
27. Ghosh, M.; Dey, A.; Kahali, S. Type-2 Fuzzy Blended Improved D-S Evidence Theory Based Decision Fusion for Face Recognition. *Appl. Soft Comput.* **2022**, *125*, 109179. [CrossRef]
28. Song, Y.; Fu, Q.; Wang, Y.-F.; Wang, X. Divergence-Based Cross Entropy and Uncertainty Measures of Atanassov's Intuitionistic Fuzzy Sets with Their Application in Decision Making. *Appl. Soft Comput.* **2019**, *84*, 105703. [CrossRef]
29. Zhang, S.; Rao, P.; Hu, T.; Chen, X.; Xia, H. A Multi-Dimensional Feature Fusion Recognition Method for Space Infrared Dim Targets Based on Fuzzy Comprehensive with Spatio-Temporal Correlation. *Remote Sens.* **2024**, *16*, 343. [CrossRef]
30. Zhang, P.; Li, T.; Wang, G.; Luo, C.; Chen, H.; Zhang, J.; Wang, D.; Yu, Z. Multi-Source Information Fusion Based on Rough Set Theory: A Review. *Inf. Fusion* **2021**, *68*, 85–117. [CrossRef]
31. Kang, B.; Deng, Y.; Hewage, K.; Sadiq, R. A Method of Measuring Uncertainty for Z-Number. *IEEE Trans. Fuzzy Syst.* **2019**, *27*, 731–738. [CrossRef]
32. Lai, H.; Liao, H. A Multi-Criteria Decision Making Method Based on DNMA and CRITIC with Linguistic D Numbers for Blockchain Platform Evaluation. *Eng. Appl. Artif. Intell.* **2021**, *101*, 104200. [CrossRef]
33. Chen, Y.-T.; Shi, J.; Ye, Z.; Mertz, C.; Ramanan, D.; Kong, S. Multimodal Object Detection via Probabilistic Ensembling. In Proceedings of the European Conference on Computer Vision, Tel Aviv, Israel, 23–27 October 2022.
34. Yager, R.R. On the Dempster-Shafer Framework and New Combination Rules. *Inf. Sci.* **1987**, *41*, 93–137. [CrossRef]
35. Sun, Q.; Ye, X.; Gu, W. A New Combination Rules of Evidence Theory. *Acta Electron. Sin.* **2000**, *28*, 117–119.
36. Razakarivony, S.; Jurie, F. Vehicle Detection in Aerial Imagery: A Small Target Detection Benchmark. *J. Vis. Commun. Image Represent.* **2016**, *34*, 187–203. [CrossRef]

remote sensing

MDPI

Article

Infrared–Visible Image Fusion through Feature-Based Decomposition and Domain Normalization

Weiyi Chen, Lingjuan Miao, Yuhao Wang *, Zhiqiang Zhou and Yajun Qiao

School of Automation, Beijing Institute of Technology, Beijing 100081, China; 3220195102@bit.edu.cn (W.C.); miaolingjuan@bit.edu.cn (L.M.); zhzhzhou@bit.edu.cn (Z.Z.); yajun@bit.edu.cn (Y.Q.)

* Correspondence: wayhao@bit.edu.cn

Abstract: Infrared–visible image fusion is valuable across various applications due to the complementary information that it provides. However, the current fusion methods face challenges in achieving high-quality fused images. This paper identifies a limitation in the existing fusion framework that affects the fusion quality: modal differences between infrared and visible images are often overlooked, resulting in the poor fusion of the two modalities. This limitation implies that features from different sources may not be consistently fused, which can impact the quality of the fusion results. Therefore, we propose a framework that utilizes feature-based decomposition and domain normalization. This decomposition method separates infrared and visible images into common and unique regions. To reduce modal differences while retaining unique information from the source images, we apply domain normalization to the common regions within the unified feature space. This space can transform infrared features into a pseudo-visible domain, ensuring that all features are fused within the same domain and minimizing the impact of modal differences during the fusion process. Noise in the source images adversely affects the fused images, compromising the overall fusion performance. Thus, we propose the non-local Gaussian filter. This filter can learn the shape and parameters of its filtering kernel based on the image features, effectively removing noise while preserving details. Additionally, we propose a novel dense attention in the feature extraction module, enabling the network to understand and leverage inter-layer information. Our experiments demonstrate a marked improvement in fusion quality with our proposed method.

Keywords: infrared and visible image fusion; unified feature space; dynamic instance normalization; non-local Gaussian filter; dense attention

1. Introduction

Recently, infrared and visible image fusion (IVIF) has gained considerable attention, owing to its extensive applications in various fields [1–3]. Single-modal images typically contain limited scene information and cannot fully reflect the true environment. Therefore, fusing information from different imaging sensors helps to enhance the informational richness of the images. Infrared and visible images have strong complementarity, i.e., infrared cameras capture thermal radiation but may not provide detailed information, while visible images are not sufficient in detecting hidden objects. Due to the complementarity and advantages of these two modalities, IVIF is widely applied in fields such as nighttime driving, military operations, and object detection.

In recent years, researchers have proposed various methods for IVIF, which can be categorized into traditional and deep learning-based methods. Traditional methods aim to design optimal representations across modalities and formulate fusion weights. These methods include multi-scale decomposition (MSD)-based methods [3–6], other transformation-based methods [7–9], and saliency-based methods [10–12]. The advancements of deep learning have significantly accelerated the evolution of IVIF. Researchers have proposed sophisticated modules or structures [13–18] for the integration of features from both infrared and visible images. Autoencoders [19–24] have also been introduced into the IVIF process due to their powerful feature extraction capabilities. Additionally, generative adversarial networks (GANs) [25–28] have been employed to enhance the fusion performance. However, existing research often neglects the differences between infrared and visible images, as well as the noise present in source images.

There are still some challenges that need to be tackled. Firstly, there is a significant difference between infrared and visible images. This difference leads to the inconsistent fusion of features when they come from these different sources. As a result, the quality of the fusion results is often affected. The differences between the infrared and visible modalities can be attributed to variations in wavelength, sources of radiation, and acquisition sensors. These modal differences lead to variations in images, such as texture, luminance, contrast, etc., subsequently affecting the fusion quality. Although decomposition representation-based methods can reduce the impact of modal differences, they often require complex decomposition and fusion rules. Secondly, low luminance may result in noisy source images. These images often impact the performance of image fusion, leading to suboptimal results. Thirdly, many methods neglect essential information from the middle layers, which are crucial in the fusion process. While dense connections [22] have been introduced into the fusion network, these connections lead to higher computational costs.

To address these challenges, we propose a novel method (UNIFusion) for IVIF, which includes cosine similarity-based image decomposition, a unified feature space, and dense attention for feature extraction. To obtain high-quality fused images, our method reduces the differences between infrared and visible features through the unified feature space, while also preserving their unique information. We first decompose the infrared and visible images into common and unique regions, respectively. Then, the features extracted from common regions are fed into the unified feature space to obtain fused features without modal differences. Specifically, we first obtain unique and common regions based on the cosine similarity between the embedded features of infrared–visible images. The unique regions contain private information that should be preserved in the fusion process, while the common regions in both infrared and visible images contain similar content. Secondly, to obtain fusion results with more information, we design a unified feature space to eliminate the differences between common features. In the space, infrared features are transformed to the pseudo-visible domain, thereby eliminating the differences between modalities. Thirdly, we propose a dense attention to enhance the feature extraction capabilities of the encoder, particularly focusing on improving the model's ability to capture important information from the input data. By applying an attention weight across all layers of the encoder, this method ensures that the model focuses on important features, which helps the model to perform fusion tasks better. Moreover, we propose the non-local Gaussian filter to enhance the fusion results. This filter can learn the shape and kernel parameters, enabling it to remove noise while retaining details.

As demonstrated in Figure 1, our method outperforms current fusion algorithms like FusionGAN [26], PMGI [29], and U2Fusion [15]. It is apparent that we can obtain better results through the unified feature space. Even the current state-of-the-art methods for IVIF cannot obtain satisfactory fused images. For example, FusionGAN generates blurred fused images, while PMGI and U2Fusion lead to fusion artifacts. Conversely, our method can improve the fusion performance by fusing multi-modal features in a consistent space.

The main contributions of this paper are summarized as follows.

- To eliminate the modal difference, we propose a domain normalization method based on the unified feature space, which enables the transformation of infrared features to the pseudo-visible domain, ensuring that all features are fused within the same domain and minimizing the impact of modal differences during the fusion process.
- We propose a feature-based image decomposition method that separates images into common and unique regions based on the cosine similarity. This approach eliminates the need to manually craft intricate decomposition algorithms, offering an adaptive solution that simplifies the process.

- We design a dense attention to allow the encoder to focus on more relevant features while ignoring redundant or irrelevant ones. Moreover, the Non-local Gaussian filter is incorporated into the fusion network to reduce the impact of noisy images on the fusion results.

Figure 1. A comparison of the fused images generated by our UNIFusion and other state-of-the-art fusion methods.

2. Related Works

In this section, we review various IVIF methods, categorizing them into traditional, AE-based, and GAN-based approaches. Additionally, related works on image-to-image translation are briefly presented to obtain a deeper understanding of the proposed models.

2.1. Traditional-Based Methods

In the study of traditional methods for IVIF, various techniques have been proposed, which include multi-scale decomposition, saliency detection, etc. Multi-scale decomposition methods [4,5,7] decompose and reconstruct the features of infrared and visible image at various levels to better fuse details, structures, etc. These approaches align the process of scale information with the human visual system. Saliency detection methods [10–12] can enhance the fusion performance on important targets by assigning higher weights to salient regions or objects. Sparse representation techniques [30] use dictionaries learned from a large set of images to encode and preserve essential information from the source images during the fusion process. These traditional approaches provide a foundation for IVIF, which can retain the image details and improve the visual effect.

2.2. CNN-Based Methods

The introduction of convolutional neural networks (CNN) has revolutionized the field of infrared and visible image fusion (IVIF). Specifically, Liu et al. [13] were pioneers in this area, applying a Siamese CNN structure to effectively generate a weight map from the source images. Over time, the architectures of CNNs in IVIF have continuously evolved. Early CNN architectures included single-branch and dual-branch configurations. For instance, Li et al. [14] incorporated residual connections to enhance the fusion capabilities. Xu et al. [31] developed a multi-scale unsupervised network based on joint attention mechanisms, significantly improving the detail preservation in the fused images. Moreover, the research by Ma et al. [17] presents a fusion technique anchored in the Transformer framework, equipped with an attention module to integrate global information. Alongside this, the impact of the lighting conditions in fusion tasks is noteworthy. PIAFusion [18] tries to improve the fusion performance based on an illumination aware module, but its model is not successful in handling complex lighting scenarios.

2.3. Autoencoder-Based Methods

Autoencoders are effective in infrared–visible image fusion as they are adept at encoding and decoding image features. This capability is essential to effectively fuse infrared and visible information. Li et al. introduced the DenseFuse method [22], which marked a significant advancement in IVIF tasks. This approach efficiently fuses visible and infrared images, paving the way for further research and development in this area. After the introduction of DenseFuse, AE-based methods for IVIF received significant development, which can be categorized as single-branch-based methods [19,20] and dual-branch-based methods [21–24]. The advancements of autoencoders have played a crucial role in improving both the efficiency and performance of the image fusion process. Additionally, the introduction of innovative modules has significantly enhanced the quality of the fused images. These modules include residual connections, channel attention, and self-attention.

Autoencoder-based methods can significantly enhance the fusion performance due to their strong capacity for feature extraction and reconstruction. This ability allows for the more comprehensive fusion of source image information, leading to superior fusion results.

2.4. GAN-Based Methods

In the IVIF task, generative adversarial networks (GANs) have been employed to generate fused images that contain rich information from the source images. Liao et al. [25] leveraged the powerful generative capabilities of GANs to produce realistic and information-rich fused images, demonstrating the advantages of GAN-based methods in infrared and visible image fusion. Furthermore, Xu et al. [27] developed a conditional GAN featuring dual discriminators, each trained on infrared and visible images. This approach effectively balances features from both types of images, thereby enhancing the fusion performance.

The architectural innovation in GAN-based methods is noteworthy. Researchers have experimented with multiple discriminators to improve the fusion performance. For example, Song et al. [28] introduced a novel GAN-based method with a triple discriminator for IVIF, which produces detailed fused images. In addition, researchers are focusing on the design of loss functions and architectures. For example, Li et al. [32] and Yuan et al. [33] used the Wasserstein distance and group convolution in GAN architectures, respectively, which led to better fusion results.

2.5. Image-to-Image Translation Methods

The objective of image-to-image (I2I) translation is to convert an image from a source domain to a target domain, ensuring that the essential characteristics of the input image are retained. Various generative adversarial network (GAN)-based frameworks have been proposed to align the output image distribution with that of the target domain. For instance, in 2016, Isola et al. introduced Pix2Pix [34], a conditional GAN model capable of translating images across domains using paired training data. Subsequently, Pix2PixHD [35] was developed to address high-resolution image translation. However, a significant challenge with these paired I2I translation methods is their dependence on paired datasets, which can be challenging and expensive to acquire, and sometimes even unattainable. Consequently, various approaches [36–39] have been explored to overcome the limitation for paired datasets. For instance, Bousmalis et al. [40] proposed an I2I translation method based on unsupervised training that applies domain adaptation in the pixel space. In our approach, we design a unified feature space to transform infrared features into the pseudo-visible domain. This ensures that all features exist within the same domain, eliminating the impact of modality differences on the fusion process.

3. Methods

3.1. Overview

Our proposed UNIFusion is an autoencoder structure, which consists of image decomposition, feature extraction, fusion, and reconstruction modules. The feature extraction module is a three-branch network based on dense attention, consisting of encoders E^{ir},

E^{vi}, and E^u, which are used to extract unique and unified features. The fusion and reconstruction module is devised to fuse features and generate fusion results, while employing a non-local Gaussian filter to reduce the adverse impact of noise on the fusion quality. The complete architecture is depicted in Figure 2, providing a detailed overview. Specifically, we decompose infrared–visible images into common regions (C^{vi} and C^{ir}) and unique regions (P^{vi} and P^{ir}). The dense attention is leveraged to effectively extract features from the common and unique regions. To eliminate modal differences, we propose the unified feature space to transform infrared features into the pseudo-visible domain. As noisy source images may degrade the fusion quality, we design a non-local Gaussian filter to minimize the impact of noise on the fusion results while maintaining the image details.

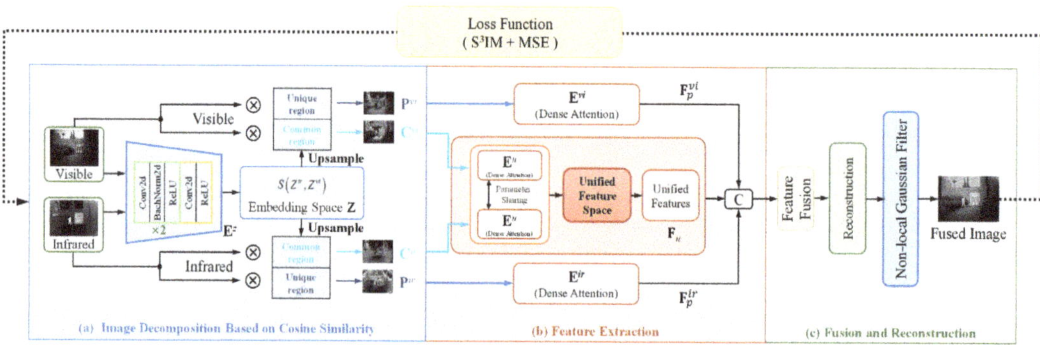

Figure 2. The overall framework of the proposed method. The method consists of (**a**) image decomposition, (**b**) feature extraction module, and (**c**) fusion and reconstruction module. (**a**) decomposes source images into common and unique regions, respectively. (**b**) is a three-branch network, consisting of encoders E^{ir}, E^{vi}, and E^u. The encoders based on dense attention are used to extract unique and unified features. (**c**) is devised to fuse features and generate fusion results, while employing a non-local Gaussian filter to reduce the adverse impact of noise on the fusion quality.

During the training phase, we use the S^{3}SIM and MSE loss functions to evaluate the similarity between the fused image and the original inputs. This helps to refine the network parameters.

3.2. Image Decomposition Based on Cosine Similarity

To obtain the common regions (C^{vi} and C^{ir}) and unique regions (P^{vi} and P^{ir}) of the source images, we embed the infrared and visible images into a shared parameter space Z to obtain consistent feature representations. By comparing the similarity of these features using cosine similarity, we can capture the directional similarity of the image features without being affected by the absolute luminance. The size of the feature map is $h \times w$ and the dimension is d, which leads to the definitions (1) and (2) for feature representation. Elements within these feature maps are denoted by the lowercase z, which are vectors in the d-dimensional space. The superscript of z indicates the modality (with vi for visible light and ir for infrared), and its subscript denotes the position of the element. The definitions are shown below:

$$Z^{vi} = \begin{bmatrix} z^{vi}_{1,1} & \cdots & z^{vi}_{1,w} \\ \vdots & \ddots & \vdots \\ z^{vi}_{h,1} & \cdots & z^{vi}_{h,w} \end{bmatrix}_{h \times w}, Z^{vi} \in \mathbb{R}^{d \times h \times w}, \quad (1)$$

$$Z^{ir} = \begin{bmatrix} z^{ir}_{1,1} & \cdots & z^{ir}_{1,w} \\ \vdots & \ddots & \vdots \\ z^{ir}_{h,1} & \cdots & z^{ir}_{h,w} \end{bmatrix}_{h\times w}, Z^{ir} \in \mathbb{R}^{d\times h\times w}, \tag{2}$$

where $z^{vi}_{i,j}$ is the element in the i-th row and j-th column of the visible feature matrix. $z^{ir}_{i,j}$ is the element in the i-th row and j-th column of the infrared feature matrix.

The cosine similarity (denoted as cs in the Equation (3)) is used to decompose infrared and visible images into common and unique regions. This is because the cosine similarity captures the structural similarity between infrared and visible images, which is more important for image fusion than absolute luminance. Two types of masks for source image decomposition are derived by computing the cosine similarity (denoted as c), namely M_c (common mask) and M_p (unique mask), as detailed in Equations (4) and (5):

$$S = cs(Z^{vi}, Z^{ir}) = \begin{bmatrix} cs(z^{vi}_{1,1}, z^{ir}_{1,1}) & \cdots & cs(z^{vi}_{1,w}, z^{ir}_{1,w}) \\ \vdots & \ddots & \vdots \\ cs(z^{vi}_{h,1}, z^{ir}_{h,1}) & \cdots & cs(z^{vi}_{h,w}, z^{ir}_{h,w}) \end{bmatrix}_{h\times w}, \tag{3}$$

$$M_c = \tfrac{1+S}{2}, \tag{4}$$

$$M_p = \tfrac{1-S}{2}, \tag{5}$$

where S is the similarity matrix of size $h \times w$, representing the cosine similarity between visible and infrared features. cs is the cosine similarity function. M_c represents the common mask, and $\frac{1+S}{2}$ normalizes the similarity scores to a range [0, 1], where 1 indicates the maximum similarity. M_p is the unique mask, and the transformation $\frac{1-S}{2}$ also normalizes the scores, with 1 indicating the maximum difference.

Next, we upsample the common mask and unique mask to align with the source image size. Element-wise multiplication is performed between the two masks (M_c and M_p) and infrared–visible images (I^{ir} and I^{vi}) to yield four decomposed outcomes (C^{ir}, P^{ir}, C^{vi}, and P^{vi}). The decomposed results are defined as followed, representing infrared–visible common regions and unique regions, respectively:

$$C^{ir} = I^{ir} \times Upsample(M_c), \tag{6}$$

$$P^{ir} = I^{ir} \times Upsample(M_p). \tag{7}$$

$$C^{vi} = I^{vi} \times Upsample(M_c), \tag{8}$$

$$P^{vi} = I^{vi} \times Upsample(M_p), \tag{9}$$

The employment of cosine similarity enables more precise decomposition, ensuring that the common regions and unique regions between the infrared and visible images are captured.

3.3. Dense Attention for Feature Extraction

Although the current fusion methods [15,22] try to utilize skip connection structures to obtain rich features, the differences between multi-scale features are not sufficiently taken into account. Specifically, low-level features capture basic input characteristics, while high-level features are more abstract, representing complex concepts and structures. Dense connections and residual connections concatenate multi-scale features directly, which can make it challenging for neural networks to differentiate important features, consequently limiting the fusion performance.

To address this limitation, we propose a dense attention-based feature extraction module to obtain multi-scale features, as shown in Figure 3. By inserting attention into every dense connection, the model can learn the significant features and relationships between different layers. Furthermore, as the network depth increases, this attention mechanism helps the model to learn long-range dependencies, improving its generalization and robustness.

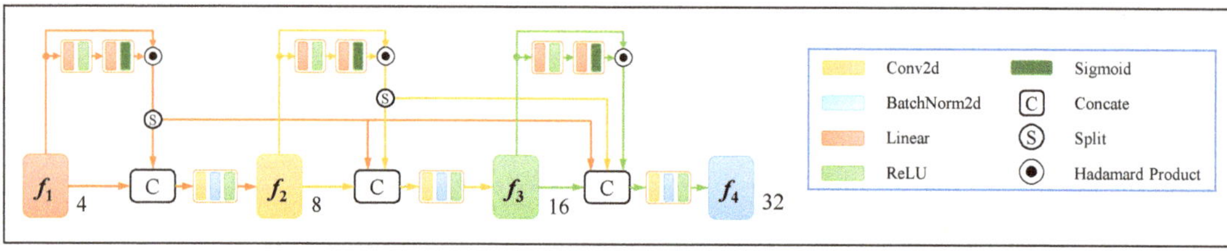

Figure 3. The structure of the feature extraction module based on dense attention.

3.4. Unified Feature Space Based on Dynamic Instance Normalization

We construct the unified feature space to eliminate the difference between infrared and visible features at the multi-scale feature level. The core components of the space include a scale-aware module, shifted patch embedding, and dynamic instance normalization (DIN), as shown in Figure 4. Specifically, the scale-aware module is trained to determine the size and shape of a patch. With the n pairs of scale and size parameters output by this module, shifted patch embedding can divide the feature map into n groups. For each group, it splits the feature map into patches according to the corresponding scale and size. DIN transforms infrared features into a pseudo-visible domain for each patch, which eliminates the differences between infrared and visible images. Subsequently, the learned confidence merges the features from the two modalities to produce the output result.

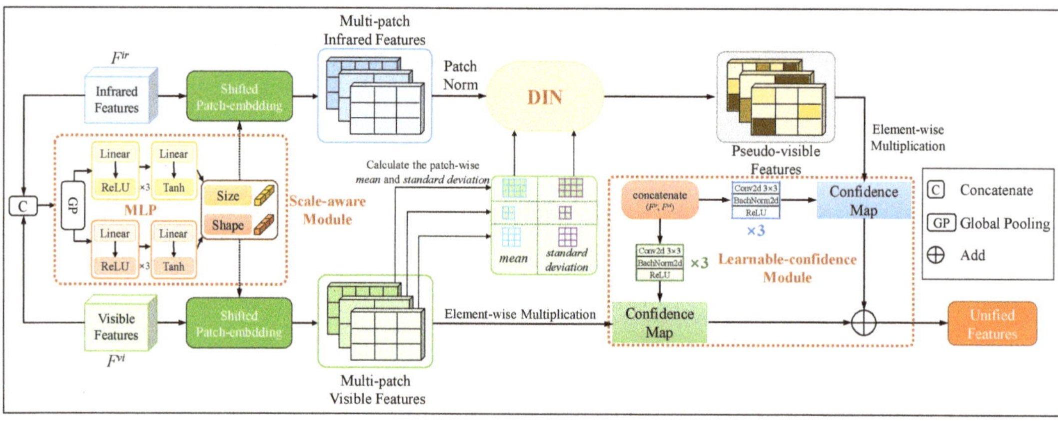

Figure 4. An illustration of the unified feature space based on dynamic instance normalization (DIN).

More specifically, the unified feature space enables the domain transformation from infrared to pseudo-visible, while also being adaptable to multi-scale targets. Dynamic instance normalization (DIN) is the core of the unified feature space, capable of transforming features from infrared features to pseudo-visible, thereby eliminating the difference between the two modalities. Moreover, we employ global pooling to concatenate features in order to enable a multilayer perceptron (MLP) to generate n pairs of size and shape parameters. The multi-patch embedding module divides the infrared and visible features into n groups along the channel dimension. Within each group, the features are segmented into patches of the same scale, determined by a set of size and shape parameters. Then, DIN transforms the infrared features to the pseudo-visible domain for each patch after shifted patch embedding. For the fusion of infrared and pseudo-visible features, we design a learnable confidence module to learn fusion weights; this method can adjust the fusion weight depending on the image content, compared with the fusion rules of addition, concatenation, and so on.

Although adaptive instance normalization (AdaIN) [41,42] plays a crucial role in image translation tasks, the core idea of AdaIN is to adjust the feature distribution of a content image to match the feature distribution of a target style image, thereby achieving style transfer. This process involves normalizing the features of the content image and then adjusting these normalized features with the statistical data (mean and variance) of the target style image. Through this method, the content image adopts the style characteristics of the style image while retaining its content structure. However, this method is not very precise due to the transformation of the domain at the level of global features. This limitation prevents independent domain transformations for each patch, restricting the effectiveness of domain transformation. To address this, we introduce dynamic instance normalization (DIN), which astutely segments the feature map into distinct subregions, as shown in Figure 5. This segmentation allows for independent domain transformations on each patch, enhancing the adaptability of the process. The DIN function is mathematically represented as

$$\mathrm{DIN}(\mathrm{X}, \mathrm{Y}) = [\mathrm{AdaIN}(x_1, y_1), \mathrm{AdaIN}(x_2, y_2), \ldots, \mathrm{AdaIN}(x_n, y_n)], \tag{10}$$

$$\mathrm{AdaIN}(x, y) = \sigma(y) \cdot \left(\frac{x - \mu(x)}{\sigma(x)}\right) + \mu(y), \tag{11}$$

where both X and Y denote global features, X represents the content input, and Y is the modal attribute input. Both X and Y are segmented into n patches, resulting in patch-wise pairs denoted as (x_i, y_i) for $i = 1, 2, \ldots, n$, where each pair corresponds to matching patches from X and Y. The terms $\mu(x)$ and $\mu(y)$ denote the means of x and y, respectively, while $\sigma(x)$ and $\sigma(y)$ denote their standard deviations.

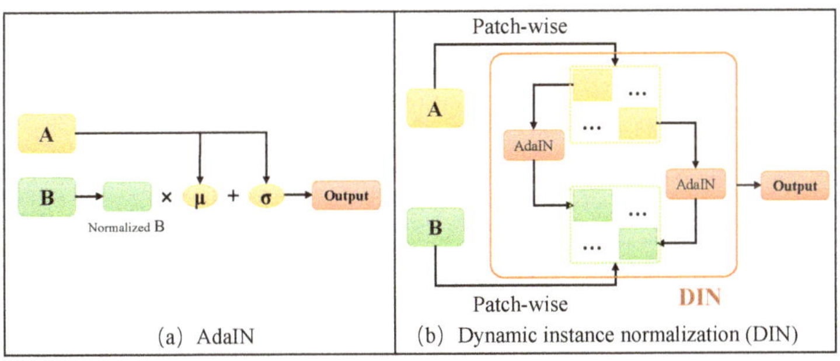

Figure 5. Different domain transformation methods. (**a**) AdaIN performs domain transformation by adjusting the global feature distribution of the content input (denoted as B), making it match the global feature distribution of the modal attribute input (denoted as A). (**b**) DIN, extended from AdaIN, adjusts the feature distribution at the patch-wise level, enabling more detailed domain adaptation.

In particular, we feed the concatenated infrared and visible features into a scale-aware module to obtain the scales and ratios. The shifted patch embedding module separately splits infrared and visible features into n groups and partitions each group of features into patches based on the scale and ratio. Infrared and visible patches can be represented as $\mathrm{X} = [x_1, x_2, \cdots, x_n]$ and $\mathrm{Y} = [y_1, y_2, \cdots, y_n]$, respectively. Applying DIN to each infrared–visible patch pair, as shown in Equation (10), we transform the infrared features into the pseudo-visible domain at the patch level. Then, we multiply them element-wise with a neural network-derived confidence metric to form the final fusion features. We obtain the final unified features by fusing pseudo-visible and visible features based on the learnable-confidence module.

3.5. Hierarchical Decoder for Fusion and Reconstruction

The hierarchical decoder does not only allow us to fuse infrared–visible features and generate fused images, but is also robust to the noise contained in source images and enhances the clarity of the fusion result. In this paper, we propose a multi-stage decoder to achieve more refined fusion, which can be divided into fusion, reconstruction, and enhancement stages.

The specific design of the hierarchical decoder is shown in Figure 6. We deploy two convolutional layers to fuse unified and unique features, receptively, in order to retain more infrared–infrared information. Then, in the reconstruction, we propose a novel module to learn the fusion strategy and obtain refined features. As every scale feature is vital to the fusion task, we not only insert a nest connection to learn the fusion strategy, but also propose a direct connection to output multi-scale features. Specifically, in the proposed architecture, features are reconstructed to match the size of the input image through a series of convolutional or transposed convolutional layers. These reconstructed features are then propagated to subsequent layers. In the final enhancement stage, we employ two distinct sets of convolutional layers to obtain a guidance feature used to obtain the filter parameters and preliminary fused images. Subsequently, we utilize a cascade of three convolutional layers to derive two-dimensional positional offsets and non-local Gaussian kernels.

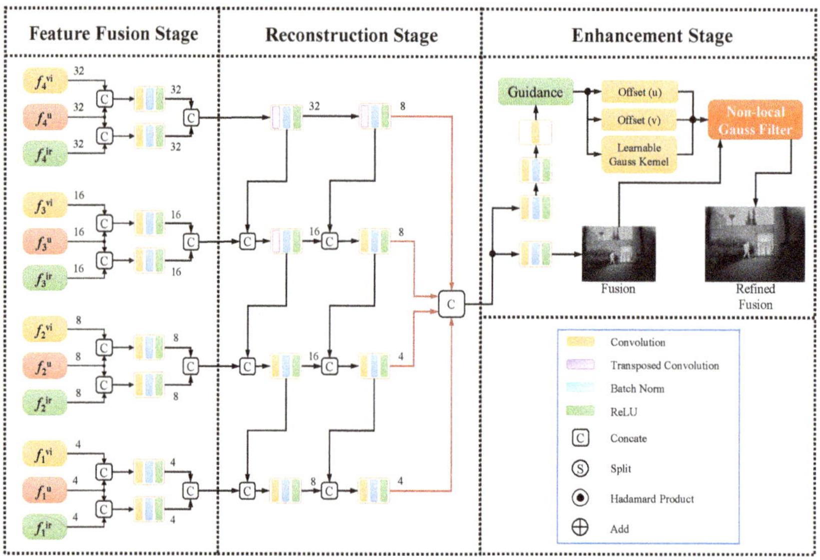

Figure 6. The structure of the hierarchical decoder.

Regarding the non-local Gaussian filter (shown in Figure 7), used for image enhancement, the process involves refining a preliminary fusion result, denoted as f. Here, $f_{i,j}$ represents the value at position (i, j) after an initial fusion step. The refined fusion outcome, $\hat{f}$, is achieved through an advanced filtering technique, mathematically formulated as

$$S_{i,j} = \sum_{n=1}^{N} w_{i,j}^{n}, \tag{12}$$

$$\hat{f}_{i,j} = \sum_{n=1}^{N} \frac{w_{i,j}^{n}}{S_{i,j}} \cdot f_{i+\Delta i_n, j+\Delta j_n}, \tag{13}$$

where $f_{i,j}$ represents the value at position (i, j), and N is the total number of neighbors, with a default value of 9. The term $w_{i,j}^{n}$ denotes learnable Gaussian kernels for the n-th neighbor of the pixel at (i,j). $S_{i,j}$ is the sum of weights for all neighbors, used to normalize the weights such that the sum of weights within the neighborhood equals 1. The terms Δi_n and Δj_n represent the positional offset values for the n-th neighbor, indicating the deviations in the row (vertical) and column (horizontal) directions, respectively, relative to the central pixel (i,j).

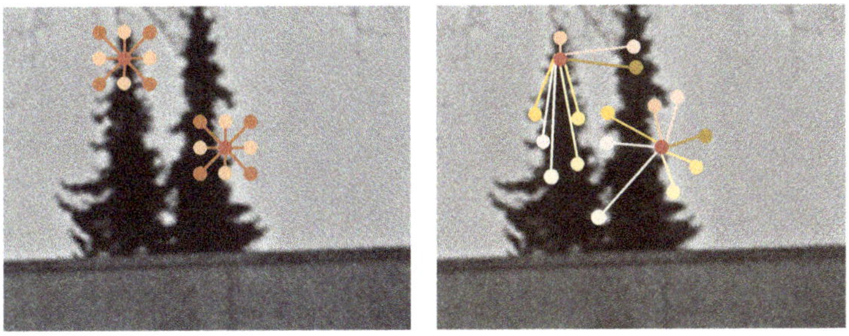
(a) Gaussian filter (b) Non-local Gaussian filter

Figure 7. An illustration of the non-local Gaussian filter, which employs a dynamic kernel to enhance the image fusion.

The non-local Gaussian filter enables the adaptive refinement of the fusion process. By dynamically adjusting the offsets and weights based on the local structures of the initial fusion result, the network can achieve a more optimized and contextually aware fusion outcome.

3.6. Loss Function

In this paper, we introduce two types of loss functions to simultaneously preserve crucial information from the source images and enhance the saliency of the fused image. Our loss functions incorporate two key components: the mean squared error (MSE) loss $\mathcal{L}mse$ and the proposed saliency structural similarity index (S^3IM) loss $\mathcal{L}_{s^3im}$. The MSE loss is used to constrain the similarity between the fusion results and the infrared–visible images. This loss focuses on maintaining fidelity to the source images by minimizing pixel-wise differences. Our proposed S^3IM loss aims to emphasize the saliency in the fused image. The total loss is calculated as follows:

$$\mathcal{L}(\theta, D) = \mathcal{L}_{mse}(\theta, D) + \lambda \mathcal{L}_{s^3im}(\theta, D), \tag{14}$$

where θ represents the parameters of the neural network, D represents the training data, and λ is the hyperparameter that balances the two losses.

Due to its efficiency and stability, the mean squared error loss $\mathcal{L}_{mse}$ can provide high accuracy and reliability in many cases. Therefore, we use it to constrain the similarity between the source images I_1, I_2, and the fused image I_f. Its definition is as follows:

$$\mathrm{MSE}(A, B) = \frac{1}{N}\sum_{i=1}^{N}(A_i - B_i)^2, \tag{15}$$

$$\mathcal{L}_{mse}(\theta, D) = \mu_1 \mathrm{MSE}(I_f, I_1) + \mu_2 \mathrm{MSE}(I_f, I_2), \tag{16}$$

where μ_1 and μ_2 are hyperparameters that balance the weights of the two MSE terms in the loss function. This allows the model to adjust the reliance on the visible image and the infrared image according to the needs of the specific task.

The structural similarity index measure (SSIM) [43] is a widely used image quality assessment metric that aims to quantify the perceptual similarity between two images. However, in infrared images, there are pixels with zero or very low intensity values, which means that the corresponding regions do not have objects with thermal radiation. In the fusion process, they should be assigned lower weights. To address this issue, we propose the saliency SSIM (S^3IM). Specifically, S^3IM can adaptively determine the loss weights based on the pixel intensity. We divide the normalized pixel values into three major regions: the low-saliency area, the linear area, and the high-saliency area, as shown in Figure 8.

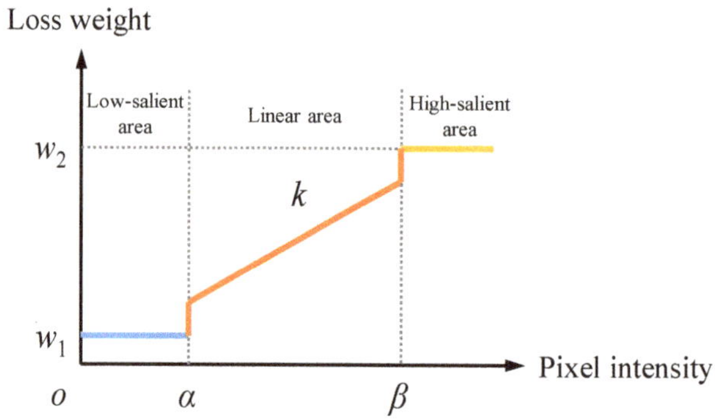

Figure 8. The schematic diagram of the s^3im weight.

The low-saliency area contains pixels with lower intensity values, which typically do not contain target information. When calculating the loss, they should be assigned a very low weight. The high-saliency region contains pixels with high intensity values, indicating objects with high thermal radiation, and they should have higher saliency in the fused image. For the remaining pixels, we adopt a linear transformation strategy to determine their loss weights, corresponding to the linear region in Figure 8. In summary, the calculation method is shown as follows:

$$h(x) = \begin{cases} w_1, x < \alpha \\ kx + b, \alpha \leq x \leq \beta \\ w_2, x > \beta \end{cases}, \tag{17}$$

$$\mathcal{L}_{s^3im}(\theta, D) = \varphi\left[1 - \text{SSIM}(I_f, I_1)\right] + h(I_2) \cdot \left[1 - \text{SSIM}(I_f, I_2)\right], \tag{18}$$

where φ is a hyperparameter used to adjust the weights of the infrared and visible images during the fusion process.

4. Experimental Results

In this section, we describe the experimental setup and the details of the network training. Following this, we perform a comparative analysis of the current fusion methods and carry out generalization experiments to highlight the benefits of our approach. Additionally, we conduct ablation studies to validate the effectiveness of our proposed methods.

4.1. Experimental Settings

We conduct experiments using four publicly available datasets. The M3FD dataset [44] is used for model training, while the TNO [45], RoadScene [15], and VTUAV [46] datasets are used to evaluate the performance of our method. The M3FD dataset contains 300 pairs of infrared and visible images for IVIF, including targets such as people, cars, buses, motorcycles, trucks, etc. These images were collected under various illuminance conditions and scenarios. The TNO dataset contains multispectral imagery from various military scenarios. The RoadScene dataset includes 221 image pairs featuring roads, vehicles, pedestrians, etc. The VTUAV dataset is used for remote sensing analysis and contains complex backgrounds and moving objects. We selected 20 pairs of infrared–visible images from both the TNO and RoadScene datasets, as well as 10 pairs from the VTUAV dataset, for the evaluation of our approach.

Our UNIFusion is compared with nine current state-of-the-art fusion methods, including a biological vision-based method, i.e., PFF [47]; an autoencoder-based method, i.e., MFEIF [48]; two generative adversarial network -based methods, i.e., FusionGAN [26] and UMF [49]; two convolutional neural network-based methods, i.e., U2Fusion [15], PMGI [29], and RFN [50]; a transformer-based method, i.e., swinfusion [17]; and a high-level task supervision-based method, i.e., PIAFusion [18].

To quantitatively evaluate the fusion performance, we utilize five key metrics: the average gradient (AG) [51], standard deviation (SD) [26], correlation coefficient (CC) [52], spatial frequency [53], and multi-scale structural similarity index (MS-SSIM) [54]. The AG measures the texture richness in the image, while the SD highlights the contrast within the fused image. The SF is indicative of the detail richness and image definition. The CC evaluates the linear relationship between the fusion results and infrared–visible images. MS-SSIM is employed to calculate the structural similarity between images. Generally, higher values in AG, SD, SF, MS-SSIM, and CC denote superior fusion performance.

4.2. Implementation Details

We trained our fusion model using the M3FD fusion dataset, which contains 300 infrared–visible pairs. During training, we randomly cropped the infrared–visible image pairs into multiple 256×256 patches, applied random affine transformations to enhance the model performance, and normalized all images to the [0, 1] range before inputting them into the fusion model. For training, we utilized the Adam optimizer with a batch size of 16. The initial learning rate was set to 5×10^{-4} and was halved every two epochs starting from epoch 30, continuing this reduction until the final epoch at 60. Additionally, we set the parameters of Equations (13)–(16) as follows: $\lambda = 1$, $\mu_1 = 1$, $\mu_2 = 1$, $\alpha = 0.2$, $\beta = 0.7$, $k = 1$, $b = 0$, $w_1 = 0.2$, $w_2 = 2$, $\varphi = 1$. The entire network was trained using the PyTorch 1.8.2 framework on an NVIDIA GeForce GTX 3080 GPU and a 3.69 GHz Intel Core i5-12600KF CPU.

4.3. Fusion Performance Analysis

In this section, we conduct a comprehensive qualitative and quantitative analysis to illustrate the advantages of our UNIFusion, comparing our method with nine state-of-the-art (SOTA) fusion approaches. In addition, we test the performance of our UNIFusion across various illumination scenarios within the VTUAV dataset.

4.3.1. Qualitative Results

The visualized comparisons of our UNIFusion with the nine SOTA methods are provided in Figures 9–11. Figures 9 and 10 present the fusion results of the different methods on the TNO and RoadScene datasets, respectively, while Figure 11 shows the color fusion results. Moreover, we evaluate our model's performance with remote sensing data collected under normal and low-light conditions, as shown in Figure 11. In our approach, we effectively transform infrared features into the pseudo-visible domain, resulting in fused images that maintain superior visual perception. This transformation process enhances the fusion of infrared and visible information, yielding more natural and clearer fusion results. Notably, our image decomposition method plays a crucial role in preserving unique information from multiple modalities, thereby highlighting salient objects in the fused images.

In Figure 9, it can be seen that FusionGAN, PMGI, RFN, U2Fusion, and UMF generate fusion results with less information and lower brightness (see the red boxes), which contain more infrared information and do not fully fuse visible image. The objects in MFEIF and PIAFusion are not salient and therefore not easily observed (see the orange boxes in Figure 9). SwinFusion suffers from overexposure and oversmoothing, resulting in some details not being clear enough (see the orange boxes in Figure 9). Although PFF can fuse more details, the results of this method contain noise (see the yellow boxes in Figure 9). On the contrary, our fused images can fuse more information through the unified feature space, which leads to rich details and structures (see the red boxes in Figure 9). Our UNIFusion can also obtain better fusion performance on small objects (see the orange boxes in Figure 9). Moreover, the results generated from our method are clear and contain less noise due to the non-local Gaussian filter (see the orange boxes in Figure 9).

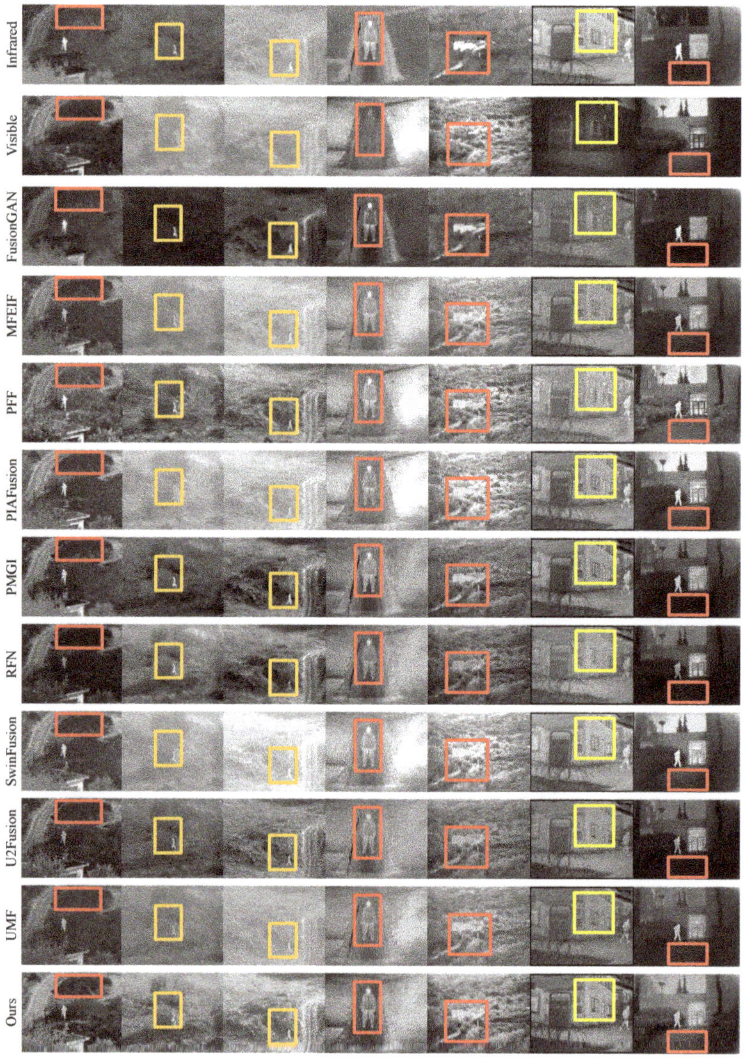

Figure 9. Qualitative comparison of the fused images from various methods on the TNO dataset.

Figures 10 and 11 show more fused images on the RoadScene dataset. In the red boxes, it can be seen that the fused images obtained from PFF contain more visible information and lees infrared information. In the fusion results obtained by FusionGAN, PMGI, and RFN, the overall brightness of the image is relatively low, leading to objects in the fused image that are not salient (see the red boxes). FusionGAN, PMGI, and RFN generate fusion results with low overall brightness, resulting in less salient objects (see the red boxes). Although MFEIF, PIAFusion, SwinFusion, and UMF produce brighter fusion results, their results appear less contrasted in Figures 10 and 11. In the orange boxes of Figure 11, the fusion result from PIAFusion and SwinFusion exhibits blurry details for the cloud, and the results of UMF and U2Fusion are unable to successfully process object edges (see the edge of the tree in orange boxes). In comparison, our method can achieve superior fusion performance in both day and night conditions. The fusion results obtained by our UNIFusion can effectively integrate the source information from infrared and visible images, and it exhibits better performance on the edges of the target.

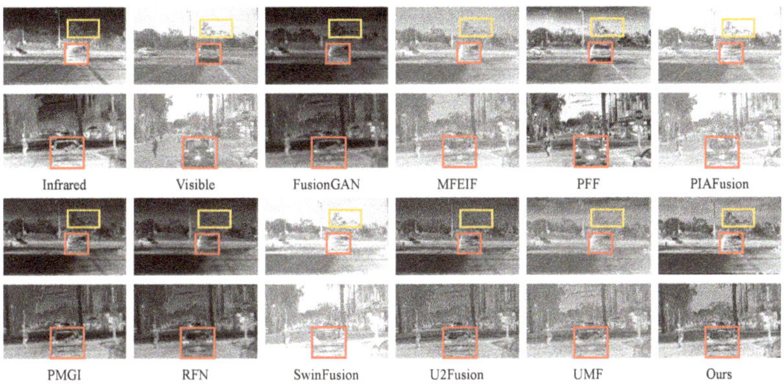

Figure 10. Qualitative comparison of the fused images from various methods on the RoadScene dataset.

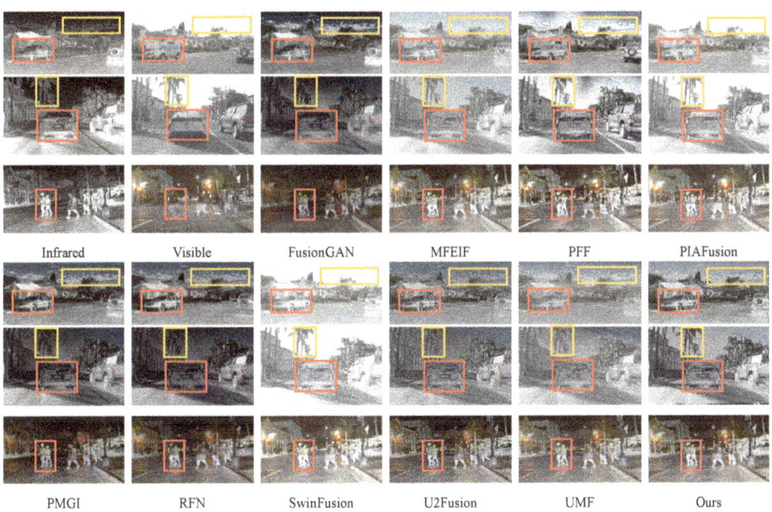

Figure 11. Qualitative comparison of the color fused images from various methods on the RoadScene dataset.

To assess the generalization of our method and its performance in low-light conditions, we conducted experiments on the VTUAV dataset. Figure 12 displays our fusion results, with Figure 12a showing the fusion results under normal-light conditions, and Figure 12b showcasing the fusion results under low-light conditions. In the normal-light scene (see the red boxes in Figure 12a), the infrared images display high thermal contrast, which our algorithm effectively integrates with the visible spectrum images, known for their rich contextual details. The resulting fusion images demonstrate the algorithm's proficiency in synthesizing the distinct attributes of each spectrum to enhance the overall image quality. Under low-light conditions (see the red boxes in Figure 12b), where visible images suffer from limited visibility, our algorithm leverages infrared imaging to accentuate thermal details otherwise obscured by darkness. The fusion process yields images that not only retain the luminance from visible light but also highlight thermal aspects, thus improving the interpretability of the scene in suboptimal lighting.

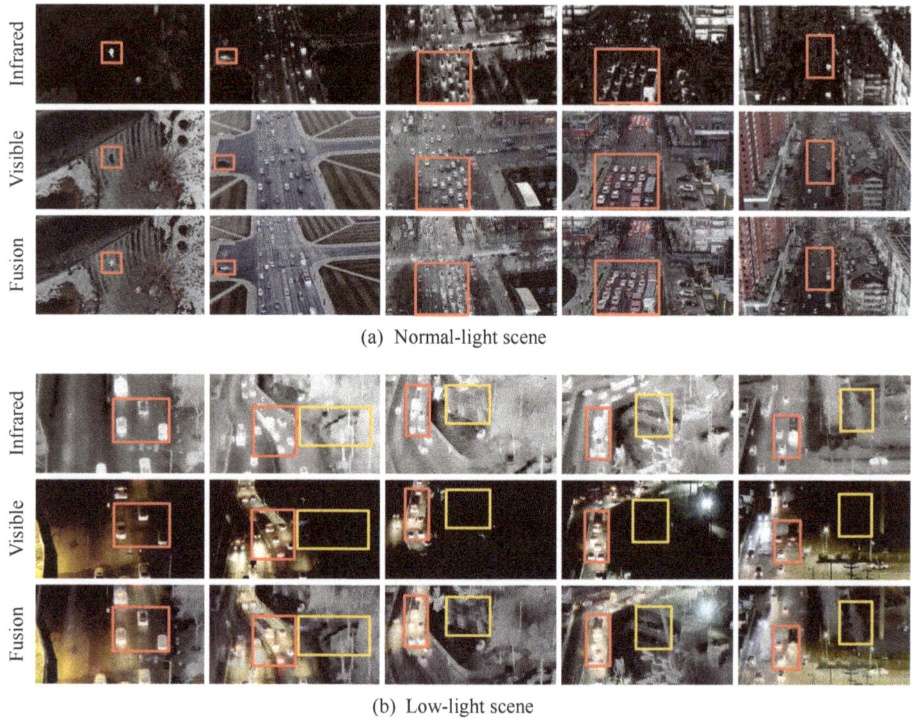

Figure 12. Fused images in normal and low-light scenes on the VTUAV dataset. The orange boxes show our fusion results in very low-light areas.

We evaluate the performance of our method using remote sensing data that include natural environments, urban landscapes, and beach scenes. Figure 13 shows our fused images in these environments. Our fusion method effectively integrates valuable information from the source images, achieving satisfactory results in terms of illumination, detail, and structural integrity. The fused images across the first, second, and third columns exhibit our method's capability to successfully fuse infrared and visible data, enhancing the clarity in details and structures, as highlighted in the red boxes. Moreover, our approach excels at retaining essential features while disregarding irrelevant information, as seen in the urban and beach scenes of the fourth and fifth columns, respectively. Despite the visible images in the fourth and fifth columns being somewhat dark and containing some details, our fusion outcome maintains these details without being affected by the abnormal illumination of the visible image. Our method is robust in preserving critical information across diverse scenes and lighting conditions.

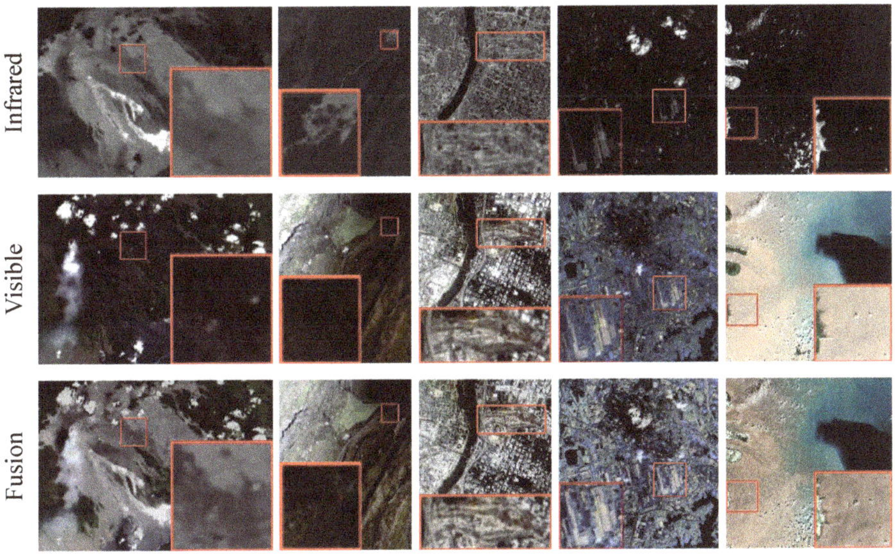

Figure 13. Fusion results in remote sensing imagery. The red boxes are enlarged to highlight the fusion performance on image details.

4.3.2. Qualitative Results

Figures 14 and 15 provide a quantitative comparison between our method and the state-of-the-art (SOTA) methods on the TNO and RoadScene datasets, respectively. The average metric values for these methods are summarized in Tables 1 and 2, respectively. Our method stands out in terms of overall performance. On the TNO dataset, our UNIFusion obtains better performance with the highest average values of SD and CC, indicating the effective integration of information from the source images while preserving the rich details in the fused images. Additionally, our method achieves the second-best results in AG and MS-SSIM, coming close to the top performer. This demonstrates our method's capability to integrate detailed information from source images effectively. In the RoadScene dataset's results, our method obtains remarkably high scores in AG, SD, and CC, further confirming its outstanding overall performance. While PFF achieves the best metrics in AG and SF by incorporating the characteristics of the human visual system, it relies on complex decomposition algorithms and faces challenges in preserving the rich information from the source images.

Table 1. Quantitative analysis on the TNO dataset. The best results are highlighted in red, the second-best in pink, and the third-best in orange.

Methods	AG	SD	SF	MS-SSIM	CC
FusionGAN [26]	3.41 ± 1.27	30.73 ± 6.10	4.32 ± 1.26	0.754 ± 0.10	0.761 ± 0.10
MFEIF [48]	4.24 ± 1.90	34.85 ± 8.26	4.86 ± 1.39	0.914 ± 0.03	0.771 ± 0.13
PFF [47]	10.02 ± 4.40	40.52 ± 7.24	8.76 ± 1.61	0.782 ± 0.09	0.722 ± 0.13
PIAFusion [18]	6.69 ± 3.27	41.95 ± 11.48	6.77 ± 1.73	0.860 ± 0.06	0.752 ± 0.13
PMGI [29]	4.86 ± 1.43	39.12 ± 4.04	5.51 ± 1.30	0.912 ± 0.07	0.750 ± 0.13
RFN [50]	3.40 ± 1.11	43.89 ± 9.63	4.24 ± 1.16	0.896 ± 0.05	0.780 ± 0.15
SwinFusion [17]	6.52 ± 2.90	39.74 ± 10.88	6.72 ± 1.62	0.890 ± 0.06	0.758 ± 0.13
U2Fusion [15]	6.91 ± 2.20	40.06 ± 7.42	7.11 ± 1.42	0.931 ± 0.03	0.779 ± 0.14
UMF [49]	4.63 ± 1.87	32.60 ± 6.83	5.15 ± 1.41	0.896 ± 0.07	0.768 ± 0.14
Ours	8.24 ± 3.58	45.20 ± 4.70	7.06 ± 0.65	0.928 ± 0.03	0.795 ± 0.14

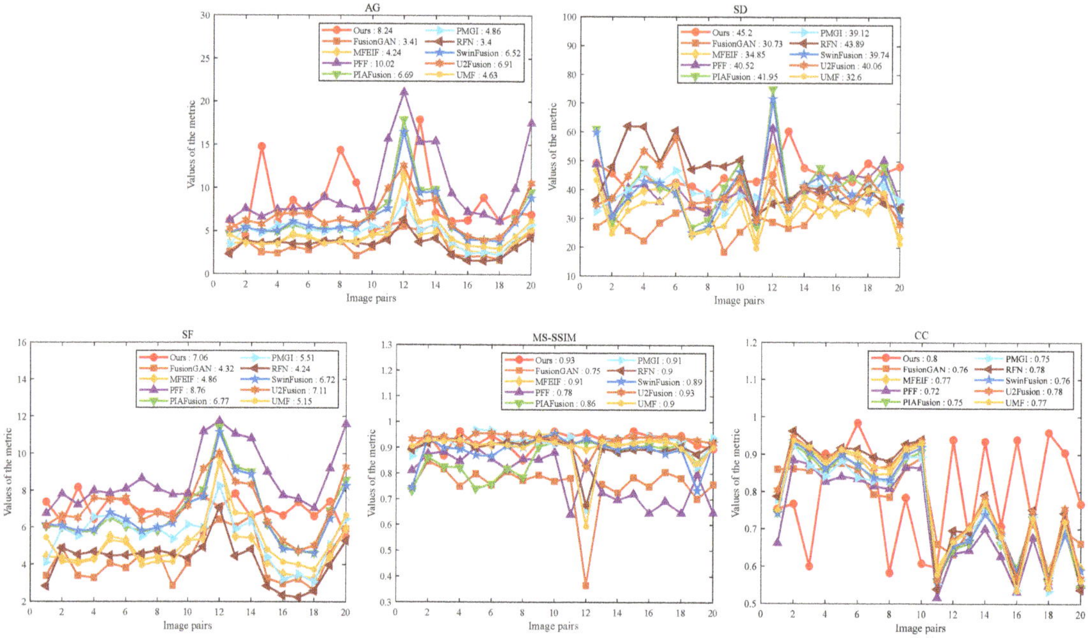

Figure 14. Comparative analysis of nine state-of-the-art methods using five metrics on the TNO dataset.

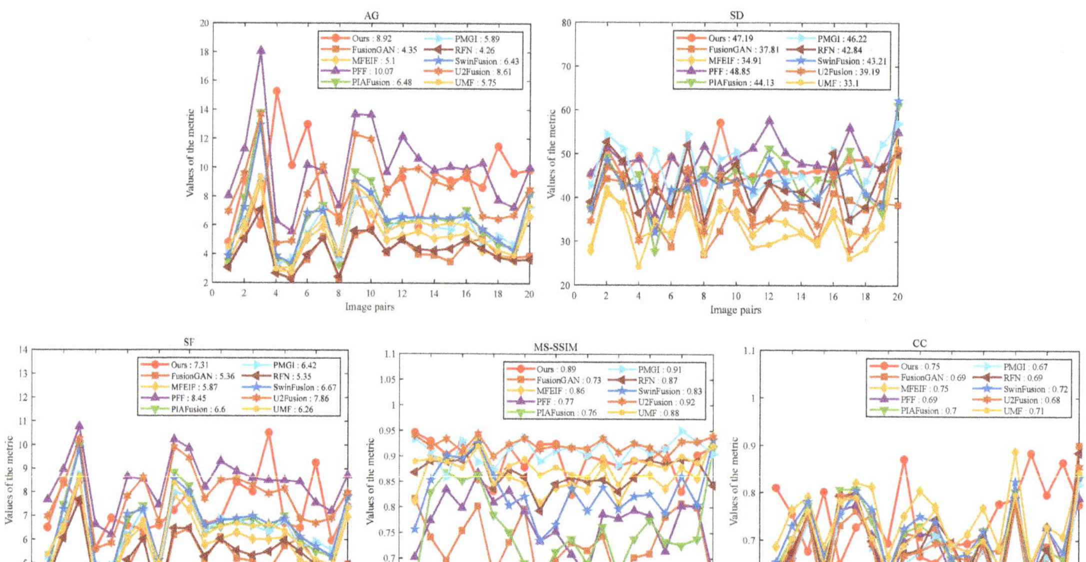

Figure 15. Comparative analysis of nine state-of-the-art methods using five metrics on the Road-Scene dataset.

Table 2. Quantitative analysis on the RoadScene dataset. The best results are highlighted in red, the second-best in pink, and the third-best in orange.

Methods	AG	SD	SF	MS-SSIM	CC
FusionGAN [26]	4.35 ± 1.41	37.81 ± 5.17	5.36 ± 1.11	0.731 ± 0.06	0.692 ± 0.07
MFEIF [48]	5.1 ± 1.58	34.91 ± 5.77	5.87 ± 1.24	0.864 ± 0.04	0.750 ± 0.06
PFF [47]	10.07 ± 2.86	48.85 ± 4.64	8.45 ± 1.13	0.770 ± 0.05	0.692 ± 0.06
PIAFusion [18]	6.48 ± 2.55	44.13 ± 6.70	6.60 ± 1.49	0.757 ± 0.08	0.701 ± 0.08
PMGI [29]	5.89 ± 1.58	46.22 ± 6.25	6.42 ± 1.14	0.911 ± 0.02	0.674 ± 0.07
RFN [50]	4.26 ± 1.18	42.84 ± 5.85	5.35 ± 1.02	0.867 ± 0.03	0.692 ± 0.08
SwinFusion [17]	6.43 ± 2.22	43.21 ± 5.88	6.67 ± 1.34	0.831 ± 0.05	0.718 ± 0.06
U2Fusion [15]	8.61 ± 2.39	39.19 ± 6.59	7.86 ± 1.23	0.923 ± 0.01	0.678 ± 0.08
UMF [49]	5.75 ± 1.72	33.10 ± 6.06	6.26 ± 1.24	0.883 ± 0.02	0.707 ± 0.07
Ours	8.92 ± 2.51	47.19 ± 3.24	7.31 ± 1.19	0.895 ± 0.02	0.752 ± 0.07

4.4. Ablation Study

We conducted experiments to analyze the effectiveness of the proposed method for infrared and visible image fusion. The fusion results with and without the unified feature space (UFS), non-local Gaussian filter (NGF), and dense attention (DA) were compared in the experiments. Figure 16 shows the fused images with and without UFS. It can be seen that the method without UFS generates blurred text on the signboard (see the red boxes in the first row of Figure 16) and does not sufficiently retain the information from the source images. In contrast, our method with UFS produces a detailed fusion result, particularly with much clearer text. From the second row of Figure 16, it can be observed that our method can retain more details of the car compared with the method without UFS. Furthermore, the red boxes in the first row of Figure 16 show that our method generates clearer edges on the signboard, indicating that the unified feature space (UFS) effectively fuses information from different modalities, thereby achieving high fusion performance. In the absence of NGF, there is an increase in noise within the fused image (see the red box in Figure 17). Compared with the method without NGF, our method not only removes more noise but also preserves image details and structures. We propose the dense attention-based feature extraction module to obtain multi-scale features, which can learn the significant features and relationships between different layers. Without dense attention, the extraction of key features becomes challenging, resulting in fusion outcomes that are lacking in detail. In Figure 18, without dense attention (DA), features such as the clouds in the sky and people on the grass appear less prominent and blurred. In contrast, our fusion results are richer in detail and clarity.

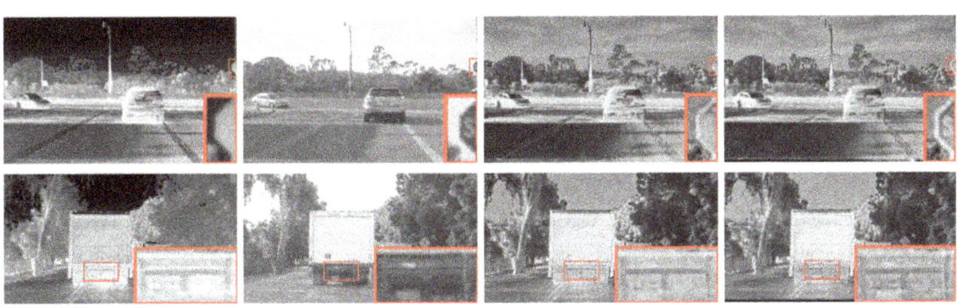

Infrared | Visible | W/O UFS | Ours

Figure 16. The fused images with and without the unified feauture space (UFS). The red boxes are enlarged to highlight the fusion performance on image details.

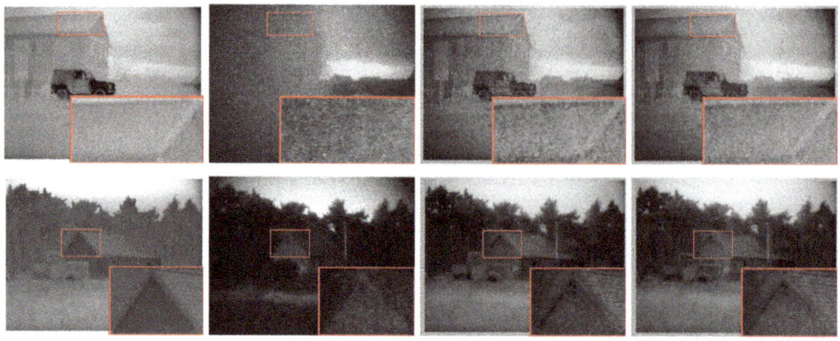

Figure 17. The fused images with and without the non-local Gaussian filter (NGF). The red boxes are enlarged to highlight the fusion performance on image details.

Figure 18. The fused images with and without the dense attention (DA). The red boxes are enlarged to highlight the fusion performance on image details.

We selected three representative metrics to demonstrate the effectiveness of each module: AG, MS-SSIM, and CC. AG indicates that the image contains rich information, while MS-SSIM and CC suggest that the fusion results retain substantial content from the source images. Table 3 presents the comparison results, which demonstrate that each component influences the overall performance. The removal of UFS lead to a marked decrease in AG, indicating its vital role in the fusion process and in maintaining rich information. The absence of NGF and DA leads to a decrease in MS-SSIM, as shown in Table 3, which shows that our proposed NGF and DA are capable of retaining more information from the source image. The absence of DA leading to a significant decrease in MS-SSIM indicates that DA captures essential features, thereby enriching the fusion results with more details from the source images. Both the qualitative and quantitative results demonstrate that the UFS, NGF, and DA are effective in removing noise while maintaining the information from the source images.

Table 3. The results of the ablation study on the TNO dataset. The best results are highlighted in red.

Methods	AG	MS-SSIM	CC
W/O UFS	7.94 ± 3.24	0.927 ± 0.03	0.79 ± 0.14
W/O NGF	8.10 ± 3.30	0.920 ± 0.03	0.79 ± 0.13
W/O DA	8.05 ± 3.17	0.908 ± 0.04	0.79 ± 0.13
Ours	8.24 ± 3.58	0.930 ± 0.03	0.80 ± 0.14

5. Conclusions

In this paper, we fuse infrared and visible images through feature-based decomposition and domain normalization. This decomposition method separates infrared and visible images into common and unique regions. We apply domain normalization to the common regions within the unified feature space to reduce modal differences while retaining unique information. The domain normalization is achieved by transforming the infrared features into a pseudo-visible domain via the unified feature space based on dynamic instance normalization (DIN). Thus, we create a consistent space for the fusion of information from diverse source images, while eliminating modal differences that affect the fusion process. To effectively extract essential features, we integrate a novel dense attention into the feature extraction process. The dense attention ensures that the network can dynamically capture key information across various layers, thereby improving the overall fusion performance in comparison to existing CNN-based methods, autoencoder-based approaches, and others. As the source images may contain noise, we propose a non-local Gaussian filter with learnable filter kernels that depend on the image content. This approach filters out noise while preserving the image details and structure. The experimental results indicate that our method can achieve fusion results of higher quality.

Author Contributions: Conceptualization, W.C. and Y.W.; methodology, Y.W. and W.C.; software, Y.W.; validation, Y.W.; formal analysis, W.C., Z.Z., and L.M.; investigation, Y.W. and Y.Q.; resources, L.M. and Z.Z.; data curation, W.C., L.M., and Z.Z.; writing—original draft preparation, Y.W.; writing—review and editing, W.C. and Y.W.; visualization, Y.W. and Y.Q.; supervision, L.M. and Z.Z.; project administration, W.C and Y.W. All authors have read and agreed to the published version of the manuscript.

Funding: This work was supported in part by the National Natural Science Foundation of China under Grant 62173040 and Grant 62071036.

Data Availability Statement: The source code of the paper is available at https://github.com/wyhlaowang/DNFusion (accessed on 28 February 2024).

Conflicts of Interest: The authors declare no conflicts of interest.

References

1. Ma, J.; Ma, Y.; Li, C. Infrared and visible image fusion methods and applications: A survey. *Inf. Fusion* **2019**, *45*, 153–178. [CrossRef]
2. Yang, Y.; Zhang, Y.; Huang, S.; Zuo, Y.; Sun, J. Infrared and visible image fusion using visual saliency sparse representation and detail injection model. *IEEE Trans. Instrum. Meas.* **2020**, *70*, 5001715. [CrossRef]
3. Hait, E.; Gilboa, G. Spectral total-variation local scale signatures for image manipulation and fusion. *IEEE Trans. Image Process.* **2018**, *28*, 880–895. [CrossRef]
4. Vishwakarma, A.; Bhuyan, M.K. Image fusion using adjustable non-subsampled shearlet transform. *IEEE Trans. Instrum. Meas.* **2018**, *68*, 3367–3378. [CrossRef]
5. Zhou, Z.; Wang, B.; Li, S.; Dong, M. Perceptual fusion of infrared and visible images through a hybrid multi-scale decomposition with Gaussian and bilateral filters. *Inf. Fusion* **2016**, *30*, 15–26. [CrossRef]
6. Zhou, Z.; Dong, M.; Xie, X.; Gao, Z. Fusion of infrared and visible images for night-vision context enhancement. *Appl. Opt.* **2016**, *55*, 6480–6490. [CrossRef]
7. Li, H.; Wu, X.J.; Kittler, J. MDLatLRR: A novel decomposition method for infrared and visible image fusion. *IEEE Trans. Image Process.* **2020**, *29*, 4733–4746. [CrossRef]
8. Bavirisetti, D.P.; Dhuli, R. Fusion of infrared and visible sensor images based on anisotropic diffusion and Karhunen-Loeve transform. *IEEE Sens. J.* **2015**, *16*, 203–209. [CrossRef]
9. Cvejic, N.; Bull, D.; Canagarajah, N. Region-based multimodal image fusion using ICA bases. *IEEE Sens. J.* **2007**, *7*, 743–751. [CrossRef]
10. Wan, T.; Canagarajah, N.; Achim, A. Segmentation-driven image fusion based on alpha-stable modeling of wavelet coefficients. *IEEE Trans. Multimed.* **2009**, *11*, 624–633. [CrossRef]
11. Han, J.; Pauwels, E.J.; De Zeeuw, P. Fast saliency-aware multi-modality image fusion. *Neurocomputing* **2013**, *111*, 70–80. [CrossRef]
12. Ellmauthaler, A.; da Silva, E.A.; Pagliari, C.L.; Neves, S.R. Infrared-visible image fusion using the undecimated wavelet transform with spectral factorization and target extraction. In Proceedings of the 2012 19th IEEE International Conference on Image Processing, Orlando, FL, USA, 30 September–3 October 2012; IEEE: Piscataway, NJ, USA, 2012; pp. 2661–2664.

13. Liu, Y.; Chen, X.; Cheng, J.; Peng, H.; Wang, Z. Infrared and visible image fusion with convolutional neural networks. *Int. J. Wavelets Multiresolution Inf. Process.* **2018**, *16*, 1850018. [CrossRef]
14. Li, Q.; Han, G.; Liu, P.; Yang, H.; Chen, D.; Sun, X.; Wu, J.; Liu, D. A multilevel hybrid transmission network for infrared and visible image fusion. *IEEE Trans. Instrum. Meas.* **2022**, *71*, 1–14. [CrossRef]
15. Xu, H.; Ma, J.; Jiang, J.; Guo, X.; Ling, H. U2Fusion: A unified unsupervised image fusion network. *IEEE Trans. Pattern Anal. Mach. Intell.* **2020**, *44*, 502–518. [CrossRef] [PubMed]
16. Jian, L.; Yang, X.; Liu, Z.; Jeon, G.; Gao, M.; Chisholm, D. SEDRFuse: A symmetric encoder–decoder with residual block network for infrared and visible image fusion. *IEEE Trans. Instrum. Meas.* **2020**, *70*, 5002215. [CrossRef]
17. Ma, J.; Tang, L.; Fan, F.; Huang, J.; Mei, X.; Ma, Y. SwinFusion: Cross-domain long-range learning for general image fusion via swin transformer. *IEEE/CAA J. Autom. Sin.* **2022**, *9*, 1200–1217. [CrossRef]
18. Tang, L.; Yuan, J.; Zhang, H.; Jiang, X.; Ma, J. PIAFusion: A progressive infrared and visible image fusion network based on illumination aware. *Inf. Fusion* **2022**, *83*, 79–92. [CrossRef]
19. Lebedev, M.; Komarov, D.; Vygolov, O.; Vizilter, Y.V. Multisensor image fusion based on generative adversarial networks. In Proceedings of the Image and Signal Processing for Remote Sensing XXV, Strasbourg, France, 9–11 Septmeber 2019; SPIE: Bellingham, WA, USA, 2019; Volume 11155, pp. 565–574.
20. Cui, Y.; Du, H.; Mei, W. Infrared and visible image fusion using detail enhanced channel attention network. *IEEE Access* **2019**, *7*, 182185–182197. [CrossRef]
21. Li, Y.; Wang, J.; Miao, Z.; Wang, J. Unsupervised densely attention network for infrared and visible image fusion. *Multimed. Tools Appl.* **2020**, *79*, 34685–34696. [CrossRef]
22. Li, H.; Wu, X.J. DenseFuse: A fusion approach to infrared and visible images. *IEEE Trans. Image Process.* **2018**, *28*, 2614–2623. [CrossRef]
23. Hou, R.; Zhou, D.; Nie, R.; Liu, D.; Xiong, L.; Guo, Y.; Yu, C. VIF-Net: An unsupervised framework for infrared and visible image fusion. *IEEE Trans. Comput. Imaging* **2020**, *6*, 640–651. [CrossRef]
24. Liu, L.; Chen, M.; Xu, M.; Li, X. Two-stream network for infrared and visible images fusion. *Neurocomputing* **2021**, *460*, 50–58. [CrossRef]
25. Liao, B.; Du, Y.; Yin, X. Fusion of infrared-visible images in UE-IoT for fault point detection based on GAN. *IEEE Access* **2020**, *8*, 79754–79763. [CrossRef]
26. Ma, J.; Yu, W.; Liang, P.; Li, C.; Jiang, J. FusionGAN: A generative adversarial network for infrared and visible image fusion. *Inf. Fusion* **2019**, *48*, 11–26. [CrossRef]
27. Ma, J.; Xu, H.; Jiang, J.; Mei, X.; Zhang, X.P. DDcGAN: A dual-discriminator conditional generative adversarial network for multi-resolution image fusion. *IEEE Trans. Image Process.* **2020**, *29*, 4980–4995. [CrossRef]
28. Song, A.; Duan, H.; Pei, H.; Ding, L. Triple-discriminator generative adversarial network for infrared and visible image fusion. *Neurocomputing* **2022**, *483*, 183–194. [CrossRef]
29. Zhang, H.; Xu, H.; Xiao, Y.; Guo, X.; Ma, J. Rethinking the image fusion: A fast unified image fusion network based on proportional maintenance of gradient and intensity. In Proceedings of the AAAI Conference on Artificial Intelligence, New York, NY, USA, 7–12 February 2020; Volume 34, pp. 12797–12804.
30. Wang, J.; Peng, J.; Feng, X.; He, G.; Fan, J. Fusion method for infrared and visible images by using non-negative sparse representation. *Infrared Phys. Technol.* **2014**, *67*, 477–489. [CrossRef]
31. Xu, D.; Zhang, N.; Zhang, Y.; Li, Z.; Zhao, Z.; Wang, Y. Multi-scale unsupervised network for infrared and visible image fusion based on joint attention mechanism. *Infrared Phys. Technol.* **2022**, *125*, 104242. [CrossRef]
32. Li, J.; Huo, H.; Li, C.; Wang, R.; Feng, Q. AttentionFGAN: Infrared and visible image fusion using attention-based generative adversarial networks. *IEEE Trans. Multimed.* **2020**, *23*, 1383–1396. [CrossRef]
33. Yuan, C.; Sun, C.; Tang, X.; Liu, R. Flgc-fusion gan: An enhanced fusion gan model by importing fully learnable group convolution. *Math. Probl. Eng.* **2020**, *2020*, 6384831 . [CrossRef]
34. Isola, P.; Zhu, J.Y.; Zhou, T.; Efros, A.A. Image-to-image translation with conditional adversarial networks. In Proceedings of the IEEE Conference on Computer Vision and Pattern Recognition, Honolulu, HI, USA, 21–26 July 2017; pp. 1125–1134.
35. Wang, T.C.; Liu, M.Y.; Zhu, J.Y.; Tao, A.; Kautz, J.; Catanzaro, B. High-resolution image synthesis and semantic manipulation with conditional gans. In Proceedings of the IEEE Conference on Computer Vision and Pattern Recognition, Salt Lake City, UT, USA, 18–23 June 2018; pp. 8798–8807.
36. Kim, T.; Cha, M.; Kim, H.; Lee, J.K.; Kim, J. Learning to discover cross-domain relations with generative adversarial networks. In Proceedings of the International Conference on Machine Learning, PMLR 2017, Sydney, Australia, 6–11 August 2017; pp. 1857–1865.
37. Liu, M.Y.; Breuel, T.; Kautz, J. Unsupervised image-to-image translation networks. In Proceedings of the 31st Conference on Neural Information Processing Systems (NIPS 2017), Long Beach, CA, USA, 4–9 December 2017; Volume 30.
38. Zhu, J.Y.; Park, T.; Isola, P.; Efros, A.A. Unpaired image-to-image translation using cycle-consistent adversarial networks. In Proceedings of the IEEE International Conference on Computer Vision, Venice, Italy, 22–29 October 2017; pp. 2223–2232.
39. Zhu, J.Y.; Zhang, R.; Pathak, D.; Darrell, T.; Efros, A.A.; Wang, O.; Shechtman, E. Toward multimodal image-to-image translation. In Proceedings of the 31st Conference on Neural Information Processing Systems (NIPS 2017), Long Beach, CA, USA, 4–9 December 2017; Volume 30.

40. Bousmalis, K.; Silberman, N.; Dohan, D.; Erhan, D.; Krishnan, D. Unsupervised pixel-level domain adaptation with generative adversarial networks. In Proceedings of the IEEE Conference on Computer Vision and Pattern Recognition, Honolulu, HI, USA, 21–26 July 2017; pp. 3722–3731.
41. Huang, X.; Belongie, S. Arbitrary style transfer in real-time with adaptive instance normalization. In Proceedings of the IEEE International Conference on Computer Vision, Venice, Italy, 22–29 October 2017; pp. 1501–1510.
42. Jing, Y.; Liu, X.; Ding, Y.; Wang, X.; Ding, E.; Song, M.; Wen, S. Dynamic instance normalization for arbitrary style transfer. In Proceedings of the AAAI Conference on Artificial Intelligence, New York, NY, USA, 7–12 February 2020; Volume 34, pp. 4369–4376.
43. Wang, Z.; Bovik, A.C.; Sheikh, H.R.; Simoncelli, E.P. Image quality assessment: From error visibility to structural similarity. *IEEE Trans. Image Process.* **2004**, *13*, 600–612. [CrossRef]
44. Liu, J.; Fan, X.; Huang, Z.; Wu, G.; Liu, R.; Zhong, W.; Luo, Z. Target-aware dual adversarial learning and a multi-scenario multi-modality benchmark to fuse infrared and visible for object detection. In Proceedings of the IEEE/CVF Conference on Computer Vision and Pattern Recognition, New Orleans, LA, USA, 18–24 June 2022; pp. 5802–5811.
45. Toet, A. The TNO multiband image data collection. *Data Brief* **2017**, *15*, 249–251. [CrossRef] [PubMed]
46. Zhang, P.; Zhao, J.; Wang, D.; Lu, H.; Ruan, X. Visible-thermal UAV tracking: A large-scale benchmark and new baseline. In Proceedings of the IEEE/CVF Conference on Computer Vision and Pattern Recognition, New Orleans, LA, USA, 18–24 June 2022; pp. 8886–8895.
47. Zhou, Z.; Fei, E.; Miao, L.; Yang, R. A perceptual framework for infrared–visible image fusion based on multiscale structure decomposition and biological vision. *Inf. Fusion* **2023**, *93*, 174–191. [CrossRef]
48. Liu, J.; Fan, X.; Jiang, J.; Liu, R.; Luo, Z. Learning a deep multi-scale feature ensemble and an edge-attention guidance for image fusion. *IEEE Trans. Circuits Syst. Video Technol.* **2021**, *32*, 105–119. [CrossRef]
49. Di, W.; Jinyuan, L.; Xin, F.; Liu, R. Unsupervised Misaligned Infrared and Visible Image Fusion via Cross-Modality Image Generation and Registration. In Proceedings of the International Joint Conference on Artificial Intelligence (IJCAI), Vienna, Austria, 23–29 July 2022.
50. Li, H.; Wu, X.J.; Kittler, J. RFN-Nest: An end-to-end residual fusion network for infrared and visible images. *Inf. Fusion* **2021**, *73*, 72–86. [CrossRef]
51. Cui, G.; Feng, H.; Xu, Z.; Li, Q.; Chen, Y. Detail preserved fusion of visible and infrared images using regional saliency extraction and multi-scale image decomposition. *Opt. Commun.* **2015**, *341*, 199–209. [CrossRef]
52. Deshmukh, M.; Bhosale, U.; et al. Image fusion and image quality assessment of fused images. *Int. J. Image Process. (IJIP)* **2010**, *4*, 484.
53. Roberts, J.W.; Van Aardt, J.A.; Ahmed, F.B. Assessment of image fusion procedures using entropy, image quality, and multispectral classification. *J. Appl. Remote Sens.* **2008**, *2*, 023522.
54. Ma, K.; Zeng, K.; Wang, Z. Perceptual quality assessment for multi-exposure image fusion. *IEEE Trans. Image Process.* **2015**, *24*, 3345–3356. [CrossRef]

remote sensing

MDPI

Article

Multi-Source T-S Target Recognition via an Intuitionistic Fuzzy Method

Chuyun Zhang [1,2], Weixin Xie [1,2], Yanshan Li [1,2] and Zongxiang Liu [1,2,*]

1 Guangdong Key Laboratory of Intelligent Information Processing, Shenzhen University, Shenzhen 518060, China; zhangchuyun2019@email.szu.edu.cn (C.Z.)
2 College of Electronics and Information Engineering, Shenzhen University, Shenzhen 518060, China
* Correspondence: liuzx@szu.edu.cn; Tel.: +86-755-26732055

Abstract: To realize aerial target recognition in a complex environment, we propose a multi-source Takagi–Sugeno (T-S) intuitionistic fuzzy rules method (MTS-IFRM). In the proposed method, to improve the robustness of the training process of the model, the features of the aerial targets are classified as the input results of the corresponding T-S target recognition model. The intuitionistic fuzzy approach and ridge regression method are used in the consequent identification, which constructs a regression model. To train the premise parameter and reduce the influence of data noise, novel intuitionistic fuzzy C-regression clustering based on dynamic optimization is proposed. Moreover, a modified adaptive weight algorithm is presented to obtain the final outputs, which improves the classification accuracy of the corresponding model. Finally, the experimental results show that the proposed method can effectively recognize the typical aerial targets in error-free and error-prone environments, and that its performance is better than other methods proposed for aerial target recognition.

Keywords: target recognition; T-S intuitionistic fuzzy rules; ridge regression; adaptive weight

1. Introduction

The complexity of the battlefield environment is enhanced significantly by high-tech equipment, which has introduced great difficulties to the acquisition of target information. As the battlefield expands to the five-dimensional space of sea, land, air, sky, and electromagnetics, the collection of target information will not only be affected by the accuracy and stability of sensor equipment, the influences of the climate environment, and the complex electromagnetic field environment, but also by other factors that lead to deviations or even errors in the collected target information. In addition, there will be interference and confusing equipment intentionally released by the enemy, which increases the uncertainty of the observation of the target. Therefore, it is difficult for a single information source to obtain accurate and complete intelligence information in such a complex environment, and also meet the requirements of actual aerial combat.

With the development of multi-source detection technology, a structure able to track multiple targets and realize target recognition is essential to a multi-sensor data fusion system. Information fusion can recognize a target from multiple dimensions and multiple directions, which data can then be comprehensively processed with the complementarity and redundancy of information, to eliminate the influence of inaccuracy and incompleteness of information obtained from a single information source. Moreover, multi-feature fusion processing is designed to obtain more accurate target features by data fusion of two or more sensors, thus breaking the limits of single-sensor detection, in which equipment generally collects the information of only one feature within a corresponding sensing range [1]. Target features obtained by different sensors are imprecise and conflict with the influence of complex environments, interference signals and so on; for example, impulsive noise may cause the collected data to deviate from the original range, leading to the drawing of the wrong conclusions in the target recognition system. Therefore, muti-feature fusion and improving the interpretability of target recognition are particularly important.

1.1. Literature Review and Motivation

For the recognition system, a series of methods have been presented, such as the Dempster–Shafer (D-S) [2–4], fuzzy set [5–7], probability statistics [8–10], the gray system [11,12], rough sets [13–15], and fractal theory [16–18]. D-S evidence theory, a general framework for information fusion, is used to combine multi-level information from multi-source environments for reasoning and dealing with uncertainty, imprecision, and incompletion [19,20]. Therefore, extended evidence theories have been well established in information fusion [21], decision analysis [22], risk assessment [23,24], pattern recognition [25], and other fields. However, traditional evidence theory has low accuracy because of the problems of constructing a basic probability assignment (BPA) and conflict management.

With regard to the framework of the BPA, some modeling approaches have been provided. Moreover, Dempster's combination method is performed to transform the BPA into probability distribution, the quality of the BPA in evidence theory will determine whether the recognition result is reasonable. Yin et al. [26] proposed a measurement model to achieve uncertainty management of the BPA via the processing of negation and the links between uncertain data and entropy. Jiang et al. [27] constructed a correlation coefficient to describe the non-intersection and the distinctions between the focal elements. Wang et al. [28] proposed a belief divergence measurement that presented the correlation of various kinds of subsets with a belief function and an appropriate probability distribution. Kaur et al. [8] processed nonnegative and symmetric divergence measures for BPA. Hu et al. [9] proposed the cross-information to change the comprehensive BPA. However, an algorithm based on decision-level data fusion needs high data preprocessing and the decision-making methods are short of general structure after obtaining the characterized distributions of basis reliability.

When coping with highly conflicting evidence, D-S evidence theory may lead to counter-intuitionistic recognition. Therefore, many methods have been proposed including Yager's combination rules method [29], Murphy's arithmetically average model of bodies of evidence [30], Li's trust-based method [31], and so on. Target recognition methods based on fuzzy set theory only need a small amount of prior knowledge to achieve more efficient and accurate recognition. Wang [32] proposed the intuitionistic fuzzy dynamic Bayesian network to transfer the outputs of intuitionistic fuzzy rules into probability. Jiang [7] established a hybrid decision-making fuzzy rough and hesitant sets model and developed a machine learning mechanism to construct the relative loss functions. Guo [33] proposed the recognition structure of UAVs based on a recurrent convolutional strategy, which influenced the degrees of super-resolution realization by setting the numbers of cycles and iterations with changes in the blur degree. Moreover, intuitionistic fuzzy sets (IFS) can conquer the inaccuracy and limitations of traditional fuzzy sets for solving specific information and eliminate the bottleneck that Bayesian models excessively rely on. Lei [34] proposed an intuitionistic fuzzy reasoning (IFR) framework to obtain the membership and non-membership degrees of the property variables of a recognition model. Dolgiy [35] combined the D-S method and Takagi–Sugeno (T-S) fuzzy system to develop the empirical process of an expert system of probability estimates based on subjective preferences of the description of typical sensors. Therefore, a novel hybrid T-S and intuitionistic fuzzy inference system are applied to target recognition in our method.

1.2. Our Contributions

In this paper, a novel MTS-IFRM is proposed for high-performance multi-target recognition in error-free and error-prone environments. The main novelties of our method include:

- Improving the robustness of the training process of the model: the features of the aerial targets are classified as inputs to the corresponding T-S target recognition model, so that features are divided into multi-level features with the target properties;
- In the T-S model algorithm, the study of premise and consequence parameter identification has been the key question. We apply an intuitionistic fuzzy C-means method based on the dynamic particle swarm optimization (DPSO) algorithm and the ridge regression model to identify the premise and consequence parameter of the T-S intuitionistic fuzzy model, respectively, which better realizes the parametric identification of the model;
- High classification accuracy can be guaranteed in error-free and error-prone environments. The adaptive weight algorithm reduces the weight corresponding to the model with a low degree of discrimination and increases the weight corresponding to the model with a high degree of discrimination, which is better distinguished from the input features.

1.3. Organization of the Article

The organization of the method is described as follows: The fuzzy target recognition model is given in Section 2. Model construction and parameter identification are presented in Section 3. The simulation results and an analysis with comparable methods are given in Section 4. Finally, the conclusions are organized in Section 5. The meanings of notation in the article are listed in Table 1.

Table 1. Notation list.

Notation	Meaning of the Notation	Notation	Meaning of the Notation
Θ	Discriminative frame	s	Scoring function set
R^l	Fuzzy rule l	N	Number of training samples
z_{CA}	Inputs of CA	X_i, V_i, P_i	Position, velocity, optimal solution of the i-th particle
E_{CA}	Universe of discourses of CA	G	Size of particle swarm
A_1^l	Intuitionistic fuzzy subsets	P_g	Current global optimal solution
p^l	Consequent parameter	$w_{\min}, w_{\max}$	Minimum, maximum inertia weights
$\mu(\bullet), v(\bullet)$	Membership, non-membership degree	c_1, c_2	Learning parameter
$\pi(\bullet)$	Intuitionistic index	T	Number of iterations
L_{RG}	Number of fuzzy rules	M	Number of label vector dimensions
y_{RG}^0	Outputs for the model	$f_{\min}, f_{\max}, f_{avg}$	Minimum, maximum, and average fitness of the particle swarm

2. Preliminaries

In this section, the preliminaries of the Dempster–Shafer evidence theory and Takagi–Sugeno intuitionistic fuzzy rules method are first introduced.

2.1. Evidence Theory

Dempster–Shafer evidence theory has flexibility and effectiveness in modeling uncertainties without prior information [19]. A discriminative frame Θ consisting of all possible propositions is defined as follows:

$$\Theta = \{\theta_1, \theta_2, \cdots, \theta_i, \cdots, \theta_n\} \quad (1)$$

Mass function mapping m from 2^Θ to [0, 1] is defined as BBA, which satisfies the following conditions:

$$m(\varnothing) = 0 \ and \ \sum_{\theta \subseteq \Theta} m(\theta) = 1 \quad (2)$$

If $m(\theta) > 0$, then θ is described as the focal element. Suppose two independent basic belief assignments m_1, m_2 construct the form $m_1 \oplus m_2$ according to Dempster's rule of combination, which can be expressed as follows:

$$m(\theta) = \begin{cases} \frac{1}{1-K} \sum\limits_{E \cap F = \theta} m_1(E) m_2(F), & \theta \neq \varnothing \\ 0 & \theta = \varnothing \end{cases} \quad (3)$$

With

$$K = \sum_{E \cap F \neq \varnothing} m_1(E) m_2(F) \quad (4)$$

where $E, F \in 2^{\theta}$ and K is the conflict coefficient of m_1 and m_2. When the evidence is highly conflicting, the evidence fusion processing will lead to counter-intuitionistic results. For a multi-source target recognition system, a degree of conflict of the information is provided by each sensor, so dealing with the conflicts between the evidence is the key to applying various evidence-based theories for accurate target recognition.

The common features of information on aerial targets, such as flight speed, acceleration, flight height and so on, can be detected by a multi-source system. Due to the problem of various forms of signal interference and other factors, a system detecting the target information will contain a lot of uncertainty. Most methods based on decision-level data fusion, such as D-S and Yager, require a high level of data preprocessing and display low interpretability. In order to improve the interpretability of the information fusion and the process of aerial target recognition, the T-S intuitionistic fuzzy model is introduced to establish mapping between the feature space and the target space. The T-S intuitionistic fuzzy model has strong learning ability and robustness, which means it can label historically detected targets with the correct categories, and input their feature information into the T-S intuitionistic fuzzy model for training after intuitionistic fuzzification, then forming a correct mapping relationship. By continuously learning target features, the final trained model can accurately obtain the mapping relationship between the features and the targets.

2.2. Takagi–Sugeno Intuitionistic Fuzzy Rules Method

When the number of input variables increases, the number of rules of the T-S model will increase exponentially, resulting in a decrease in training performance. For typical aerial targets, we divide the features of the aerial targets into two or three groups for modeling. Figure 1 illustrates the classification and the process of target recognition.

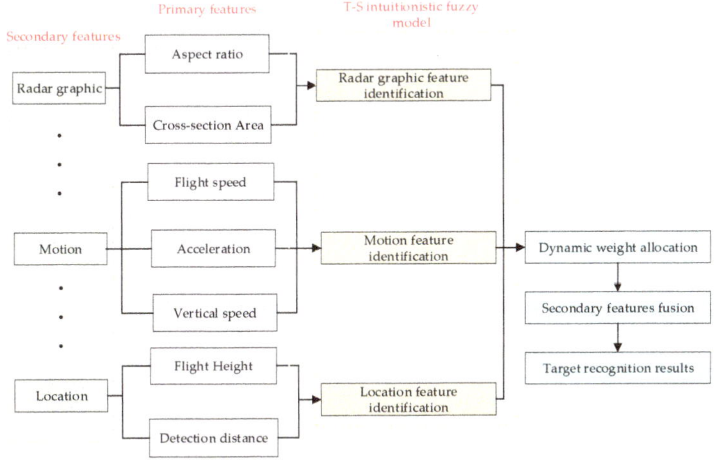

Figure 1. Target recognition T-S intuitionistic fuzzy model.

First, the features are divided into primary features and secondary features with the target properties, and each secondary feature contains two or three primary features. Then, the model is trained by the training data to obtain the premise and consequence parameters, and the primary features are fused and judged by the trained MTS-IFRM. Finally, the identity estimation results of the target are fused with secondary features to obtain the final recognition result of the target.

The main difficulty of aerial target recognition lies in the fusion of multiple features. Achieving accurate recognition of targets from imprecise and conflicting feature data is the key. This section will mainly introduce the proposed aerial target recognition algorithm. The MTS-IFRM is designed by taking the radar graphic (RG) as an example, the inputs of the model are the feature values of aspect ratio (AR) and cross-sectional area (CA) after intuitionistic fuzzification, then we define the MTS-IFRM based on a fuzzy set:

$$\begin{array}{c} Rule\ R^l\text{: } If\ z_{CA} \text{ is } A_1^l,\ and\ z_{AR}\ is A_2^l,\ then: \\ f_{RG}^l(z_{RG}) = p_{RG0}^l + p_{RG}^l S(z_{CA}) + p_{RG}^l S(z_{AR}), l = 1, 2, \ldots, L_{RG} \end{array} \tag{5}$$

where the part after "*if*" denotes the premise and the part after "*then*" denotes the consequence of the rule. $z_{CA} = \{\langle CA, \mu(CA), v(CA)\rangle | CA \in E_{CA}\}$ and $z_{AR} = \{\langle AR, \mu(AR), v(AR)\rangle | AR \in E_{AR}\}$ denote the inputs of the CA and AR after intuitionistic fuzzification, respectively. $\mu(\bullet)$ and $v(\bullet)$ are the degrees of membership and the non-membership, respectively, which represent the intuitionistic fuzzy number. Then, $0 \leq \mu(\bullet) + v(\bullet) \leq 1$. $\pi(\bullet) = 1 - \mu(\bullet) - v(\bullet)$ denotes the intuitionistic index of the intuitionistic fuzzy number. The specific process can be referenced in [36]. E_{CA} and E_{AR} denote the universe of discourses of the CA and AR, respectively. A_1^l and A_2^l denote the intuitionistic fuzzy subsets corresponding to the inputs z_{CA} and z_{AR} of rule l, respectively. The input vector $z_{RG} = [z_{CA}, z_{AR}]$ denotes the premise variable of the model. $p_{RG}^l = \left[p_{RG0}^l, p_{RG1}^l, p_{RG2}^l\right]$ denotes the consequence part. $S(\bullet)$ denotes the scoring function with the abilities of sequencing and decision-making, which converts an intuitionistic fuzzy set into a definite numerical value [37]. L_{RG} denotes the RG number of fuzzy rules. Therefore, the weighted average y_{RG}^0 of the final outputs for each rule $f_{RG}^l(z_{RG})$ are obtained by:

$$y_{RG}^0 = \sum_{l=1}^{L_{RG}} \frac{\mu^l(z_{RG}) f_{RG}^l(z_{RG})}{\sum_{\tilde{l}=1}^{L_1} \mu^{\tilde{l}}(z_{RG})} = \sum_{l=1}^{L_{RG}} \tilde{\mu}^l(z_{RG}) \cdot f_{RG}^l(z_{RG}) \tag{6}$$

where $\mu^l(z_{RG})$ denotes the fuzzy membership degree of fuzzy rule l to input z_{RG}. The normalization method is defined as:

$$\tilde{\mu}^l(z) = \frac{\mu^l(z_{RG})}{\sum_{\tilde{l}=1}^{L_{RG}} \mu^{\tilde{l}}(z_{RG})} \tag{7}$$

where

$$\mu^l(z_{RG}) = \mu_{A_1^l}(z_{CA}) \cdot \mu_{A_2^l}(z_{AR}) \tag{8}$$

$$\tilde{\mu}^l(z_{RG}) = \frac{\mu^l(z_{RG})}{\sum_{\tilde{l}=1}^{L_{RG}} \mu^{\tilde{l}}(z_{RG})} \tag{9}$$

$$\mu_{A_1^l}(z_{CA}) = \lambda_1 \mu_{A_1^l}(z_{CA}) + \lambda_2 v_{A_1^l}(z_{CA}) + \lambda_3 \pi_{A_1^l}(z_{CA}) \tag{10}$$

$$\mu_{A_1^l}(z_{AR}) = \lambda_1 \mu_{A_1^l}(z_{AR}) + \lambda_2 v_{A_1^l}(z_{AR}) + \lambda_3 \pi_{A_1^l}(z_{AR}) \tag{11}$$

Here, $\mu_{A_i^l}(\bullet)$, $v_{A_i^l}(\bullet)$ and $\pi_{A_i^l}(\bullet)$ are calculated by the premise parameter identification. $\mu_{A_i^l}(\bullet)$ can be expressed by using a suitable index λ (generally setting $\lambda_1 = 1$, $\lambda_2 = 0$, and $\lambda_3 = 0.5$).

Similarly, the MTS-IFRM based on the feature of motion (M) and location (L) can be established. The output results of the corresponding model are defined as follows:

$$y_{MF}^{0} = \sum_{l=1}^{L_M} \frac{\mu^{l}(z_M) f_M^{l}(z_M)}{\sum_{\tilde{l}=1}^{L_M} \mu^{\tilde{l}}(z_M)} = \sum_{l=1}^{L_M} \tilde{\mu}^{l}(z_M) \cdot f_M^{l}(z_M) \quad (12)$$

$$y_{L}^{0} = \sum_{l=1}^{L_L} \frac{\mu^{l}(z_3) f_L^{l}(z_L)}{\sum_{\tilde{l}=1}^{L_3} \mu^{k}(z_L)} = \sum_{l=1}^{L_L} \tilde{\mu}^{l}(z_L) \cdot f_{LF}^{l}(z_L) \quad (13)$$

where $z_M = [z_{FS}, z_A, z_{VS}], z_L = [z_{FH}, z_{DD}], z_{FS}, z_A, z_{VS}, z_{FH}$ and z_{DD} denote flight speed, acceleration, vertical speed, flight height, and detection distance features after intuitionistic fuzzification, respectively.

3. Aerial Target Recognition Methods Based on the MTS-IFRM

According to the above analysis, parameter identification is a central role of a T-S rule-based system, which evaluates the quality of the rule modeling. Therefore, the related work of the MTS-IFRM contains the structure identification of consequent parameters based on the ridge regression method, the identification of the premise part with a novel intuitionistic fuzzy C-means (IFCM) clustering model, and the adaptive weight algorithm.

3.1. Construction of MTS-IFRM

In this section, we take the training of the RG consequence parameters as an example. First, according to Equations (5) and (6), let:

$$s_e = \left(1, s^T\right)^T \quad (14)$$

where $s = [S(z_{CS}), S(z_{AR})]$ denotes the scoring function set of the input z. So that:

$$\tilde{s}_{RG}^{l} = \tilde{\mu}^{l}(z_{RG}) s_e \quad (15)$$

$$s_{g,RG} = \left(\left(\tilde{s}_{RG}^{1}\right)^T, \left(\tilde{s}_{RG}^{2}\right)^T, \ldots, \left(\tilde{s}_{RG}^{L_{RG}}\right)^T\right)^T \quad (16)$$

$$p_{RG}^{l} = \left(p_{RG}^{l}, p_{RG}^{l}, p_{RG}^{l}\right)^T \quad (17)$$

$$p_{g,RG} = \left(\left(p_{RG}^{1}\right)^T, \left(p_{RG}^{2}\right)^T, \ldots, \left(p_{RG}^{L_{RG}}\right)^T\right)^T \quad (18)$$

where $\tilde{\mu}^{l}(z_{RG})$ is acquired in Equation (7). Next, the output of the model is denoted as:

$$y_{RG}^{0} = \left(p_{g,RG}\right)^T s_{g,RG} \quad (19)$$

In Equation (19), we obtain the RG output of the MTS-IFRM. To solve the target recognition problem, each secondary feature needs to have the corresponding output. Therefore, the MTS-IFRM is constructed. The ridge regression method, a modified analysis of the least-squares estimation, can deal with multicollinearity by operating the unbiased estimator. To obtain a more reliable estimate of the consequent parameter, ridge regression analysis is constructed to train the model:

$$\min_{p_{g,m,RG}} J(p_{g,m,RG}) = \frac{1}{2} \sum_{m=1}^{M} \sum_{n=1}^{N} \left(\left(p_{g,m,RG}\right)^T s_{g,n,RG} - \tilde{y}_{n,m}\right)^2 + \gamma_1 \sum_{m=1}^{M} \sum_{n=1}^{N} \left(p_{g,m,RG}\right)^T p_{g,m,RG} \quad (20)$$

Equation (20) contains the minimization of empirical risk and structure risk. Where $p_{g,m,RG}$ denotes the consequent parameter of the m-th aerial target, N denotes the number of training samples, $\tilde{y}_{n,m}$ denotes the M-dimensional label vector of the n-th training sample,

γ_1 represents the regularization parameter. To adjust the consequent parameter $p_{g,m,RG}$, the final optimization result is calculated by the first-order necessary condition:

$$\frac{\partial J\left(p_{g,m,RG}\right)}{\partial p_{g,m,RG}} = \sum_{m=1}^{M}\sum_{n=1}^{N}\left(s_{g,n,RG}(s_{g,n,RG})^T + \gamma_1 I_{l\times l}\right)\cdot p_{g,m,RG} - \sum_{m=1}^{M}\sum_{n=1}^{N}\left(s_{g,n,RG}\tilde{y}_{n,m}\right) = 0 \quad (21)$$

In Equation (21), $p_{g,m,RG}$ is as follows:

$$p_{g,m,RG} = \sum_{n=1}^{N}\left(\gamma_1 I_{l\times l} + s_{g,n,RG}(s_{g,n,RG})^T\right)^{-1}\sum_{m=1}^{M}\sum_{n=1}^{N}\left(s_{g,m,RG}\tilde{y}_{n,m}\right) \quad (22)$$

Therefore, a new MTS-IFRM of RF for aerial target recognition can be expressed as follows according to Equations (5) and (22):

$$\begin{array}{r}\text{Rule } R^{l'}\text{: } If\ z'_{CS}\ is\ A_1^{l'},\ \text{and } z'_{AR}\ is A_2^{l'},\ then: \\ f_{RG}^{l'}(z'_{RG}) = p_{g,j,RG0}^{l'} + p_{g,j,RG1}^{l'}S(z'_{CS}) + p_{g,j,RG2}^{l'}S(z'_{AR}), l' = 1,2,\ldots,L'_{RG}\end{array} \quad (23)$$

where z'_{CS} and z'_{AR} are the CS features and AR features after intuitionistic fuzzification, respectively. $p_{g,m,RGi}^{l'}$ denotes the consequent parameter corresponding to rule l' of model m, here, $i = 0,1,2$ and L'_{RG} denotes the number of rules. Similarly, the corresponding rules of the MTS-IFRM for the motion feature (MF) and location feature (LF) can be established in the same construction procedures.

3.2. Premise Identification

IFCM and the FCM clustering are very sensitive to the initial clustering center position and are prone to converging to the local optimal solution in a noisy environment. Moreover, the variation factors of dynamic evolution theory are introduced into the PSO algorithm to improve the clustering optimization model [38].

Suppose the position of the i-th particle is $X_i = (x_{i,1}, x_{i,2}, \ldots, x_{i,d})$, the velocity is $V_i = (v_{i,1}, v_{i,2}, \ldots, v_{i,d})$ and $P_i = (p_{i,1}, p_{i,2}, \ldots, p_{i,d})$ is the optimal solution in d-dimensional space, where $i = 1,2,\cdots,G$, G is the size of the particle swarm, then the velocity and position updated in the j-th dimension at an iteration are:

$$v_{i,j}(t+1) = \omega v_{i,j}(t) + c_1 r_1(p_{i,j} - x_{i,j}(t)) + c_2 r_2(p_{g,j} - x_{i,j}(t)) \quad (24)$$

$$x_{i,j}(t+1) = x_{i,j}(t) + v_{i,j}(t+1), \quad j = 1,2,\cdots,d \quad (25)$$

where $P_g = (p_{g,1}, p_{g,2}, \ldots, p_{g,d})$ denotes the current global optimal solution, w denotes the inertia weight. r_1 and r_2 are random numbers in the interval [0, 1], respectively. c_1 and c_2 denote the learning parameter of the DPSO, respectively, and are defined as follows:

$$c_1 = 2.5 - 2\times\frac{t}{T} \quad (26)$$

$$c_2 = 2.5 + 2\times\frac{t}{T} \quad (27)$$

where t denotes the number of iterations in this round, and T denotes the maximum number of iterations. c_1 and c_2 change dynamically to meet the changing rule with the increase in the number of iterations. Therefore, the algorithm can adaptively expand the local search range in the early stage of iteration and accelerate the global convergence speed in the late iteration. This learning mechanism is used to accelerate the overall convergence.

In the iterative process, the inertia weight can affect the search range of the current round according to the speed of the previous round. At the end of each round of iterations, the fitness function of the selected particle swarms should be obtained. Moreover, the inertia weights can be dynamically adjusted based on the results of the fitness values, which will make the selected particle swarms in this round of iterations have a more balanced position. The nonlinear adaptive inertia weight strategy is used to calculate the inertia weight, and the method is as follows:

$$w = \begin{cases} w_{\min} + \frac{(w_{\max} - w_{\min}) \times (f_i - f_{\min})}{f_{avg} - f_{\min}}, f_i \leq f_{avg} \\ w_{\max}, f_i > f_{avg} \end{cases} \tag{28}$$

where $w_{\max}$ and $w_{\min}$ are the maximum and minimum inertia weights set, respectively, and $f_{\min}$ and $f_{\max}$ represent the minimum and maximum fitness values of the particle swarm in this round, respectively. f_{avg} represents the average fitness of a particle swarm. At this point, the speed of the particle swarm mainly refers to the speed of the previous round to increase the activity of the particle swarm. Conversely, the speed of the particle swarm at this time mainly refers to the local optimal position and the global optimal position to accelerate the particle swarm to move closer to the dominant space.

Suppose $Z = \{\mathbf{z}_1, \mathbf{z}_2, \ldots, \mathbf{z}_N\}$ is the dataset, where $\mathbf{z}_n = [z_1, z_2, \ldots, z_d]^T$ and $z_i = \{\langle x_i, \mu(x_i), v(x_i)\rangle | x_i \in E\}$, $1 \leq i \leq d$. N is the number of data items. m is the number of clusters. Here, $V = \{v_1, v_2, \ldots, v_m\}$, $v_m \in R^d$, is a set of M clustering centers where $M \geq 2$. Each clustering center vector can be expressed as $v_m = [c_1^m, c_2^m, \ldots, c_d^m]$, where $c_i^m = \{\langle c_i^m, \mu_{v_m}(c_i^m), v_{v_m}(c_i^m)\rangle | 1 \leq i \leq d, 1 \leq m \leq M\}$. The objective function is given below:

$$\begin{gathered} J_m(U,V) = \sum_{n=1}^{N} \sum_{m=1}^{M} \mu_{nm}{}^{c_0} d_{nm}{}^2(z_n, v_m) \\ \mu_{nm} \in [0,1],\ 1 \leq m \leq M,\ 1 \leq n \leq N \\ \sum_{m=1}^{M} \mu_{nm} = 1,\ \forall n, m \end{gathered} \tag{29}$$

where μ_{nm} is the membership degree of the sample data in the m-th class. $U = [\mu_{nm}]_{N \times M}$ denotes the fuzzy membership matrix of $X.c_0 \in [1, +\infty)$ denotes the fuzzification index. $d_{nm}{}^2(z_n, v_m)$ denotes the ordinary Euclidean distance between the measurement point z_n and the clustering center v_m, which is defined as:

$$d_{nm}{}^2(z_n, v_m) = \frac{1}{2}\sum_{i=1}^{d} p_i \Big\{ [\mu_{z_n}(x_i) - \mu_{v_m}(c_i^m)]^2 + [v_{z_n}(x_i) - v_{v_m}(c_i^m)]^2 + [\pi_{z_n}(x_i) - \pi_{v_m}(c_i^m)]^2 \Big\} \tag{30}$$

where $p_i = (1/p, 1/p, \ldots, 1/p)$, $\mu_{z_n}(x_i)$, $v_{z_n}(x_i)$ and $\pi_{z_n}(x_i)$ are the fuzzy membership degree, non-membership degree, and intuitionistic index of input data z_n, respectively. $\mu_{v_k}(c_i^m)$, $v_{v_k}(c_i^m)$ and $\pi_{v_k}(c_i^m)$ are the fuzzy membership degree, non-membership degree, and intuitionistic index of clustering center v_m, respectively.

Therefore, to obtain the optimal objective function by DPSO, it can be considered that the smaller the result of the objective function $J_m(U,V)$, the better the fitness of the particles, so the particle fitness can be expressed by the following:

$$f(x_i) = \frac{\lambda}{J(U,V)} = \frac{\lambda}{\sum_{n=1}^{N} \sum_{m=1}^{M} \mu_{nm}{}^2 d_{nm}{}^2(z_n, v_{i,m})} \tag{31}$$

where $v_{i,m}$ denotes the intuitionistic fuzzy number of the m-th dimension of particle x_i and also denotes m-th clustering center. λ is a constant, which can be manually adjusted according to the specific situation. The main steps for DPSO-IFCM are summarized in Figure 2.

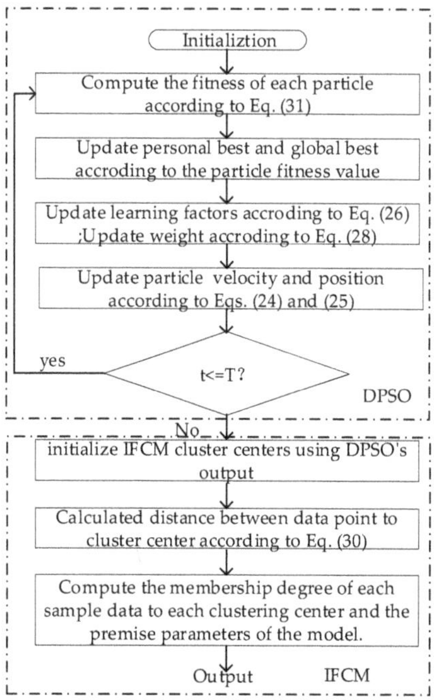

Figure 2. DPSO-IFCM algorithm processes.

In Figure 2, it is shown that the proposed DPSO-IFCM clustering algorithm includes the following steps:

1. Initialization: Initialize G particles to form G first-generation particles, where each particle randomly generates M clustering centers. The fitness value is calculated by Equation (31) and determines the current optimal position of each particle i by the fitness value, and the position of the current particle swarm with the highest fitness is p_g;
2. Compute the velocity and position of each particle in the new particle swarm using Equations (24) and (25);
3. Compute the fitness value of each particle in the new particle swarm using Equation (31) and compare it with the previous generation. For the same individual, if the individual fitness in the new population is larger than the corresponding individual in the previous generation, replace the individual of the previous generation and this becomes the optimal position of particle i, otherwise, it remains unchanged;
4. Compare the fitness value of the optimal individual of the new particle swarm with the optimal individual of the previous generation, if the fitness is greater than the previous generation, update the optimal position of the population to the optimal position of the new particle swarm, otherwise, it remains unchanged, then $t = t + 1$.
5. Repeat Steps 2–4 until a criterion is met that is usually of a sufficiently good fitness or a maximum number of iterations;
6. Obtain the individual position with the highest fitness value as the initial clustering center of the IFCM algorithm;
7. Compute the membership degree μ_{nm} of each sample dataset to each clustering center and the premise parameters $\mu_{A_i^m}(x_i), v_{A_i^m}(x_i), \pi_{A_i^m}(x_i)$ of the model. A detailed method can be found in Ref. [36].

Finally, we input the intuitionistic fuzzy features into the trained MTS-IFRM. The output of the j-th model is:

$$\tilde{y}_j = \sum_{l'=1}^{L'} \frac{\mu^{l'}(\mathbf{z}) f^{l'}(\mathbf{z})}{\sum_{i=1}^{L'} \mu^{i}(\mathbf{z})} = \sum_{l'=1}^{L'} \tilde{\mu}^{l'}(\mathbf{z}) \cdot f^{l'}(\mathbf{z}) \tag{32}$$

3.3. Adaptive Weight Algorithm

From Equation (32), we know that every target has a corresponding MTS-IFRM, then each model is trained and obtains the corresponding label vector output. If the features of the input data are more similar to a certain class, then the value of the corresponding class in the label vector output will be closer to one, otherwise, the value will be closer to zero. When the values of more than one class are relatively close, the class cannot be well distinguished from the input features; that is, the degree of discrimination is not obvious. At this point, we can focus on other models to realize the classification and recognition of the target; that is, reduce the weight corresponding to the model with a low degree of discrimination, and increase the weight corresponding to the model with a high degree of discrimination. First, the initial weight of each model is $1/h$, h denotes the number of secondary features, the weight distribution is also related to the following two points:

1. For a certain secondary feature, in the output result of the corresponding model, if all the values in the output vector are less than 0.5, the possibility of the feature belonging to the target being classified is too low. Therefore, the secondary feature should be reduced according to the impact of the secondary features on the classification results, the weight corresponding to the secondary features is reduced and assigned to other features. Suppose that the maximum value of the label vector output by the model is $x_{\max}$, the weight of the corresponding model can be expressed as:

$$S_1(x_{\max}) = \frac{1}{1 + e^{h_1 \cdot (h_2 - x_{\max})}} \tag{33}$$

The final output matrix can be obtained:

$$f_1(x_{\max}) = \frac{1}{h} \cdot \frac{1}{1 + e^{h_1 \cdot (h_2 - x_{\max})}} \tag{34}$$

where h_1 and h_2 are two constants to control the speed of weight change. Figure 3 shows the weight change under $h_1 = 20, h_2 = 0.25$.

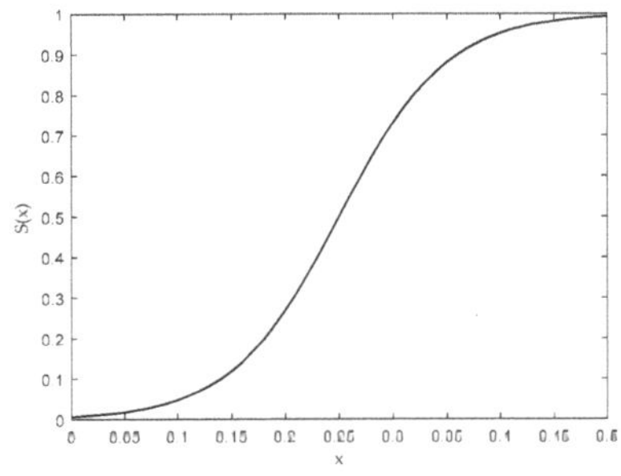

Figure 3. Weight adjustment in case 1.

In Figure 3, when $x_{\max}$ is less than 0.5, the weight of the corresponding model will gradually decrease. When $x_{\max} = 0.25$, the weight of the corresponding model will decrease rapidly. When the weight is below 0.1, the corresponding model weight is close to 0 and the larger weight will be allocated to the model that can be better identified, which can obtain a higher recognition accuracy.

2. For a certain secondary feature to the corresponding T-S IFM output, if the maximum value in the label vector is greater than 0.5, and the difference between the maximum value and the second large value is less than 0.3, then the classification ability of the secondary features for all of the targets to be classified is weak. However, because the maximum value in the label vector is greater than 0.5, the feature has a certain classification ability for a certain type or several types of targets, but it cannot determine which type the input feature data belongs to. Therefore, the corresponding weight can be appropriately reduced and assigned to other features.

Suppose that the difference between the maximum value and the sub-maximum value in the label vector output by the model is $x_{dif} = x_{first} - x_{\text{second}}$, $0 \leq x_{dif} \leq 0.3$. Finally, in case 2, Figure 4 shows the weight adjustment under $h_1 = 20, h_2 = 0$. Different from case 1, case 2 cannot clearly distinguish which category the target belongs to, because there is a value in the label vector, only the weight is appropriately reduced. From Figure 4, the weight is reduced to at most half of the original.

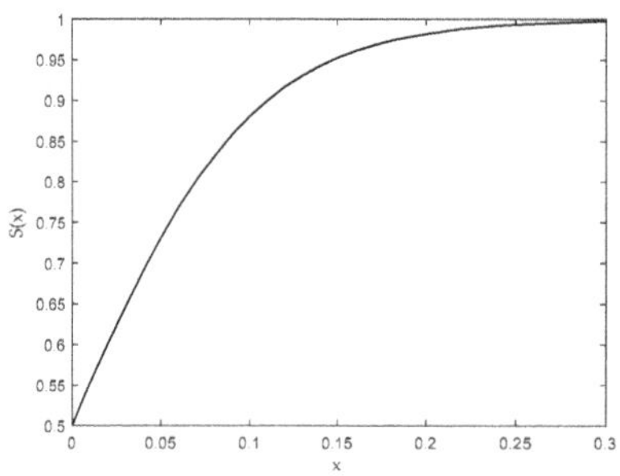

Figure 4. Weight adjustment in case 2.

According to the above two points of analysis, the final weight allocation method of each model is designed, and the process is as follows:

To assign the reduced weight portion of the model of cases 1 and 2 equally to the other models, first, the number of secondary features that do not satisfy the above two cases can be expressed as:

$$num = \begin{cases} num, & x_{\max} \leq 0.5 \ \ or \ \ x_{dif} \leq 0.3 \\ num + 1, & 0.5 < x_{\max} \leq 1 \ \ and \ \ 0.3 < x_{dif} \leq 1 \end{cases} \tag{35}$$

Equation (35) denotes the number of models with obvious classification effects. Then, the final weight adjustment of each model can be expressed as:

$$W_i = \begin{cases} f_{\max}(x_{\max}), & x_{\max} \leq 0.5 \\ f_2(x_{dif}), & 0.5 < x_{\max} \leq 1 \ \ and \ \ x_{dif} \leq 0.3 \\ \frac{1}{n} + \frac{1}{num} \sum\limits_{j=1, j \neq i}^{n} (1 - f(x_j)), & 0.5 < x_{\max} \leq 1 \ \ and \ 0.3 < x_{dif} \leq 1 \end{cases} \tag{36}$$

where W_i denotes the final weight of the *i-th* model. $f(x_j)$ denotes the weight of the corresponding model when case 1 or case 2 occurs. Therefore, the final fusion results are calculated as follows:

$$y^0 = W_{RF} y^0_{RF} + W_{MF} y^0_{MF} + W_{LF} y^0_{LF} \tag{37}$$

3.4. Computational Complexity Analysis

In the proposed MTS-IFRM, the main program includes the implementation of the DPSO-IFCM algorithm and the structure identification of consequence parameters based on the ridge regression method. The total computational complexity of the ridge regression is calculated as $N(L \cdot N \cdot M^2)$, where L is the number of intuitionistic T-S fuzzy rules, N is the number of samples, and M is the number of label vector dimensions. In the DPSO-IFCM algorithm, the total computational complexity of the main loop of DPSO is $N(G \cdot T \cdot d)$, where G is the size of the particle swarm, T is the maximum number of iterations, d is the dimension of the solution space, and the calculation time of the IFCM is mainly used for the fuzzy membership μ_{nm} and the computational complexity is $N(L \cdot N \cdot M)$. In summary, the computational cost of the proposed algorithm is determined by L, N, M, G, T, and d.

4. Simulation Results and Analysis

To evaluate the performance of the MTS-IFRM approach to the problem of recognizing aerial targets in a complex environment, two examples were used to show the recognition performance of MTS-IFRM compared to that of the standard forms of the D-S [19], Yager [29], Murphy [30], multi-sensor data fusion algorithm (MSDF) [32], Kaur [8], and Hu [9] in a complicated environment. Table 2 presents the feature ranges of five typical aerial targets (bomber (Br), fighter (Fr), helicopter (Hr), air-to-ground missile (AGM), and tactical ballistic missile (TBM)).

Table 2. Feature ranges of five aerial targets.

	Br	Fr	Hr	AGM	TBM
Flight height (km)	25–35	7–13	1.6–2.5	3.8–5.2	55–80
Detection distance (km)	350–450	250–350	130–180	100–140	130–180
Flight speed (m/s)	300–500	500–700	70–130	1000–1500	1700–2300
Acceleration (m/s^2)	0–20	0–50	0–30	150–250	200–400
Vertical speed (m/s)	0–50	0–300	0–50	800–1200	1600–2300
Cross-section area (m^2)	0.25–0.35	0.17–0.23	0.08–0.12	0.05–0.08	0.06–0.11
Aspect ratio	1.2–2.0	2.6–3.6	3.2–4.8	6.7–9.3	8.5–11.5

Table 2 shows the complete discernment frame is $\Theta = \{\mathrm{Br}, \mathrm{Fr}, \mathrm{HG}, \mathrm{AGM}, \mathrm{TBM}\}$, and the target recognition feature set is $\mathbf{E} = \{E_A, E_B, E_C, E_D, E_E, E_F, E_G\}$, which represents the credibility of the evidence of the flight altitude (FH), detection distance (DD), flight speed (FS), acceleration (A), vertical speed (VS), cross-section area (CA), and aspect ratio (AR), respectively. The training data is generated within the scope of feature ranges, the experiment uses 125 sets of target feature data within the appropriate range as the training phase with the rules of nine sets. Table 3 presents seven training datasets from the training datasets.

Table 3. The feature data of aerial targets.

Serial Number	FH (km)	DD (km)	FS (m/s)	A (m/s^2)	VS (m/s)	CA (m^2)	AR	Target
1	58.6	135.6	1845.5	210.4	1685.6	0.08	9.5	TBM
2	4.2	125.8	1250.7	211.9	952.2	0.06	8.4	AGMM
3	8.3	344	612.3	42.6	258.4	0.22	2.7	Fr
4	4.5	132.8	1252.1	158.7	958.8	0.06	6.9	AGM
5	31.6	377.4	315.3	14.6	25.3	0.31	1.6	Br
6	1.7	136.6	88.6	11.3	29.3	0.09	3.8	Hr
7	56.6	179.1	2200.6	365.6	1936.7	0.10	11.5	TBM

Table 3 shows that the collected 13–14 d historical feature datasets with results are obtained as the training datasets and the testing target is recognized according to the trained MTS-IFRM, then the feature datasets and model parameters of the target are updated with the recognition result.

The fuzzy membership function is very important for the initial recognition process because of the uncertainty in the feature data. By analyzing the features of aerial targets, the Gaussian membership function is used to recognize the target in Equation (38) and

Table 4 presents δ and x of five typical aerial targets with difference features, showing the fuzzy membership functions corresponding to the detection distance.

$$\mu(x_i) = \exp(-\frac{\|x - x_i\|}{\delta}) \tag{38}$$

Table 4. Five typical aerial targets with different features.

	Br	Fr	Hr	AGM	TBM
FH (km)	(30,7.5)	(10,4.5)	(2,1)	(4.5,1)	(65,15)
DD (km)	(400,80)	(300,80)	(200,60)	(120,45)	(150,60)
FS (m/s)	(400,150)	(600,150)	(100,50)	(1200,500)	(2000, 500)
A(m/s^2)	(10,10)	(25,25)	(15,15)	(200,60)	(300,100)
VS(m/s)	(25,25)	(150,150)	(25,25)	(1000,300)	(1950,600)
CA (m^2)	(0.3,0.08)	(0.2,0.06)	(0.1,0.03)	(0.06,0.02)	(0.08,0.03)
AR	(1.5,0.5)	(3,0.6)	(4,0.8)	(8,1.3)	(10,1.5)

Table 4 shows the appropriate membership function $\mu(x_i)$ can be designed by adjusting δ and x with the different features of the targets x_i by analyzing the various feature attributes of each target in Table 2. Then, take the feature of detection distance as an example. Figure 5 presents the fuzzy membership functions corresponding to the detection distance.

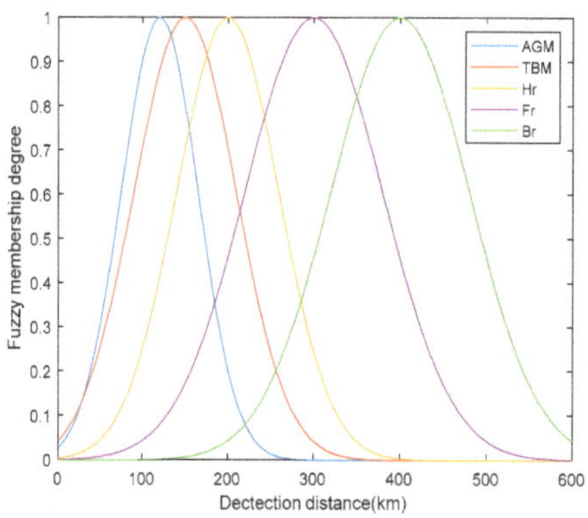

Figure 5. Fuzzy membership functions of detection distance.

From Figure 5, the fuzzy membership degree of each target will be different with different values of primary features. When the detection distance is 450 km, the fuzzy membership degree belonging to target Br is the highest, which is 0.8226, and the fuzzy membership degree belonging to the target AGM is the lowest, approaching zero. When the target features obtained by the radar system are inaccurate and uncertain, the features are calculated by the membership function, thus effectively recognizing the target initially. Figure 6 shows the target recognition framework based on fuzzy membership degree and evidence theory.

In Figure 6, the supporting information of the target obtained by the fuzzy membership function may not be consistent. We use the recognition result of the target obtained by the fuzzy membership function as the confidence degree, and evidence theory is used to fuse the confidence degree and obtain a target recognition result.

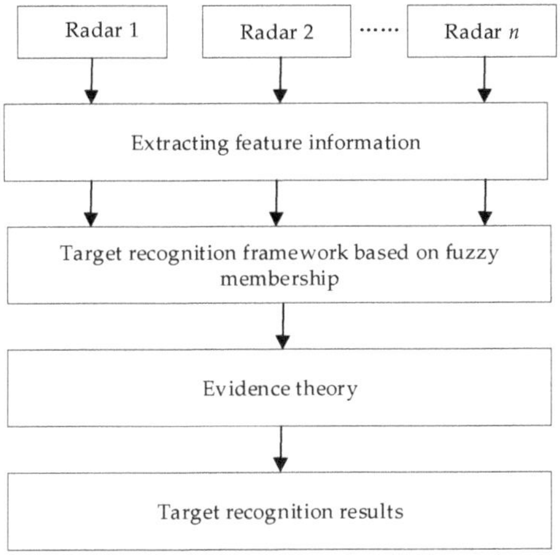

Figure 6. The target recognition framework based on fuzzy membership and evidence theory.

4.1. Example 1: The Data Does Not Contain Fault Features

In this example, data without fault features is employed to show the performance of the methods, that is, all target features support a certain target. Suppose the radar detects a suspicious target, the target features are: $A = 23$ km, $B = 450$ km, $C = 350$ m/s, $D = 10$ m/s^2, $E = 40$ m/s, $F = 0.31$ m^2, and $G = 1.0$. Table 5 presents the BPA example of multi-source information fusion.

Table 5. The BPA example of the multi-source information fusion.

Evidence	Br	Fr	Hr	AGM	TBM	X
E_A	0.4185	2.37×10^{-4}	0	0	3.93×10^{-4}	0.5809
E_B	0.6766	0.0297	2.88×10^{-4}	0	0	0.2936
E_C	0.8837	0.0297	0	0.0549	1.84×10^{-5}	0
E_D	0.3857	0.0614	0.3451	1.70×10^{-5}	8.59×10^{-5}	0
E_E	0.3525	0.2691	0.3525	1.80×10^{-5}	2.01×10^{-5}	0
E_F	0.9660	0.0340	0	0	0	0
E_G	0.3679	1.49×10^{-5}	7.81×10^{-5}	0	0	0.6321

Table 5 shows the corresponding BPA functions and X denotes the unknown term. The features are expressed with fuzzy membership for the unknown targets detected by radar, all the features of the unknown target have high credibility for the target Br, and no feature opposes the Br. Tables 6–9 show the recognition results of the target with different numbers of evidence in an error-free environment.

Table 6. Comparison of algorithms with E_A and E_B in an error-free environment.

Method	m(Br)	m(Fr)	m(Hr)	m(AGM)	m(TBM)	m(X)	Target
D-S	0.8095	0.0176	0	0	0	0.1728	Br
Yager	0.2832	0	0	0	0	0.7168	X
Murphy	0.7905	0.0705	0.0715	0.0047	0	0.0627	Br
MSDF	0.7946	0.0692	0.0694	0.0047	0	0.0621	Br
Kaur	0.8056	0.0862	0.0891	0.0056	0	0.0135	Br
Hu	0.8134	0.0923	0.0451	0.0026	0	0.0466	Br
MTS-IFRM	0.9894	0.0064	0.0042	0	0	0	Br

Table 7. Comparison of algorithms with E_C, E_D and E_E in an error-prone environment.

Method	m(Br)	m(Fr)	m(Hr)	m(AGM)	m(TBM)	m(X)	Target
D-S	0.9610	0.0390	0	0	0	0	Br
Yager	0.1201	0.0049	0	0	0	0.8750	X
Murphy	0.9020	0.0385	0.0391	0.0021	0	0.0183	Br
MSDF	0.9050	0.0374	0.0375	0.0021	0	0.0180	Br
Kaur	0.9156	0.0395	0.0357	0.0035	0	0.0057	Br
Hu	0.9265	0.0402	0.0315	0.0018	0	0	Br
MTS-IFRM	0.9342	0.0187	0.0472	0	0	0	Br

Table 8. Comparison of algorithms with E_F and E_G in an error-free environment.

Method	m(Br)	m(Fr)	m(Hr)	m(AGM)	m(TBM)	m(X)	Target
D-S	0.9782	0.0218	0	0	0	0	Br
Yager	0.3554	0	0	0	0	0.6446	X
Murphy	0.7905	0.0705	0.0715	0.0047	0	0.0627	Br
MSDF	0.7946	0.0692	0.0694	0.0047	0	0.0621	Br
Kaur	0.8165	0.0712	0.0718	0.0056	0	0.0349	Br
Hu	0.8564	0.0522	0.0559	0.0062	0	0.0293	Br
MTS-IFRM	0.9790	0.0210	0	0	0	0	Br

Table 9. Comparison of algorithms with **E** in an error-free environment.

Method	m(Br)	m(Fr)	m(Hr)	m(AGM)	m(TBM)	m(X)	Target
D-S	0.9998	1.75×10^{-4}	0	0	0	0	Br
Yager	0.0121	0	0	0	0	0.9879	X
Murphy	0.9970	0.0014	0.0014	0	0	0	Br
MSDF	0.9973	0.0013	0.0013	0	0	0	Br
Kaur	0.9981	0.0010	9×10^{-4}	0	0	0	Br
Hu	0.9985	0.0008	0.0011	0	0	0	Br
MTS-IFRM	0.9813	0.0015	0.0172	0	0	0	Br

From Tables 6–9 when the quantity of evidence increases, the recognition accuracy of the other six methods steadily improves except for Yager. The reason is that Yager assigns all the conflicts between evidence to X, which leads to cumulative conflicts between pieces of evidence in the synthetic evidence, and the value of X will increase as the quantity of fusing conflicting evidence increases. When the quantity of evidence is small, the MTS-IFRM maintains better target recognition performance and faster convergence because it can deal with the uncertainty well. Regardless of whether fewer features or more features are available, the MTS-IFRM has higher accuracy when recognizing the targets.

4.2. Example 2: The Data Contains Fault Features

The dataset simulated in this paper contains one or more fault features obtained by the equipment, so that the multiple features do not all support a certain target. Suppose the radar detects a suspicious target, the obtained target features are: $A = 23$ km, $B = 450$ km, $C = 350$ m/s, $D = 10$ m/s^2, $E = 40$ m/s, $F = 0.31$ m^2, and $G = 4.1$. Except for the target aspect ratio, other features are the same as in example 1. Due to the influence of factors such as noise and the working status of the sensor device, the target aspect ratio feature is abnormal, and the BPA of the aspect ratio can be expressed as:

$$E_G : m_G(\mathrm{Br}) = 0,\ m_G(\mathrm{Fr}) = 0.0340,\ m_G(\mathrm{Hr}) = 0.9658,$$
$$m_G(\mathrm{AGM}) = 0.0001, m_G(\mathrm{TBM}) = 0, m_G(\mathrm{X}) = 0.$$

The aspect ratio has a high degree of support for target Hr, while the support degree for Br is 0. Therefore, E_G shows significant conflict with the other evidence. Tables 10 and 11 compare the target recognition performance of the algorithms.

Table 10. Comparison of algorithms with E_F and E_G in an error-free environment.

Method	m(Br)	m(Fr)	m(Hr)	m(AGM)	m(TBM)	m(X)	Target
D-S	0	1	0	0	0	0	Fr
Yager	0	0.0012	0	0	0	0.9988	X
Murphy	0.7060	0.0631	0.2003	0.0035	0	0.0270	Br
MSDF	0.7746	0.0702	0.1257	0.0037	0	0.0258	Br
Kaur	0.7945	0.0642	0.1254	0.0034	0	0.0125	Br
Hu	0.8563	0.0281	0.1043	0.0021	0	0.0092	Br
MTS-IFRM	0.9639	0.0345	1.23×10^{-4}	0	0	0	Br

Table 11. Comparison of algorithms with E in an error-prone environment.

Method	m(Br)	m(Fr)	m(Hr)	m(AGM)	m(TBM)	m(X)	Target
D-S	0	1	0	0	0	0	Fr
Yager	0	0	0	0	0	1	X
Murphy	0.9830	6.31×10^{-4}	0.0163	0	0	0	Br
MSDF	0.9965	5.89×10^{-4}	0.0029	0	0	0	Br
Kaur	0.9905	0.0084	0.0011	0	0	0	Br
Hu	0.9942	0.0049	0.0009	0	0	0	Br
MTS-IFRM	0.9811	0.0184	0.0019	0	0	0	Br

Tables 10 and 11 show that because of the conflicting evidence E_G, D-S finally determines that Fr is the final result, which is counter-intuitionistic. Meanwhile, the Yager is also unable to correctly recognize the target because it assigns the high-conflict part of the evidence to X. Murphy, MSDF, Kaur, Hu, and the MTS-IFRM can process the conflicting evidence and realize reasonable results. The Murphy method has lower convergence because it calculates the averages without considering the correlations between the evidence, the MSDF method modifies the entropy method to calculate the weight of the evidence, and the Kaur and Hu methods comprehensively improve the credibility of evidence by analyzing the discrepancy in different aspects. Moreover, the accuracy of the MTS-IFRM is higher compared to other methods in the case of fewer features. The MTS-IFRM establishes a higher stability and reliability structure when confronting uncertainty.

The reasons why the MTS-IFRM shows better performance for aerial target recognition can be explained as follows. First, the MTS-IFRM is constructed according to intuitionistic fuzzy theory, which deals with uncertainty data of aerial targets using DPSO-IFCM clustering. Second, the adaptive weight algorithm is used to further improve the classification accuracy of the model, which is crucial for addressing the target recognition problem in an error-free or error-prone environment.

To further verify the effectiveness of the method, a dataset of 10,000 target features is randomly generated within the range given in Table 12 as the test dataset of the simulation.

Table 12. Range of the test dataset.

	Br | δ	Fr | δ	Hr | δ	AGM | δ	TBM | δ
FH (km)	30 | 15	10 | 5	2 | 1	4.5 | 2	70 | 30
DD (km)	400 | 200	300 | 150	150 | 75	120 | 60	150 | 75
FS (m/s)	400 | 200	600 | 300	100 | 50	1200 | 600	2000 | 1000
A(m/s^2)	10 | 10	25 | 25	15 | 15	200 | 100	300 | 150
VS(m/s)	25 | 25	100 | 100	20 | 20	1000 | 500	2000 | 1000
CA (m^2)	0.30 | 0.15	0.20 | 0.1	0.10 | 0.05	0.05 | 0.02	0.10 | 0.05
AR	1.5 | 0.75	2.5 | 1.0	4.0 | 2.0	8.0 | 4.0	10.0 | 5.0

The data model for the simulation feature parameters is:

$$F_{ij} = f_{ij} \pm randn \times \delta_{ij} \tag{39}$$

where f_{ij} denotes the j-th feature of the target i corresponding to the deviation δ_{ij}, *randn* denotes a normal random number with a mean of 0 and a variance of 1. Six algorithms with higher recognition rate methods are employed in the experiment.

In Table 13, a( ) represents the recognition rate of the target "·", which is obtained by dividing the number of correctly recognized samples by the total number of testing samples, and in bold is the best simulation result under the same conditions. After fusing the seven features, Figure 7 shows the final recognition rates of six algorithms.

Table 13. Recognition rates for five algorithms.

	E_A, E_B	E_C,E_D,E_E	E_F,E_G	E
D-S	a(Br) = 0.4970	a(Br) = 0.6085	a(Br) = 0.7044	a(Br) = 0.8381
	a(Fr) = 0.6394	a(Fr) = 0.8698	a(Fr) = 0.3802	a(Fr) = 0.9347
	a(Hr) = 0.6663	a(Hr) = 0.9133	a(Hr) = 0.5227	a(Hr) = 0.9997
	a(AGM) = 0.6102	a(AGM) = 0.8389	a(AGM) = 0.4165	a(AGM) = 0.9160
	a(TBM) = 0.4207	a(TBM) = 0.7138	a(TBM) = 0.2678	a(TBM) = 0.9041
Murphy	a(Br) = 0.5976	a(Br) = 0.7861	a(Br) = 0.5915	a(Br) = 0.9174
	a(Fr) = 0.7350	a(Fr) = 0.8473	a(Fr) = 0.7326	a(Fr) = 0.9064
	a(Hr) = 0.7956	a(Hr) = 0.9602	a(Hr) = 0.7904	a(Hr) = 0.9990
	a(AGM) = 0.6463	a(AGM) = 0.7983	a(AGM) = 0.6228	a(AGM) = 0.9044
	a(TBM) = 0.4862	a(TBM) = 0.6907	a(TBM) = 0.4968	a(TBM) = 0.8611
MSDF	a(Br) = 0.6173	a(Br) = 0.7892	a(Br) = 0.6141	a(Br) = 0.9087
	a(Fr) = 0.7709	a(Fr) = 0.8557	a(Fr) = 0.7776	a(Fr) = 0.8951
	a(Hr) = 0.8553	a(Hr) = 0.9749	a(Hr) = 0.8489	a(Hr) = 0.9990
	a(AGM) = 0.6977	a(AGM) = 0.8217	a(AGM) = 0.7023	a(AGM) = 0.9212
	a(TBM) = 0.5365	a(TBM) = 0.7119	a(TBM) = 0.5362	a(TBM) = 0.8549
Kaur	a(Br) = 0.6215	a(Br) = 0.8021	a(Br) = 0.6042	a(Br) = 0.9213
	a(Fr) = 0.7821	a(Fr) = 0.8566	a(Fr) = 0.7511	a(Fr) = 0.9155
	a(Hr) = 0.8163	a(Hr) = 0.9713	a(Hr) = 0.8224	a(Hr) = 0.9990
	a(AGM) = 0.7062	a(AGM) = 0.8078	a(AGM) = 0.6634	a(AGM) = 0.9156
	a(TBM) = 0.6035	a(TBM) = 0.7256	a(TBM) = 0.5264	a(TBM) = 0.8744
Hu	a(Br) = 0.7654	a(Br) = 0.8156	a(Br) = 0.6317	a(Br) = 0.9315
	a(Fr) = 0.7905	a(Fr) = 0.8557	a(Fr) = 0.7812	a(Fr) = 0.9213
	a(Hr) = 0.8632	a(Hr) = 0.9812	a(Hr) = 0.8497	a(Hr) = 0.9992
	a(AGM) = 0.7256	a(AGM) = 0.8247	a(AGM) = 0.7123	a(AGM) = 0.9336
	a(TBM) = 0.6636	a(TBM) = 0.7311	a(TBM) = 0.5546	a(TBM) = 0.8639
MTS-IFRM	a(Br) = 0.8834	a(Br) = 0.7145	a(Br) = 0.8345	a(Br) = 0.9354
	a(Fr) = 0.7341	a(Fr) = 0.8844	a(Fr) = 0.5341	a(Fr) = 0.9555
	a(Hr) = 0.7589	a(Hr) = 0.9253	a(Hr) = 0.8835	a(Hr) = 0.9952
	a(AGM) = 0.7954	a(AGM) = 0.9051	a(AGM) = 0.4954	a(AGM) = 0.9862
	a(TBM) = 0.8795	a(TBM) = 0.9493	a(TBM) = 0.7101	a(TBM) = 0.9899

In Figure 7, the MTS-IFRM algorithm has better performance than the other five methods and is slightly inferior to other algorithms for the Hr. The main reasons for this: in other methods, the preliminary recognition of the target with the fuzzy membership function will have high accuracy, and the results will be fused by the evidence theory method. Moreover, Table 2 shows that the features of flight height and speed for Hr have a large difference from those of other targets, for example, suppose the radar detects a suspicious target, the target features are: A = 1.7 km, B = 135 km, C = 75 m/s, D = 10 m/s^2, E = 25 m/s, F = 0.09 m^2 and G = 3.8, in our proposed method, the existing target features are used to construct the T-S intuitionistic fuzzy training model, and the feature datasets and model parameters of the target are updated with the recognition result, which has higher requirements for training data. If the number of Hr in the training data is insufficient, the suspicious target may be recognized as an AGM or TBM with E_A and E_B, because Hr, AGM, and TBM have the similar feature ranges of detection distance. In addition, due to the similar feature ranges of acceleration and vertical speed, the suspicious target may be recognized as a Br or Fr with E_C, E_D and E_E. However, the final recognition rate of the MTS-IFRM is more than 99% for the Hr with abundant training datasets. Overall, if a richer and more effective training dataset can be obtained, the recognition accuracy of the proposed MTS-IFRM can be improved.

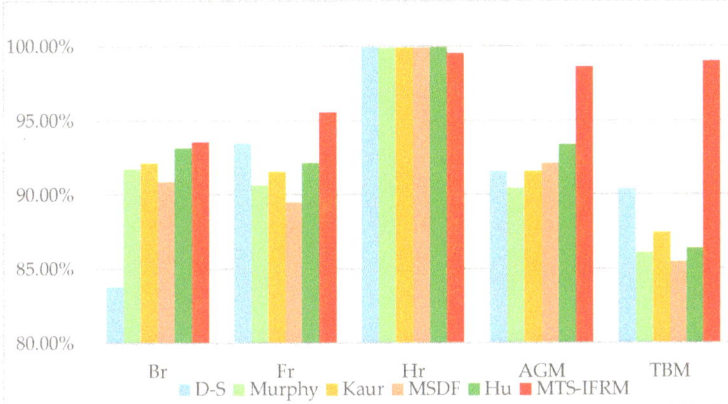

Figure 7. The recognition rates of six algorithms.

5. Conclusions

In this paper, a target recognition approach based on MTS-IFRM is proposed, which constructs a fuzzy classification model to enhance the robustness of the recognition process. The intuitionistic fuzzy theory and ridge regression method are employed in the consequent identification, the intuitionistic fuzzy C-regression clustering based on dynamic optimization can realize the premise identification. Then, the adaptive weight algorithm improves the classification accuracy of the corresponding model. The experimental results show that the MTS-IFRM can effectively recognize aerial targets in error-free and error-prone environments, and its performance is better than the methods proposed for aerial target recognition.

Although the proposed MTS-IFRM can show encouraging results for target recognition, many issues remain. For example, when fusing the outputs of multiple models, the method of the weight distribution is still relatively rough. As the features of the target increase, a more complete weight allocation algorithm needs to fuse the outputs of multiple models accurately. In the future, further methods can be proposed to improve accuracy by extending the models to adjust to different types of datasets and by developing more efficient objective functions for the MTS-IFRM using specific samples.

Author Contributions: Conceptualization, C.Z., W.X. and Z.L.; methodology, C.Z., W.X. and Z.L.; software, Y.L. and Z.L.; formal analysis, C.Z., Y.L. and Z.L.; investigation, Y.L. and Z.L. resources, Y.L. and Z.L.; data curation, C.Z.; writing—original draft preparation, C.Z.; writing—review and editing, C.Z., W.X. and Z.L.; visualization, C.Z.; supervision, C.Z.; project administration, C.Z.; funding acquisition, Z.L. All authors of the article have provided substantive comments. All authors have read and agreed to the published version of the manuscript.

Funding: This work was supported by the National Natural Science Foundation of China (62171287, 62076165), the Science & Technology Program of Shenzhen (No.JCYJ20220818100004008), the Innovation Team Project of the Department of Education of Guangdong Province (No. 2020KCXTD004) and the Science and Technology on Information Systems Engineering Laboratory (No. 05202206).

Data Availability Statement: The data presented in this study are partly available on request from the corresponding author. The data are not publicly available due to their current restricted access.

Conflicts of Interest: The authors declare no conflict of interest.

References

1. Zhao, K.; Sun, R.; Li, L. An improved evidence fusion algorithm in multi-sensor systems. *Appl. Intell.* **2021**, *51*, 7614–7624. [CrossRef]
2. Chen, A.; Tang, X.; Cheng, B.C. Multi-source monitoring information fusion method for dam health diagnosis based on Wasserstein distance. *Inform. Sci.* **2023**, *632*, 378–389. [CrossRef]
3. Liang, Q.; Liu, Z.; Chen, Z. A networked method for multi-evidence-based information fusion. *Entropy* **2022**, *25*, 69. [CrossRef] [PubMed]
4. Wang, Y.C.; Wang, J.; Huang, M.J. An evidence combination rule based on a new weight assignment scheme. *Soft Comput.* **2022**, *26*, 7123–7137. [CrossRef]
5. Meng, L.Y.; Li, L.Q. Time-sequential hesitant fuzzy set and its application to multi-attribute decision making. *Complex Intell. Syst.* **2022**, *8*, 4319–4338. [CrossRef]
6. Meng, L.Y.; Li, L.Q.; Xie, W.X. Time-sequential hesitant fuzzy entropy, cross-entropy and correlation coefficient and their application to decision making. *Eng. Appl. Artif. Intell.* **2023**, *123*, 106455. [CrossRef]
7. Jiang, H.; Hu, B.Q. A decision-theoretic fuzzy rough set in hesitant fuzzy information systems and its application in multi-attribute decision-making. *Inform. Sci.* **2021**, *579*, 103–127. [CrossRef]
8. Kaur, M.; Srivastava, A. A new divergence measure for belief functions and its applications. *Int. J. Gen. Syst.* **2023**, *52*, 455–472. [CrossRef]
9. Hu, Z.; Su, Y.; Hou, W. Multi-sensor data fusion method based on divergence measure and probability transformation belief factor. *Appl. Soft Comput.* **2023**, *145*, 110603. [CrossRef]
10. Lv, Y.W.; Yang, G.H. Centralized and distributed adaptive cubature information filters for multi-sensor systems with unknown probability of measurement loss. *Inform. Sci.* **2023**, *630*, 173–189. [CrossRef]
11. Wang, W.; Zhang, Y. A method based on grey theory toward the multi-sensor information fusion of human centered robots. In Proceedings of the 12th International Convention on Rehabilitation Engineering and Assistive Technology, 13 July 2018; pp. 293–296.
12. Akshaya, T.G.; Sreeja, S. Multi-sensor data fusion for aerodynamically controlled vehicle based on FGPM. *IFAC—PapersOnLine* **2020**, *53*, 591–596.
13. Fu, C.; Qin, K.; Yang, L. Hesitant fuzzy β-covering (T, I) rough set models: An application to multi-attribute decision-making. *J. Intell. Fuzzy Syst.* **2023**. *preprint*. [CrossRef]
14. Zhang, L.; Zhan, J.; Xu, Z. Covering-based generalized IF rough sets with applications to multi-attribute decision-making. *Inform. Sci.* **2019**, *478*, 275–302. [CrossRef]
15. Zhang, L.; Zhu, P. Asymmetric models of intuitionistic fuzzy rough sets and their applications in decision-making. *Int. J. Mach. Learn. Cyb.* **2023**, *14*, 3353–3380. [CrossRef]
16. Zhou, Z.; Zhao, C.; Cai, X. Three-dimensional modeling and analysis of fractal characteristics of rupture source combined acoustic emission and fractal theory. *Chaos Soliton. Fract.* **2022**, *160*, 112308. [CrossRef]
17. Lai, Y.; Zhao, K.; He, Z. Fractal characteristics of rocks and mesoscopic fractures at different loading rates. *Geomech. Energy Environ.* **2023**, *33*, 100431. [CrossRef]
18. Wang, H.; Liu, S.V.; Qu, X. Field investigations on rock fragmentation under deep water through fractal theory. *Measurement* **2022**, *199*, 111521. [CrossRef]
19. Dempster, A.P. Upper and lower probabilities induced by a multivalued mapping. In *Classic Works of the Dempster-Shafer Theory of Belief Functions*; Springer: Berlin/Heidelberg, Germany, 2008; pp. 57–72.
20. Chen, L.; Deng, Y. A new failure mode and effects analysis model using Dempster–Shafer evidence theory and grey relational projection method. *Eng. Appl. Artif. Intell.* **2018**, *76*, 13–20. [CrossRef]
21. He, Y.; Guo, H.; Li, X. A Collaborative Relay Tracking Method Based on Information Fusion for UAVs. *IEEE Trans. Aerosp. Electron. Syst.* **2023**, *59*, 6894–6906. [CrossRef]
22. Xiao, F. GEJS: A generalized evidential divergence measure for multisource information fusion. *IEEE Trans. Syst. Man Cybern. Syst.* **2022**, *53*, 2246–2258. [CrossRef]
23. Sezer, S.I.; Akyuz, E.; Arslan, O. An extended HEART Dempster–Shafer evidence theory approach to assess human reliability for the gas freeing process on chemical tankers. *Reliab. Eng. Syst. Safe* **2022**, *220*, 108275. [CrossRef]
24. Pan, Y.; Zhang, L.; Li, Z.W. Improved fuzzy Bayesian network-based risk analysis with interval-valued fuzzy sets and D–S evidence theory. *IEEE Trans. Fuzzy Syst.* **2019**, *28*, 2063–2077. [CrossRef]
25. Zhu, C.; Xiao, F. A belief Rényi divergence for multi-source information fusion and its application in pattern recognition. *Appl. Intell.* **2023**, *53*, 8941–8958. [CrossRef]
26. Yin, L.; Deng, X.; Deng, Y. The negation of a basic probability assignment. *IEEE Trans. Fuzzy Syst.* **2018**, *27*, 135–143. [CrossRef]
27. Jiang, W. A correlation coefficient for belief functions. *Int. J. Approx. Reason.* **2018**, *103*, 94–106. [CrossRef]
28. Wang, H.; Deng, X.; Jiang, W. A new belief divergence measure for Dempster–Shafer theory based on belief and plausibility function and its application in multi-source data fusion. *Appl. Artif. Intell.* **2021**, *97*, 104030. [CrossRef]
29. Yager, R.R. On the Dempster-Shafer framework and new combination rules. *Inform. Sci.* **1987**, *41*, 93–137. [CrossRef]
30. Murphy, C.K. Combining belief functions when evidence conflicts. *Decis. Support Syst.* **2000**, *29*, 1–9. [CrossRef]

31. Li, X.; Zhao, Y.; Fan, C. Multi-sensor Data Fusion Algorithm Based on Dempster-Shafer Theory. In Proceedings of the 7th International Conference on Computer and Communications (ICCC), Chengdu, China, 10–13 December 2021; pp. 288–293.
32. Wang, L.; Yang, H.Y.; Zhang, S.W. Intuitionistic Fuzzy Dynamic Bayesian Network and its application to terminating situation assessment. *Procedia Comput. Sci.* **2019**, *154*, 238–248. [CrossRef]
33. Guo, L.; Yang, R.; Zhong, Z. Target recognition method of small UAV remote sensing image based on fuzzy clustering. *Neural Comput. Appl.* **2022**, *34*, 12299–12315. [CrossRef]
34. Lei, Y.; Kong, W. Technique for target recognition based on intuitionistic fuzzy reasoning. *IET Signal Process.* **2012**, *6*, 255–263. [CrossRef]
35. Dolgiy, A.I.; Kovalev, S.M.; Kolodenkova, A.E. Processing heterogeneous diagnostic information on the basis of a hybrid neural model of Dempster-Shafer. In Proceedings of the Artificial Intelligence: 16th Russian Conference, RCAI 2018, Moscow, Russia, 24–27 September 2018; Proceedings 16, pp. 79–90.
36. Zhang, C.Y.; Li, L.Q.; Huang, S. Multiple target data-association algorithm based on Takagi-Sugeno intuitionistic fuzzy model. *Neurocomputing* **2023**, *536*, 114–124. [CrossRef]
37. Chen, S.M.; Liu, A.Y. Multiattribute decision making based on nonlinear programming methodology, novel score function of interval-valued intuitionistic fuzzy values, and the standard deviations of the score values in the score matrix. *Inform. Sci.* **2023**, *607*, 119381. [CrossRef]
38. Dhanachandra, N.; Chanu, Y.J. An image segmentation approach based on fuzzy c-means and dynamic particle swarm optimization algorithm. *Multimed. Tools Appl.* **2020**, *79*, 18839–18858. [CrossRef]

remote sensing

MDPI

Article

Comparative Analysis of Pixel-Level Fusion Algorithms and a New High-Resolution Dataset for SAR and Optical Image Fusion

Jinjin Li [1], Jiacheng Zhang [1], Chao Yang [1], Huiyu Liu [1], Yangang Zhao [2] and Yuanxin Ye [1,*]

1 Faculty of Geosciences and Environmental Engineering, Southwest Jiaotong University, Chengdu 611756, China; swjtuljj@163.com (J.L.); swjtu_zjc@163.com (J.Z.); yc18483685462@my.swjtu.edu.cn (C.Y.); huiyu20200131@163.com (H.L.)

2 The Second Topographic Surveying Brigade of Ministry of Natural Resources, Xian 710054, China; zyg15926@163.com

* Correspondence: yeyuanxin110@163.com

Abstract: Synthetic aperture radar (SAR) and optical images often present different geometric structures and texture features for the same ground object. Through the fusion of SAR and optical images, it can effectively integrate their complementary information, thus better meeting the requirements of remote sensing applications, such as target recognition, classification, and change detection, so as to realize the collaborative utilization of multi-modal images. In order to select appropriate methods to achieve high-quality fusion of SAR and optical images, this paper conducts a systematic review of current pixel-level fusion algorithms for SAR and optical image fusion. Subsequently, eleven representative fusion methods, including component substitution methods (CS), multiscale decomposition methods (MSD), and model-based methods, are chosen for a comparative analysis. In the experiment, we produce a high-resolution SAR and optical image fusion dataset (named YYX-OPT-SAR) covering three different types of scenes, including urban, suburban, and mountain. This dataset and a publicly available medium-resolution dataset are used to evaluate these fusion methods based on three different kinds of evaluation criteria: visual evaluation, objective image quality metrics, and classification accuracy. In terms of the evaluation using image quality metrics, the experimental results show that MSD methods can effectively avoid the negative effects of SAR image shadows on the corresponding area of the fusion result compared with CS methods, while model-based methods exhibit relatively poor performance. Among all of the fusion methods involved in the comparison, the non-subsampled contourlet transform method (NSCT) presents the best fusion results. In the evaluation using image classification, most experimental results show that the overall classification accuracy after fusion is better than that before fusion. This indicates that optical-SAR fusion can improve land classification, with the gradient transfer fusion method (GTF) yielding the best classification results among all of these fusion methods.

Keywords: synthetic aperture radar (SAR); optical image; image fusion; image classification

1. Introduction

With the rapid development of different types of sensors that obtain information from the Earth, various remote sensing images have become available for users. Among them, optical images and synthetic aperture radar (SAR) images are two of the most commonly used data in remote sensing applications. SAR images have unique characteristic structure and texture information, making them adaptable for collection at any time without being affected by weather conditions. However, due to the special measurement method of SAR systems (i.e., side-looking imaging), the gray values of SAR images are different from the spectral reflectance of the Earth's surface, which brings difficulties for the interpretation of SAR images in certain scenarios. As is shown in Figure 1, considering that optical images contain rich spectral information, they can directly reflect the colors and textural details of ground objects. Therefore, optical and SAR images are fused to obtain fusion results containing complementary information, thus enhancing the performance of subsequent remote sensing applications [1].

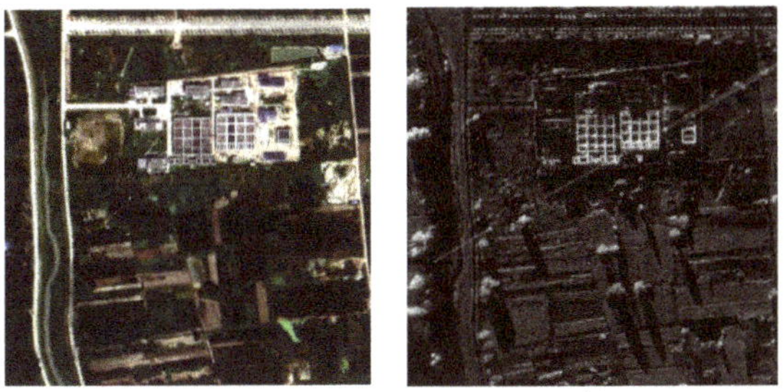

(a) (b)

Figure 1. Optical image and SAR image of the same scene: (**a**) Optical image. (**b**) SAR image.

According to the stage of data integration, the fusion technology can be divided into three categories: pixel-level, feature-level, and decision-level [2]. Compared with feature-level and decision-level methods, pixel-level fusion methods involve higher computational complexity. Pixel-level fusion methods, despite their higher computational complexity compared to feature-level and decision-level approaches, are widely employed in remote sensing image fusion due to their superior accuracy. These methods have the properties of effective retention of original data, limited information loss, and abundant and accurate image information [3]. As more and more algorithms and their improved versions have been used to fuse optical and SAR images, researchers have compared the performance of these methods for improving ground object interpretation. For instance, Battsengel et al. compared the performance of intensity–hue–saturation (IHS) transform, Brovey transformation, and principal components analysis (PCA) in urban feature enhancement [4]. The analysis revealed that the images transformed through IHS have better characteristics in spectral and spatial separation of different urban levels. However, in a comparative experiment conducted by Sanli et al., IHS showed the worst results [5].

As mentioned above, the performance of the same fusion method can exhibit significant variations across different scenes owing to the special imaging mechanism of SAR and its distinct image content. The fusion quality is affected not only by the quality of the input image, but also by the performance of the fusion method. Accordingly, it is worth considering the selection of a suitable method among many fusion methods and the choice of appropriate metrics for evaluation. In order to compare the performance of various fusion methods objectively, some researchers quantitatively evaluate the effect of fusion methods through objective fusion quality evaluation metrics [6–9], but there are few fusion methods and evaluation metrics involved in experiments, which fail to cover all of the categories of pixel-level fusion methods.

In addition to evaluating fusion quality based on traditional image quality evaluation metrics, it is worth exploring how to use classification accuracy to evaluate fusion quality, particularly in the context of improving image interpretation and land classification. The quality of the input images will affect the accuracy of the classification results [10–13]. Radar can penetrate clouds, rain, snow, haze, and other weather conditions, thus obtaining the reflection information from the target surface. As a result, SAR data can be collected at nearly any time and under any environmental conditions. However, these data are susceptible to speckle noise, thus resulting in poor interpretability, and they lack spectral information. In contrast, optical images contain rich spectral information. In the application of land cover classification, the fusion of optical and SAR data is beneficial to distinguish ground object types that might be indistinguishable due to their similar spectral characteristics. Thus, in order to improve the image classification results, numerous researchers have used SAR and optical image fusion for land cover classification [14–18].

Gaetano et al. deal with the fusion of optical and SAR data for land cover monitoring. Experiments show that the fusion of optical and SAR data can greatly improve the classification accuracy compared with raw data or even multitemporal filtering data [15]. Hu et al. propose a fusion approach for the joint use of SAR and hyperspectral data, which is used for land use classification. The classification results show that the fusion method can improve the classification performance of hyperspectral and SAR data, and it can collect the complementary information of the two datasets well [17]. Kulkarni et al. present a hybrid fusion approach to integrate information from SAR and MS imagery to improve land cover classification [18]. Dabbiru et al. investigate the impact of an oil spill in an ocean area. The main purpose of that study was to apply fusion technology to SAR and optical images and explore the application value of fusion technology in the classification of oil-covered vegetation in coastal zones [19]. However, few researchers take classification accuracy as an evaluation metric to evaluate the performance of different fusion methods.

Optical and SAR image fusion has garnered significant attention owing to the special complementary advantages. However, many existing methods borrow migrations of fusion models from other fields (e.g., optical and infrared images, multi-focused images), with a lack of algorithmic exploration for the study of optical and SAR specificity. In recent years, deep learning has greatly driven the applied research on image fusion, but the studies are mostly focused on specific application scenarios, such as target extraction, cloud removal, land classification, etc. [20–22], in which the algorithms mainly deal with local feature information rather than global pixel information. In most of the latest research articles on pixel-level fusion of optical and SAR images based on deep learning, no specific code files have been published to objectively verify the advantages and disadvantages of the algorithms. Therefore, in this paper, in order to better experimentally verify the algorithms within the field of pixel-level image fusion, several types of traditional algorithms that are well-established and publicly available are selected for comparative analysis.

This paper makes the following three contributions:

1. We systematically review the current pixel-level fusion algorithms for optical and SAR image fusion, and then we select eleven representative fusion methods, including CS methods, MSD methods, and model-based methods for comparison analysis.
2. Based on the evaluation indicators of low-level visual tasks, we combine these with the evaluation indicators of subsequent high-level visual tasks to analyze the advantages and disadvantages of existing pixel-level fusion algorithms.
3. We produce a high-resolution SAR and optical image fusion dataset, including 150 pairs of images of urban, suburban, and mountain settings, which can provide data support for relevant research. The download link for the dataset is https://github.com/yeyuanxin110/YYX-OPT-SAR (accessed on 21 January 2023).

This paper extends our early work [23] by adding two datasets, including a self-built high-resolution dataset named YYX-OPT-SAR and a publicly medium-resolution dataset named WHU-OPT-SAR, to evaluate the fusion methods. In order to evaluate the performance of different fusion methods in subsequent classification applications, we also employ classification accuracy as an evaluation criterion to assess the quality of different fusion methods.

2. Pixel-Level Methods of Optical–SAR Fusion

As an important branch of information fusion technology, the pixel-level fusion of images can be traced back to the 1980s. With the increasing maturity of SAR technology, researchers have explored the fusion of optical and SAR images to enhance the performance of remote sensing data across various applications. In the multi-source remote sensing data fusion competition held by the IEEE Geoscience and Remote Sensing Society (IEEE GRSS) in 2020 and 2021, the theme of SAR and multispectral image fusion has been consistently included, which underscores the growing significance of optical and SAR image fusion in recent years. At present, research on pixel-level fusion algorithms of optical and SAR images based on deep learning remains relatively limited in depth, so the pixel-level fusion algorithms selected in this paper are relatively mature, traditional algorithms. Generally speaking, traditional pixel-level fusion methods can be divided into CS methods, MSD methods, and model-based methods [2]. Because of their different data processing strategies, these three methods have their own advantages and disadvantages in optical and SAR image fusion.

2.1. CS Methods

The fusion process of CS methods is shown in Figure 2. CS methods aim to obtain the final image fusion result by replacing a certain component of the positive transformation of the optical image with the SAR image and then applying the corresponding inverse transformation. In this way, the obtained image fusion result incorporates the spectral information from the optical image and the texture information from the SAR image. For instance, Chen et al. utilize the IHS transform to fuse hyperspectral and SAR images. The fusion results not only have a high spectral resolution but also contain the surface texture features of SAR images, which enhances the interpretation of urban surface features [24]. The conventional PCA method is improved by Yin and Jiang, and the fusion result demonstrates better performance in preserving both spatial and spectral contents [25]. Yang et al. use the Gram–Schmidt algorithm to fuse GF-1 images with SAR images and successfully improve the classification accuracy of coastal wetlands by injecting SAR image information into the fusion results [26].

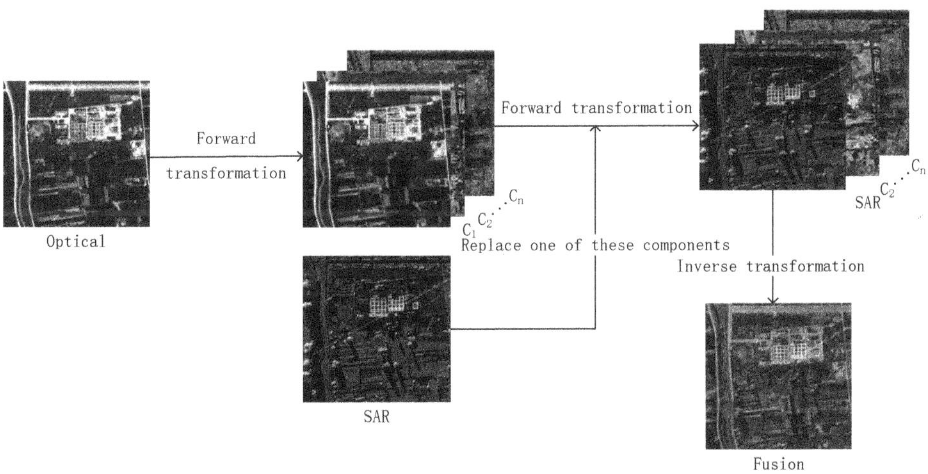

Figure 2. Fusion process of the CS method.

With the characteristics of simplicity and low computational complexity, CS methods can obtain fusion results with abundant spatial information in pan-sharpening and other fusion tasks. However, in multi-sensor and multi-modal image fusion, such as SAR–optical image fusion, serious spectral distortions occur frequently in partial areas because of low correlation. Recently, the research on pixel-level fusion algorithms of optical and SAR images has developed toward the multi-scale decomposition method.

2.2. MSD Methods

MSD methods divide the original image into the main image and the multilayer detail image according to the decomposition strategy, and each image encapsulates distinct potential information from the original image [27]. While the number of subbands decomposed by different methods varies, these methods share a similar process framework, which is shown in Figure 3. According to the decomposition strategies, MSD methods can be divided into three categories: wavelet-based methods, pyramid-based methods, and multi-scale geometric analysis (MGA)-based methods [27].

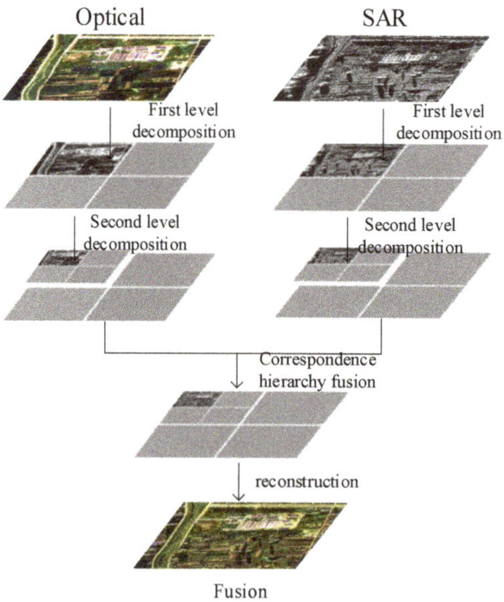

Figure 3. Schematic diagram of fusion process of multi-scale decomposition methods.

Kulkarni and Rege use the wavelet transform to fuse SAR and multispectral images, and they apply the activity level measurement method based on local energy to merge detail subbands, which not only enhances spatial information but also avoids spectral distortions [1]. Eltaweel and Helmy apply the Non-subsampled shearlet transform (NSST) for multispectral and SAR image fusion. The fusion rules based on local energy and the dispersion index are used to integrate the low-frequency coefficients decomposed through NSST, and the multi-channel pulse coupled neural network (m-PCNN) is utilized to guide the fusion process of bandpass subbands. The fusion results show good object contour definition and structural details [28].

The primary goal of MSD methods is to extract multiplex features of the input image into different scales of subbands, and thus to realize the optimal selection and integration of diverse pieces of salient information through specifically designed fusion rules. Activity-level measurement and coefficient combination are essential steps in MSD methods. As a critical factor affecting the quality of the fused image, activity-level measurement is used to express the salience of each coefficient and then provide the evaluation criterion and calculation basis for the weight assignment in the coefficient combination process. And the activity-level measurement methods can be divided into three categories: the coefficient-based, window-based, and region-based measures. Equally important are coefficient combination rules, which involve various operations, such as weighted average, maximum value, and consistency verification, that help to control the contribution of different frequency bands to the merging results with predefined or adaptive rules [29].

2.3. Model-Based Methods

Model-based fusion methods relate to the fusion of optical and SAR images as an image generation problem. The final fusion result is derived by establishing a mathematical model that describes the mapping relationship from the source image to the fusion result, or by establishing a constraint relationship between the fusion result and the source image.

In addition, in order to enhance the fusion effect, a probability model and a priori constraint can be introduced into the model, albeit at the expense of increased solution complexity. Representative methods within this category include variational model methods and sparse representation (SR) methods. Variational model methods establish an energy functional consisting of different terms based on prior constraint information. The fusion result is obtained by minimizing the energy functional under the premise that the existence of a minimum for the energy functional is proved. On the other hand, the methods based on SR select different linear combinations from overcomplete dictionaries to describe image signals. Yang and Li are pioneers in employing SR for the image fusion task, and they propose an SR-based image fusion method using the sliding window technique [30]. The schematic diagram of the method is shown in Figure 4.

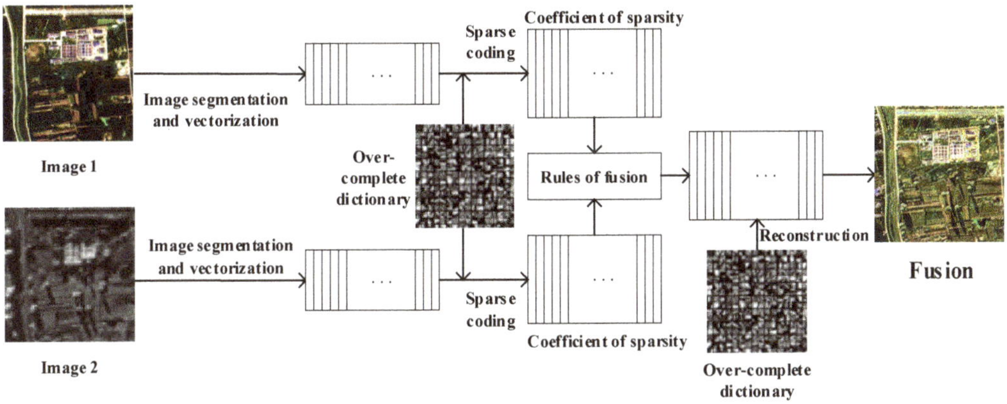

Figure 4. Flow chart of image fusion based on sparse representation.

Wei Zhang and Le Yu introduce the variational model for pan-sharpening into the fusion process of SAR and multispectral images, which obtains the final fusion result by minimizing the energy functional composed of linear combination constraints, color constraints, and geometric constraints. The experiment demonstrates that a variational model-based fusion method is acceptable for SAR and multispectral image fusion in terms of spectral preservation [31]. Additionally, Huang proposes a cloud removal method for optical images based on sparse representation fusion, which uses SAR and low-resolution optical images to provide high-frequency and low-frequency information for reconstructing the cloud occlusion area and achieves good visual effect and radiation consistency [20].

2.4. Method Selection

Generally speaking, the traditional pixel-level fusion methods can be divided into CS methods, MSD methods, and model-based methods. According to the decomposition strategy, MSD methods can be divided into three categories: wavelet-based methods, pyramid-based methods, and multi-scale geometric analysis (MGA) methods. In order to compare the differences between fusion methods in different categories, we choose some classical methods for the following two points in each category. First, there are publicly available algorithms with dependable performance to conduct comparative experiments. Second, they have been used in optical and SAR image fusion fields. Table 1 shows a list of investigated methods.

For MSD methods, the "averaging" rule is selected to merge low-pass bands, while the "max-absolute" rule is employed to merge high-pass MSD bands. Two instances of the "max-absolute" rule are applied, one being the conventional rule and the other incorporating a local window-based consistency verification scheme [32]. These are denoted by the numbers "1" and "2" appended to the corresponding abbreviation, as shown in Table 2, to explore their respective impacts on the final fusion results. The decomposition levels and decomposition filters presented in Table 3 are chosen according to the research conclusion of the literature [33]. In the sparse representation based on the sliding window, the step size and the window size are fixed to one and eight, respectively [30], the K-means generalized singular value decomposition (K-SVD) algorithm [34] is used to build an overcomplete dictionary, and the orthogonal matching pursuit (OMP) algorithm [35] is utilized for sparse coding. The parameter selection of other methods adopts the recommended values from the corresponding literature.

Table 1. Pixel-level fusion methods participating in comparison.

Category		Method
CS		Intensity–Hue–Saturation (IHS) transform [36] Principal Component Analysis (PCA) [37] Gram–Schmidt (GS) transform [38]
MSD	Pyramid-based	Laplacian pyramid (LP) [39] Gradient pyramid (GP) [40]
	Wavelet-based	Discrete wavelet transform (DWT) [41] Dual tree complex wavelet transform (DTCWT) [42]
	MGA	Curvelet transform (CVT) [43] Non-subsampled contourlet transform (NSCT) [44]
Model-based		SR [30] Gradient Transfer Fusion (GTF) [45]

Table 2. The fusion rule of the high-frequency component, represented by different serial numbers.

Method	Rule
XX_1	max-absolute
XX_2	"max-absolute" rule with a local window-based consistency verification scheme

(XX denotes a fusion method).

Table 3. Filters and number of decomposition layers in MSD methods.

Category	Method	Filters	Levels
Pyramid-based	LP	/	4
	RP	/	4
Wavelet-based	DWT	Daubechies (db1)	4
	DTCWT	First: LeGall 5-3 Other: Q-shift_06	4
MGA	CVT	/	4
	NSCT	Pyramid: pyrexc Orientation: 7–9	{4, 8, 8, 16}

3. Evaluation Criteria for Image Fusion Methods

3.1. Visual Evaluation

The visual evaluation is conducted to assess the quality of the fused image based on human observation. Observers judge the spectral fidelity, the visual clarity, and the amount of information in the image according to their subjective feelings. Although visual evaluation has no technical obstacles in implementation and directly reflects the visual quality of images, its reliability is influenced by various factors, such as the observer's self-experience, display variations in hardware, and ambient lighting conditions, leading to lower reproducibility and stability. Generally, the visual evaluation serves as a supplement in combination with statistical evaluation methods.

3.2. Statistical Evaluation

The statistical evaluation of image quality is a fundamental aspect of digital image processing encompassing various fields, such as image enhancement, restoration, and compression. Numerous conventional image quality evaluation metrics, like standard deviation, information entropy, mutual information, and structural similarity, have been widely applied. These metrics can objectively evaluate the quality of fusion results and provide quantitative numerical references for the comparative analysis of fusion methods. In addition to these conventional metrics, researchers have proposed some quality metrics specially designed for image fusion, such as the weighted fusion quality index Q_W and the edge-dependent fusion quality index Q_E [46], as well as the objective quality metric based on structural similarity Q_Y [47].

The objective quality evaluation of image fusion can be carried out in two ways [48]. The first way is to compare the fusion results with a reference image, which is commonly used in pan-sharpening and multi-focus image fusion. However, in multimodal image fusion tasks, such as SAR–optical image fusion, obtaining an ideal reference image is challenging. Therefore, this paper uses the non-reference metrics to objectively evaluate the quality of the fusion image. The fusion results are comprehensively compared from different aspects through nine representative fusion evaluation metrics: information entropy (EN), peak signal-to-noise ratio (PSNR), mutual information (MI), standard deviation (SD), the metric $Q^{AB/F}$ based on edge information preservation [49], the universal image quality index Q_o [50], the weighted fusion quality index Q_W, the edge-dependent fusion quality index Q_E, the similarity-based image fusion quality index Q_Y, and the human visual system (HVS)-model-based quality index Q_{CB} [51]. Based on the different emphases of these evaluation indexes, they can be divided into four categories [52,53]. Table 4 presents the definitions and characteristics of the selected nine quality metrics.

3.3. Fusion Evaluation According to Classification

Most of the subsequent applications of remote sensing images focus on image classification and object detection. At present, there have been researches on object detection of remote sensing images [54], but there are few traditional methods. Therefore, this paper chooses image classification as an index to evaluate the performance of image fusion in subsequent applications. In the evaluation of image classification, three classic methods, including Support Vector Machine (SVM) [55], Random Forests (RF) [56], and Convolutional Neural Network (CNN) [57], are used to perform image classification. It is crucial to evaluate the accuracy of classification results. According to the results of the accuracy evaluation, we can judge whether the classification method is accurate and whether the classification degree meets the needs of the subsequent analysis. This information enables us to identify which fusion method yields the best classification result. The commonly used method to evaluate the accuracy of classification results is the confusion matrix, also known as the error matrix. It reflects the correct and incorrect classification of the corresponding classification results of each category in the validation data. The confusion matrix is a square matrix with a side length of c, where c is the total number of categories and the values on the diagonal are the number of correctly classified pixels in each category.

Overall accuracy (OA) refers to the ratio of the total number of pixels correctly classified to the total number of pixels in the verified sample. It provides the overall evaluation of the quality of the classification results. User accuracy (UA) represents the degree to which a class is correctly classified in the classification results. It is calculated as the ratio of the number of correctly classified pixels in each class to the total number of pixels sorted into that class by the classifier (the sum of row elements corresponding to that class).

Table 4. Definition and significance of nine quality indices.

Category	Definition	Range	Characteristic
Information-theory-based	$EN = -\sum_{l=0}^{L-1} p_l \log_2 p_l$, L is the number of the gray level and p_l is the normalized histogram of an image	$[0, \log_2 L]$	Reflects the amount of information contained in the image
	$PSNR = 10\log_{10}\frac{L^2}{MSE}$ where $MSE = \frac{MSE_{AF}+MSE_{BF}}{2}$, $MSE_{XF} = \frac{1}{MN}\sum_{i=0}^{M-1}\sum_{j=0}^{N-1}(X(i,j)-F(i,j))^2$	$[0, +\infty]$	Reflects the distortion
Image-feature-based	$SD = \sqrt{\sum_{i=1}^{M}\sum_{i=1}^{N}(F(i,j)-\mu)^2}$	$[0, +\infty]$	Reflects the distribution and contrast of the image
	$Q^{AB/F} = \frac{\sum_{i=1}^{N}\sum_{j=1}^{M}\left(Q^{AF}(i,j)w^A(i,j)+Q^{BF}(i,j)w^B(i,j)\right)}{\sum_{i=1}^{N}\sum_{j=1}^{M}\left(w^A(i,j)+w^B(i,j)\right)}$ $Q^{XF}(i,j) = Q_g^{XF}(i,j)Q_\alpha^{XF}(i,j)$, $Q_g^{XF}(i,j)$ and $Q_\alpha^{XF}(i,j)$ denote the edge strength and orientation preservation values at pixel (i,j); w is the weighting factor	$[0, 1]$	Evaluates the edge information preserved in the fused image
Structural-similarity-based	$Q_Y = \begin{cases}\lambda(w)\mathrm{SSIM}(A,B\mid w)+(1-\lambda(w))\mathrm{SSIM}(B,F\mid w), & \mathrm{SSIM}(A,B\mid w)\geqslant 0.75\\ max\{\mathrm{SSIM}(A,B\mid w),\mathrm{SSIM}(A,B\mid w)\}, & \mathrm{SSIM}(A,B\mid w)<0.75\end{cases}$ $\mathrm{SSIM}(X,Y\mid w)$ is local structural similarity, $\lambda(w)$ is the weighting factor	$[0, 1]$	Reflects the structural similarity between two images
	$Q_0 = (Q_0(A,F)+Q_0(B,F))/2$, where $Q_0(X,Y) = \frac{2\sigma_{XY}}{\sigma_X^2+\sigma_Y^2}\cdot\frac{2\mu_X\mu_Y}{\mu_X^2+\mu_Y^2}$	$[-1, 1]$	Reflects the loss of correlation, luminance distortion, and contrast distortion of the fused image
	$Q_W = \sum_{w\in W} c(w)(\lambda(w)Q_0(A,F\mid w)+(1-\lambda(w)Q_0(B,F\mid w))$ $c(w)$ is normalized salience, $\lambda(w)$ is saliency weight, and $Q_0(X,Y\mid w)$ is Wang–Bovik image quality index	$[-1, 1]$	Indicates the amount of salient information transferred into the fused image
	$Q_E = Q_W(A,B,F)\cdot Q_W\left(\acute{A},\acute{B},\acute{F}\right)^\alpha$ $\acute{A},\acute{B},\acute{F}$ are edge images of A, B, and F; α is the adjustable parameter	$[-1, 1]$	Evaluates the edge information preserved in the fused image
Human-perception-inspired	$Q_{CB} = \overline{\lambda_A(x,y)Q_{AF}(x,y)+\lambda_B(x,y)Q_{BF}(x,y)}$, λ_A, λ_B is the saliency map, Q_{AF}, Q_{BF} is the information preservation value	$[0, 1]$	Assesses the image quality of the fused image

(M, N are the width and height of the image; X and Y represent any image; A and B represent the source image; F represents the fusion result).

4. Datasets

To promote the development of optical–SAR data fusion methods, access to a substantial volume of high-quality optical and SAR image data is essential. SAR and optical images with a sub-meter resolution provide abundant shape structure and texture information of landscape objects. Accordingly, their fusion results are beneficial for accurate image interpretation, and they reflect the specific performance of the used algorithm, thus enabling a persuasive assessment of the fusion methods.

To facilitate research in optical and SAR image fusion technology, we constructed a dataset named YYX-OPT-SAR. This dataset comprises 150 pairs of optical and SAR images covering urban, suburban, and mountain settings, and it is characterized by scene diversity with sub-meter resolution. This dataset can also provide data support for the study of optical and SAR image fusion technology.

The SAR images were collected around Weinan City, Shaanxi Province, China. In order to form high-resolution SAR and optical image pairs, we downloaded optical images of the corresponding areas from Google Earth. The exact location of data collection is shown in Figure 5.

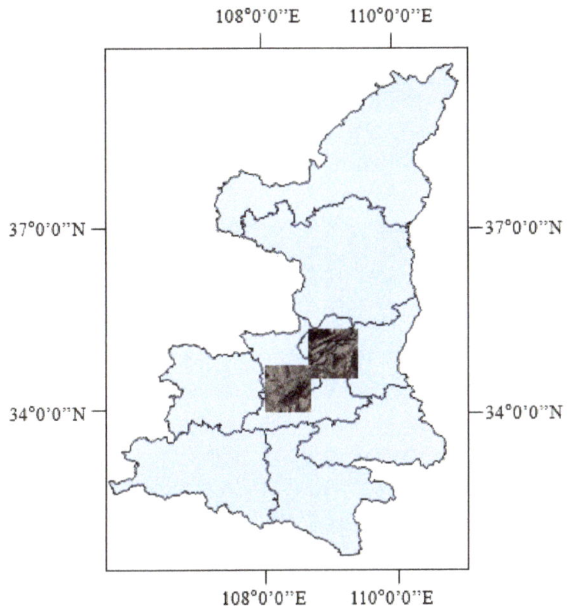

Figure 5. YYX-OPT-SAR dataset: Geographic location of the dataset in Shaanxi Province, China.

After the acquisition of heterogeneous image data, image registration is required to carry out the subsequent phase of fusion. At present, there are many excellent heterologous image registration methods [58,59]. In this paper, an efficient matching algorithm named channel features of orientated gradients (CFOG) [60] is utilized to achieve high accuracy registration with a match error of less than one pixel. In order to maximize the use of available scenes and ensure that each pair of cropped images can fully express the features of optical and SAR images, so as to facilitate the visual evaluation and the subsequent fusion result analysis, we crop the registered optical and SAR image pairs into non-overlapping image blocks with a size of 512 $\times$ 512 pixels. Then, according to different image coverage scenes, we categorize the obtained image pairs into three types: urban, suburban, and mountain. Each type comprises 50 pairs of images, resulting in a total of 150 pairs of images. Some samples are shown in Figure 6.

Another large ground object fusion dataset used in this paper named WHU-OPT-SAR [61] contains medium-resolution optical and SAR images. This dataset, with a resolution of 5 m, covers 51,448.56 square kilometers in Hubei Province, including 100 pairs of 5556 $\times$ 3704 (pixel) images. The exact location and coverage of these images on the map are shown in Figure 7. The optical images in the dataset were obtained from the GF-1 satellite (2 m resolution), while the SAR images were obtained from the GF-3 satellite (5 m resolution), and a unified resolution of 5 m was achieved through bilinear interpolation. Some samples of this dataset are shown in Figure 8.

In the experiment, we produce a high-resolution SAR and optical image dataset covering three different types of scenes: urban, suburban, and mountain. Such a dataset and a publicly available medium-resolution dataset named WHU-OPT-SAR are used collectively to evaluate these fusion methods using three different kinds of evaluation criteria. Detailed specifications of the two datasets are given in Table 5.

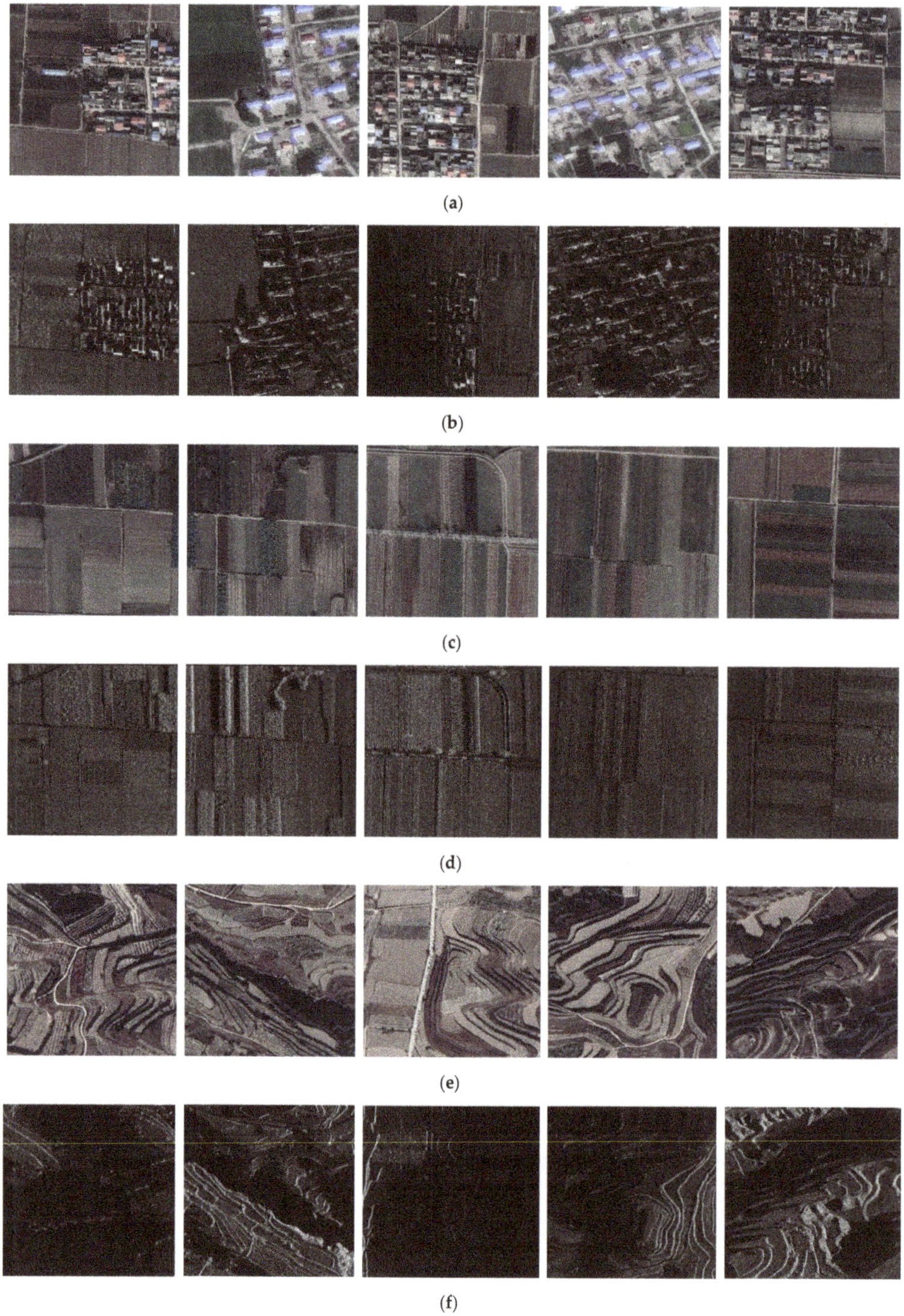

Figure 6. Three types of images for the experiment: (a) Optical images covering the urban setting. (b) SAR images covering the urban setting. (c) Optical images covering the suburban setting. (d) SAR images covering the suburban setting. (e) Optical images covering the mountains. (f) SAR images covering the mountains.

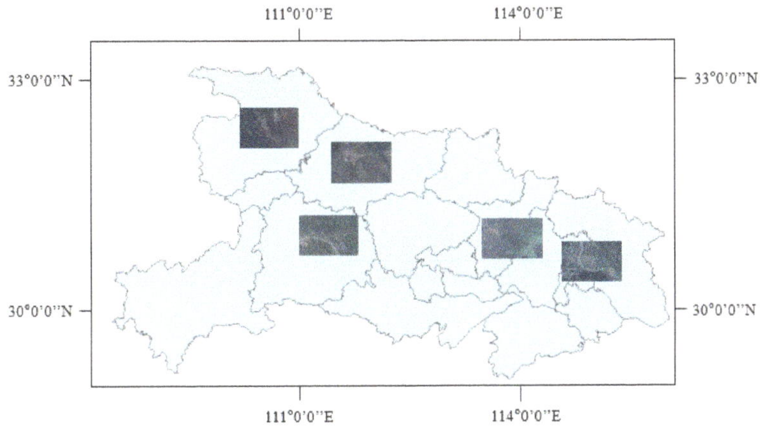

Figure 7. WHU-OPT-SAR dataset: Geographic location of the dataset in Hubei Province, China.

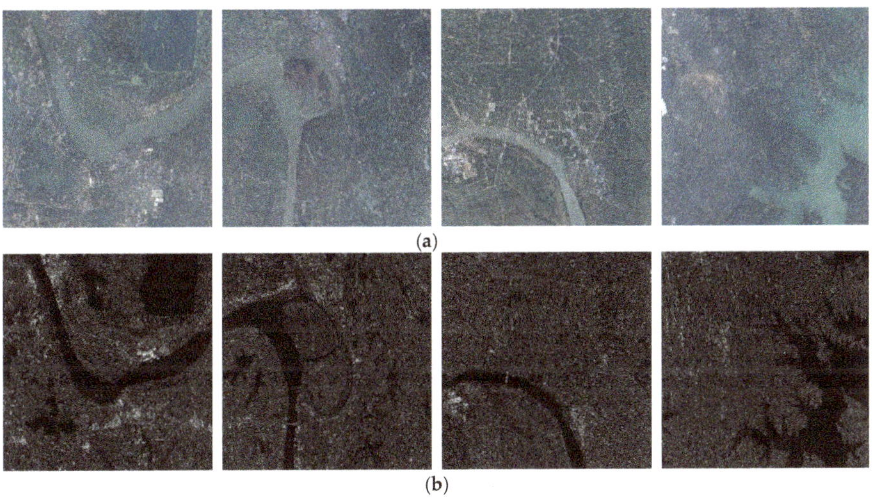

Figure 8. Some samples from WHU-OPT-SAR: (**a**) Optical images. (**b**) SAR images.

Table 5. Specifications of the datasets.

	YYX-OPT-SAR	WHU-OPT-SAR
Number of images (pairs)	150	100
Image pixel size	512 × 512	5556 × 3704
Ground resolution (m)	0.5	5
The surrounding areas	Weinan City, Shaanxi Province in China	Wuhan City, Hubei Province in China

5. Experimental Analysis

5.1. Visual Evaluation

5.1.1. Visual Evaluation of High-Resolution Images

The datasets proposed in the previous section are fused using the 11 fusion methods given in the second section (Table 1) to generate the corresponding fusion results. CS methods select a specific component from the forward transform and replace it with the SAR image for inverse transformation, which makes full use of SAR image information.

Compared with other types of fusion methods, this strategy makes the fusion results include the texture feature of SAR images and introduce shadows in SAR images. Figure 9 illustrates the fusion results of CS methods (including IHS and PCA) and MSD methods (including GP and NSCT). It can be clearly seen that the fusion results of CS methods introduce shadows in the SAR images, which makes image interpretation challenging and fails to achieve the purpose of fusing complementary information. Compared with the results of MSD methods, those of CS methods present worse global spectral quality, often manifesting as color distortion in the areas of roads and vegetation.

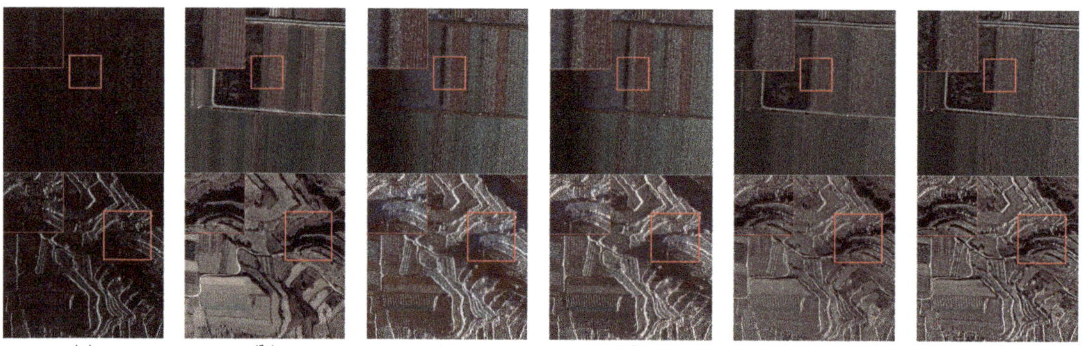

(a) (b) (c) (d) (e) (f)

Figure 9. Fusion results of component substitution methods and partial multiscale decomposition methods. (**a**) SAR. (**b**) Optical image. (**c**) IHS. (**d**) PCA. (**e**) GP_1. (**f**) NSCT_1. A larger version of the red square is shown in the upper left corner.

MSD methods exhibit lower overall color distortion compared to CS methods. Visual observation of the results obtained by applying the two different high-frequency component fusion rules is basically consistent, as shown in Figure 10. Therefore, in the rest of the qualitative evaluation, we only select the fusion results obtained based on one high-frequency fusion rule in each MSD method. As a result, we select the "max-absolute" rule for experimental analysis. On the one hand, by merging the separated low-frequency components, MSD methods effectively retain the spectral information of optical images; on the other hand, by using specific fusion rules for the integration of high-frequency components, the bright textures and edge features of the SAR images are combined into the fusion results to effectively filter out the shadows (as seen in the last two columns of Figure 9). Figure 11 shows the fusion results of different MSD methods. It is apparent that the fusion results of LP and DWT combine more SAR image information, thus introducing the brighter edge features and noise information from SAR images. However, these two methods have color distortion in some areas, such as the edges of houses (the first row of Figure 11) and trees (the second row of Figure 11).

The model-based fusion methods, including SR and GTF, showcase their advantages and disadvantages due to their different fusion strategies. SR tends to make an either–or choice between optical and SAR images, which is consistent with the sparse coefficient selection rule (specifically, the "choose-max" fusion rule with the L1-norm activity level measure). Consequently, the fusion result of SR resembles that of the SAR image in the non-shaded part and that of the optical image in the shaded part, with a rough transition between the two regions. In comparison, GTF integrates the optical image information effectively, but the texture details of SAR are not well introduced, resulting in fuzzy object edges. Figure 12 shows the selected fusion results of these two model-based methods.

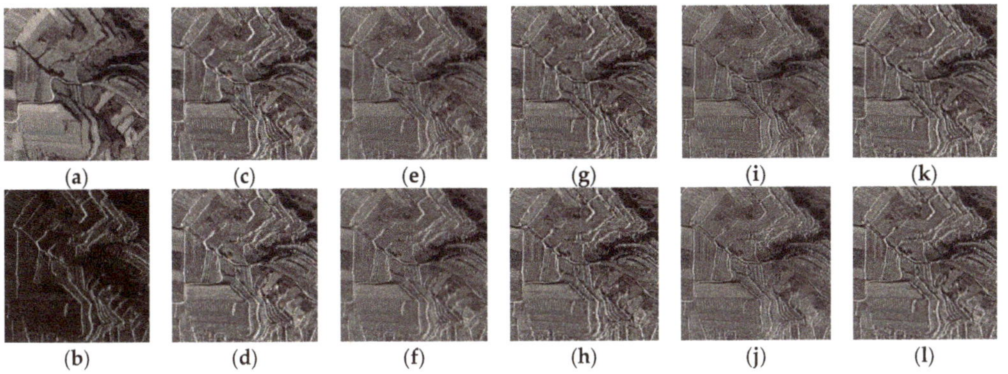

Figure 10. The fusion results obtained using two high-frequency component fusion rules of multiscale decomposition. (**a**) Optical. (**b**) SAR. (**c**) LP_1. (**d**) LP_2. (**e**) GP_1. (**f**) GP_2. (**g**) DWT_1. (**h**) DWT_2. (**i**) CVT_1. (**j**) CVT_2. (**k**) NSCT_1. (**l**) NSCT_2. XX_1 denotes "max-absolute" rule; XX_2 denotes "max-absolute" rule with a local window-based consistency verification scheme.

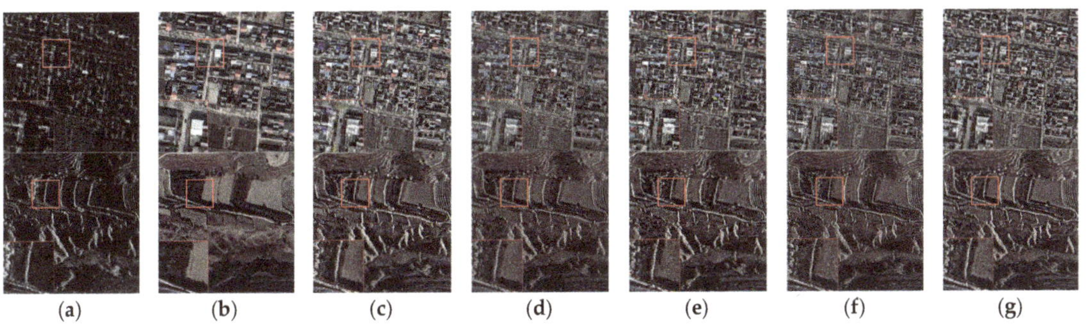

Figure 11. Fusion results of different multiscale decomposition methods. (**a**) SAR. (**b**) Optical. (**c**) LP_1. (**d**) GP_1. (**e**) DWT_1. (**f**) CVT_1. (**g**) NSCT_1. A larger version of the red square is shown in the lower left corner.

The fusion results of all fusion methods under the same image are shown in Figure 13, and the enlarged image of the selected area is displayed in the upper left corner. From the perspective of visual effect, the fusion results of different types of fusion methods are obviously different. CS methods take the SAR image as a component to participate in inverse transformation and effectively use the pixel intensity information of the SAR image. Compared with other fusion methods, CS methods combine more SAR image information, thus introducing more SAR image texture features and shadows. From the box selection area, it can also be seen that the image texture features and shadows are more similar to the SAR image. MSD methods have advantages in preserving the spectral information of optical images by combining the separated low frequency components. At the same time, the high-frequency component is selected through specific fusion rules, and the bright and edge features of the SAR image are fused into the fusion result effectively, while the shadows are filtered. Among them, LP, DWT, and DTCWT combine more SAR image information and introduce brighter edge features and noise information in the SAR image, while the color transition is not natural, such as the roads in the figure. GTF can retain spectral information better, but the boundary of ground objects is fuzzy.

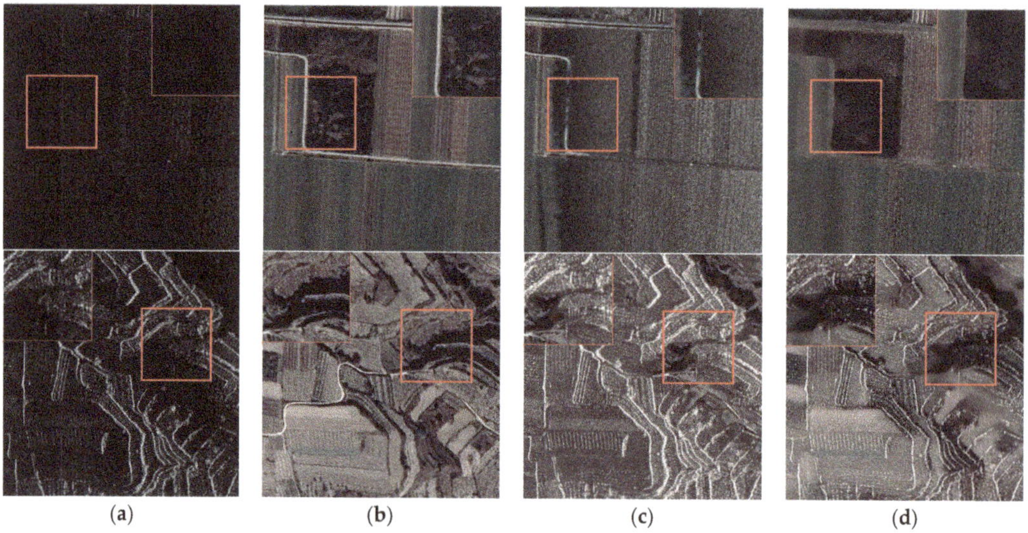

Figure 12. Fusion results of different model-based methods: (a) SAR. (b) Optical. (c) SR. (d) GTF. A larger version of the red square is shown in the corner.

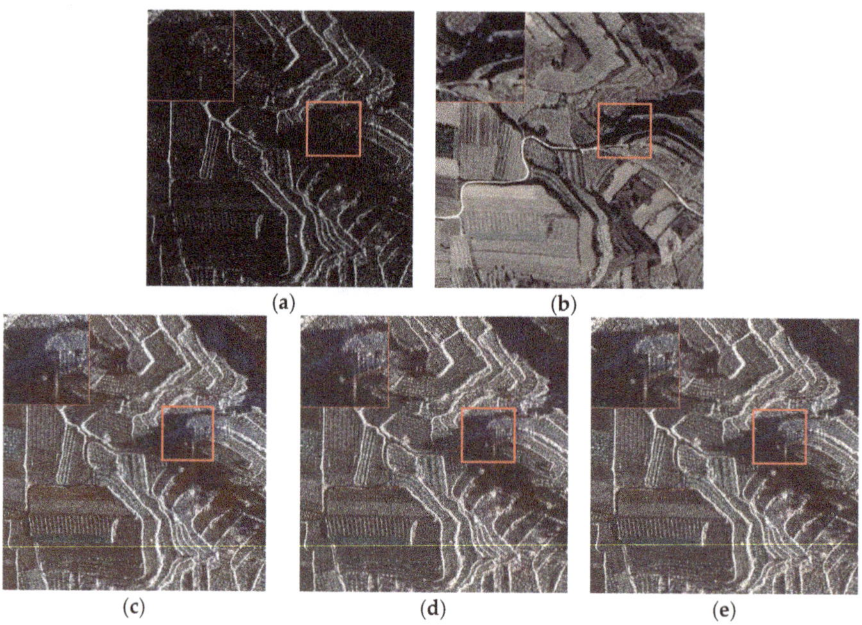

Figure 13. Cont.

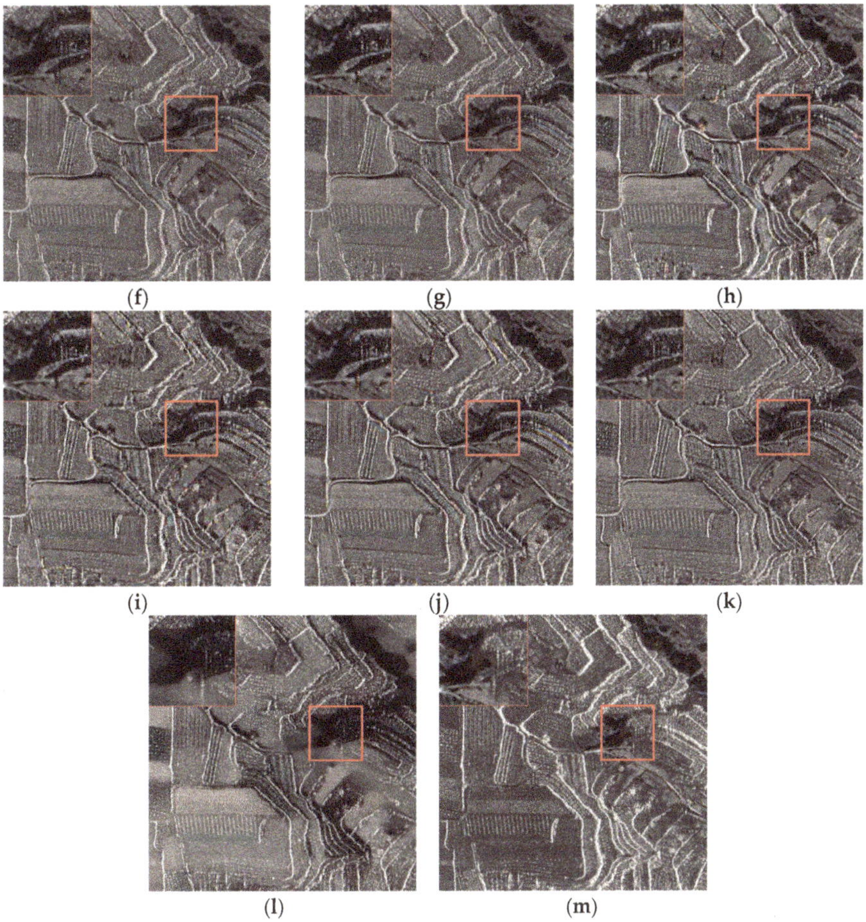

Figure 13. Fusion results of different methods for the high-resolution images: (**a**) SAR. (**b**) Optical. (**c**) IHS. (**d**) PCA. (**e**) GS. (**f**) GP. (**g**) NSCT. (**h**) LP. (**i**) DWT. (**j**) DTCWT. (**k**) CVT. (**l**) SR. (**m**) GTF. A larger version of the red square is shown in the upper left corner.

5.1.2. Visual Evaluation of Medium-Resolution Images

From the perspective of visual effect, the fusion results of medium-resolution images exhibit obvious differences among different types of fusion methods. As is shown in Figure 14, CS methods take the SAR image as a component and participate in the inverse transformation, utilizing the pixel intensity information of the SAR image effectively. Compared with other fusion methods, CS methods combine more SAR image information, which introduces more texture features and shadows from SAR images.

By combining the separated low-frequency components, MSD methods excel in preserving the spectral information of optical images. At the same time, by using specific fusion rules to select high-frequency components, they effectively incorporate the bright point features and edge features of SAR images into the fusion results while filtering out shadows.

GTF can retain the spectral information better. The same kind of ground objects share the same color in the optical images, but the boundaries of ground objects appear blurred. The fusion results of GTF contain less texture information from SAR images, and they only contain the brighter edge information.

5.2. Statistical Evaluation

5.2.1. Statistical Evaluation of High-Resolution Images

The nine quality assessment metrics shown in Table 4 are used for quantitative analysis of the fusion methods. Because each fused image has three bands, we calculate the average value of the metrics of these three bands and use it as the final assessment metric. In addition, we analyze the fusion results of different scenes, including urban, suburban, and mountain scenes, separately. Considering that each scene contains 50 fused images, the average value of their metrics is taken as the result of each method in such a scene. Figure 15 depicts the assessment metric values of the fusion results of each compared method. The higher the metric value, the better the fusion quality.

The EN index reflects the amount of image information, and the PSNR index can measure the ratio of signal to noise and then reflect the degree of image distortion. Figure 15a shows that the fusion result of the mountain scene contains more information than the other two types of ground objects, indicating that more SAR image information is combined. However, when SAR image information is introduced, the noise information of the SAR image is also introduced. Therefore, as shown in Figure 15c, the PSNR value corresponding to the mountain scene is lower than that of the other two types of ground objects.

From the perspective of different types of fusion methods, the fusion quality of CS methods (such as IHS, PCA, and GS) is generally at the same level. For the images in suburban and mountain areas, his achieves the maximum values on most metrics (such as EN, SD, $Q^{AB/F}$, and Q_y). In the images covering the urban scene, most metrics of the fusion results obtained through PCA achieve the maximum values (such as PSNR, Q_w, Q_e, and Q_{cb}).

In the MSD methods, LP has the highest EN and SD values in all images of the three different scenes, which proves that the fusion results of LP contain higher contrast and richer information content than those of other MSD methods. In the three types of images, NSCT achieves the highest values in most metrics (such as Q_w, Q_e, and Q_o), indicating a better fusion effect for NSCT.

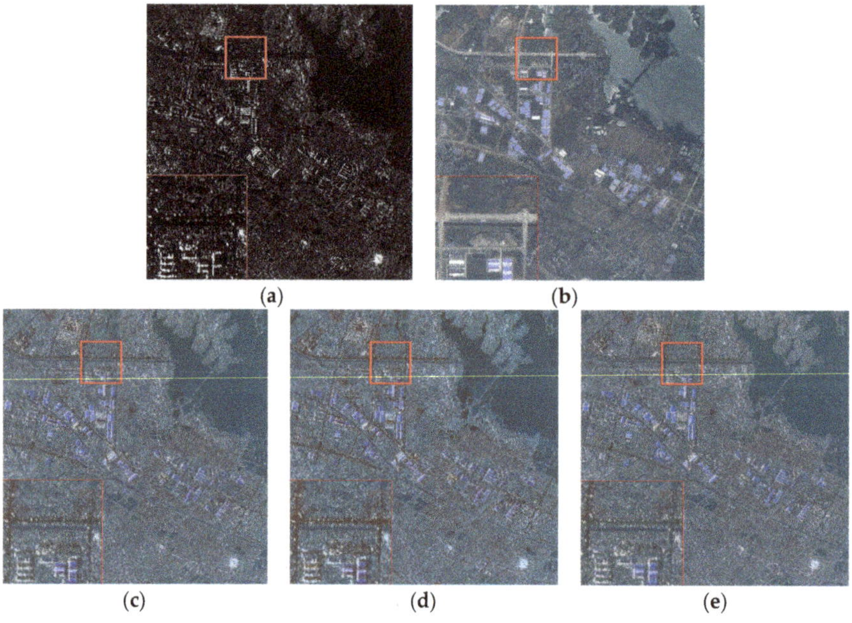

Figure 14. *Cont.*

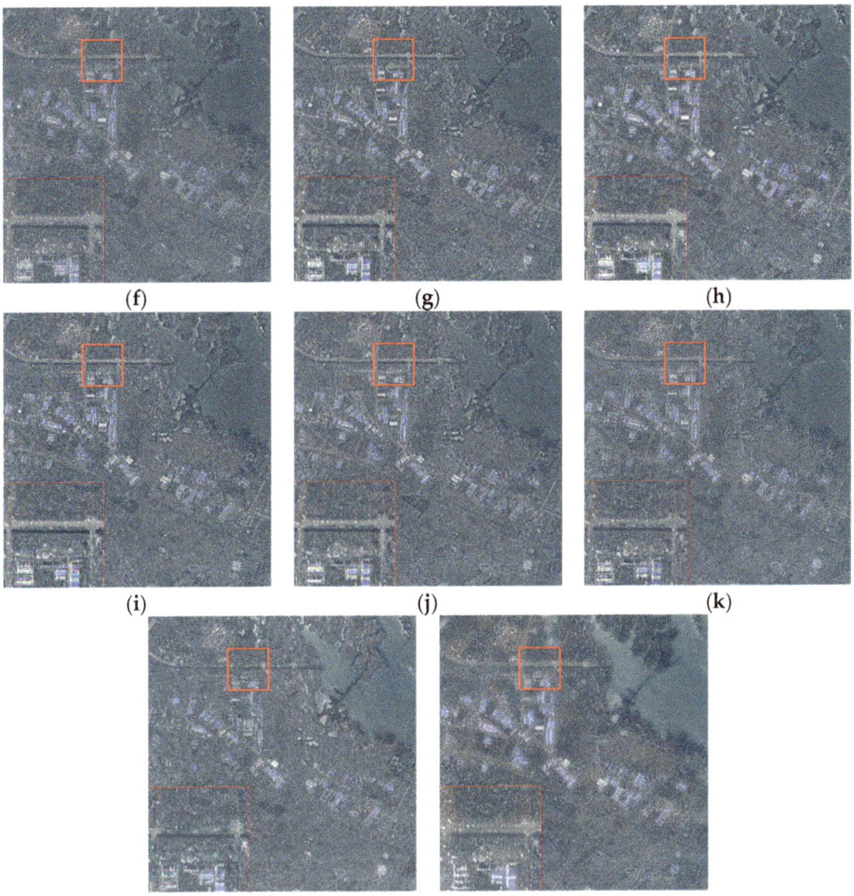

(f) (g) (h) (i) (j) (k) (l) (m)

Figure 14. Fusion results of different methods for the medium-resolution images: (**a**) SAR. (**b**) Optical. (**c**) IHS. (**d**) PCA. (**e**) GS. (**f**) GP. (**g**) NSCT. (**h**) LP. (**i**) DWT. (**j**) DTCWT. (**k**) CVT. (**l**) SR. (**m**) GTF. A larger version of the red square is shown in the lower left corner.

In the model-based methods, GTF obtains higher EN, SD, and PSNR for the three types of images, revealing that the fusion results of GTF contain more information. From the image quality assessment metrics, the fusion results of SR are worse than those of GTF and MSD methods on the whole.

Considering that the visual interpretation of SR is poor and the spectral distortion is serious, the fusion methods that obtain the highest values on each metric except SR are listed in Table 6. GTF has the highest EN and SD values, indicating that the fusion result contains more information. MSD methods obtain the highest values in most of the image quality assessment metrics in the three types of images; in particular, NSCT has the highest Q_w, Q_e, and Q_o values, which indicates that NSCT presents the best fusion result in all of these fusion methods.

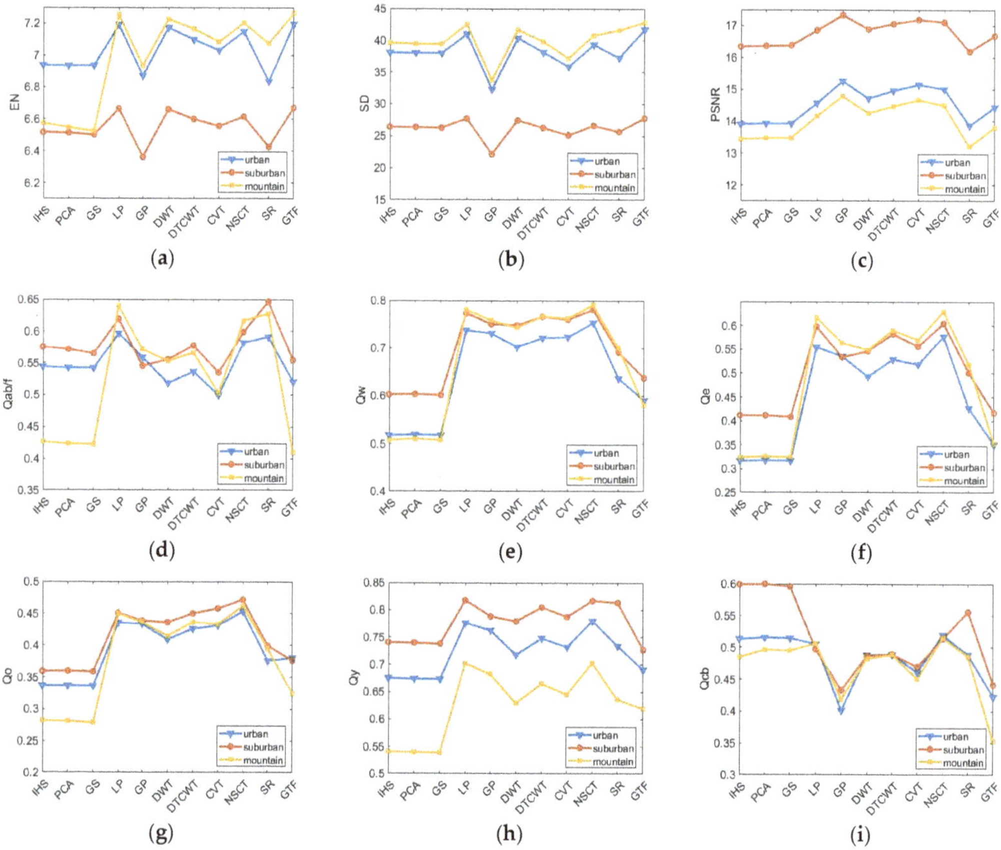

Figure 15. Nine indexes obtained using each method for the high-resolution images of the three types of features: (**a**) EN. (**b**) SD. (**c**) PSNR. (**d**) $Q^{AB/F}$. (**e**) Q_w. (**f**) Q_e. (**g**) Q_o. (**h**) Q_y. (**i**) Q_{cb}.

Table 6. Fusion methods for obtaining the highest index value for various types of high-resolution images (excluding SR).

	EN	SD	PSNR	$Q^{AB/F}$	Q_o	Q_w	Q_e	Q_y	Q_{cb}
Urban	GTF	GTF	GP	LP	NSCT	NSCT	NSCT	NSCT	NSCT
Suburban	GTF	GTF	GP	LP	NSCT	NSCT	NSCT	LP	PCA
Mountain	GTF	GTF	GP	LP	NSCT	NSCT	NSCT	NSCT	NSCT

Based on the above subjective comparison and objective analysis, NSCT performs best when dealing with the fusion of optical and SAR images, mainly including urban and mountain scenes. In terms of statistical evaluation metrics, NSCT and LP have their own advantages in optical and SAR image fusion of suburban areas. However, from the perspective of visual effect, the fusion images obtained through LP have color distortion. Therefore, combining visual effect and statistical evaluation metrics, NSCT can obtain the best fusion effect in the image fusion of these three types of ground objects.

5.2.2. Statistical Evaluation of Medium-Resolution Images

In addition to the quantitative analysis of the fusion results of the high-resolution images, we also select the dataset of the medium-resolution images for image fusion and quantitatively analyze the performance of different fusion methods on this dataset. Figure 16 shows the assessment metric values of the fusion results of all of the compared methods.

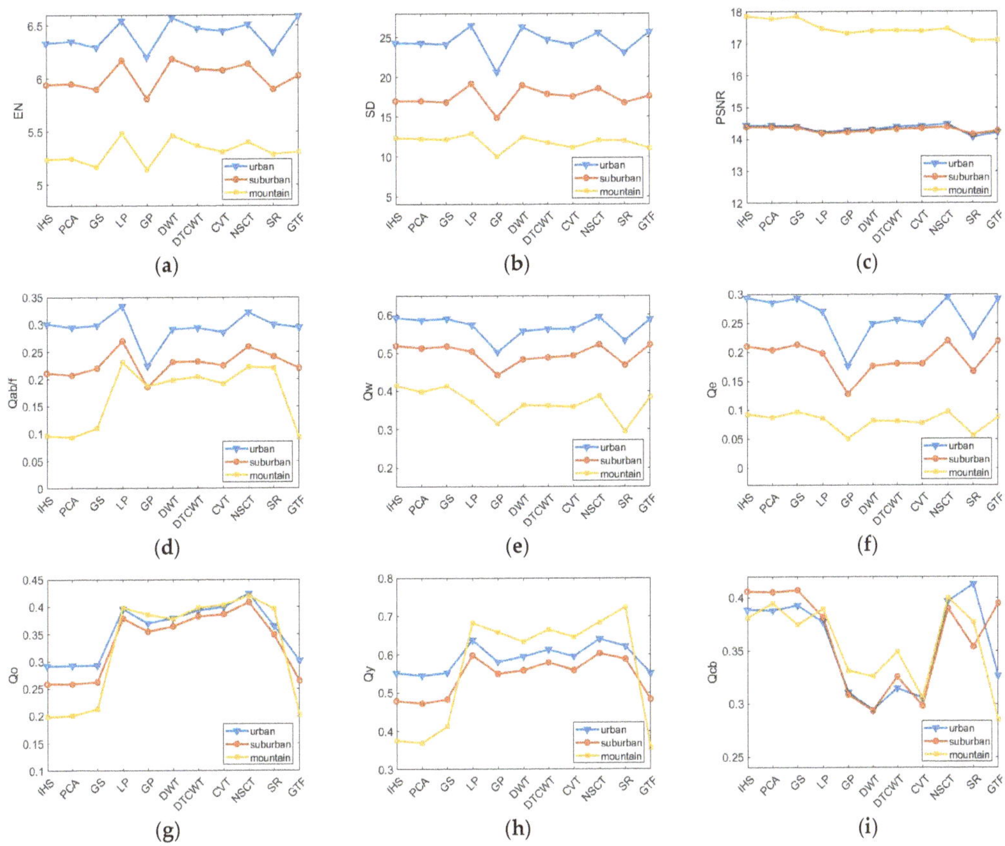

Figure 16. Nine indexes obtained using each method for the medium-resolution images of the three types of features: (**a**) EN, (**b**) SD, (**c**) PSNR, (**d**) $Q^{AB/F}$, (**e**) Q_w, (**f**) Q_e, (**g**) Q_o, (**h**) Q_y, (**i**) Q_{cb}.

From the perspective of different types of fusion methods, the fusion results of the medium-resolution images present a similar law to those of the high-resolution images. The fusion quality of CS methods (such as IHS, PCA, and GS) is almost at the same level, among which IHS achieves the highest PSNR and Q_W values in mountain scenes.

Most of the quality assessment metrics of MSD methods are higher than those of CS methods. Among them, LP has the highest SD and $Q^{AB/F}$ values among the three types of images. NSCT obtains the highest Q_O, Q_y, and Q_e values for the three types of images, indicating that NSCT can obtain better fusion results.

The fusion result obtained through GTF for the medium-resolution images is worse than that for the high-resolution images. This discrepancy is because the quality of the fusion results of GTF depends on the information richness of the original optical and SAR images, and the information of the medium-resolution images is less than that of the high-resolution images.

The fusion methods (excluding SR) that obtain the highest value on each metric are listed in Table 7. It can be observed that LP has the highest SD and $Q^{AB/F}$ and NSCT has the highest Q_y, Q_e, and Q_o for the three types of images. Therefore, NSCT demonstrates the best performance in image fusion across various metrics.

Table 7. Fusion methods for obtaining the highest index value for various types of medium-resolution images (excluding SR).

	EN	SD	PSNR	$Q^{AB/F}$	Q_o	Q_w	Q_e	Q_y	Q_{cb}
Urban	GTF	LP	NSCT	LP	NSCT	NSCT	NSCT	NSCT	NSCT
Suburban	DWT	LP	NSCT	LP	NSCT	NSCT	NSCT	NSCT	GS
Mountain	LP	LP	IHS	LP	NSCT	IHS	NSCT	NSCT	NSCT

Based on the subjective comparison and objective metric analysis of the two groups of data, the conclusions can be drawn as follows. The fusion results of CS methods combine more SAR image information, like texture features and shadows. Nevertheless, the visual effect is worse than that of MSD methods. The surface boundary of GTF is fuzzy, and the visual effect is not as good as that of MSD methods. In the MSD methods, LP and NSCT are at the forefront of most metrics, indicating that these two methods obtain better fusion results. However, considering the color distortion of LP, NSCT performs best among all of the compared methods.

5.3. *Fusion Evaluation According to Classification*

5.3.1. Fusion Evaluation of High-Resolution Images According to Classification

The datasets mentioned earlier are fused using the 11 fusion methods outlined in the second section (Table 1) to generate the corresponding fusion results. CS methods select the specific component of the forward transform and replace it with the SAR image for inverse transformation, thus maximizing the utilization of SAR image information, like texture feature. Figure 9 represents the fusion results of CS methods (including IHS and PCA) and MSD methods (including GP and NSCT). It is evident that the fusion results of the CS methods introduce shadows in the SAR images, thus complicating image interpretation and failing to achieve the purpose of fusing complementary information. In comparison to MSD methods, CS methods present worse global spectral quality, with noticeable color distortion in road and vegetation areas. In this section, we evaluate the 11 fusion methods through image classification for high-resolution images. In the experiment, 50 pairs of fused images, including some typical ground objects, such as bare ground, low vegetation, trees, houses, and roads, are classified by SVM, RF, and CNN, respectively. Given that the dataset contains multiple fused images, the average of their measurements is taken as the result of each method. Tables 8–10 show the classification accuracy results. For instance, the classification results of a pair of optical and SAR images are shown in Figure 17. From the classification accuracy table, it is obvious that CNN achieves higher overall accuracy compared to SVM and RF. Simultaneously, the bare ground is more prone to be misclassified, while the houses and roads exhibit lower misclassification rates. This is because the spectral characteristics of the bare ground are highly uncertain, and the spectral characteristics of the houses and roads are obviously different from those of other categories. From Figure 17, it can be seen that CNN produces classification results more similar to the labels, indicating superior performance compared to SVM and RF, which show more instances of misclassification.

From Figure 17 and the classification accuracy table, it can be concluded that the fused images obtain better classification accuracy compared with single optical or SAR image, which demonstrates that image fusion effectively integrates complementary information of multimodal images and improves classification accuracy. Compared with the single SAR image, the optical image yields better classification accuracy. The special imaging mechanism of SAR leads to the inherent multiplicative speckle noise in SAR images, seriously affecting the interpretation of SAR images. As a result, the classification accuracy of SAR images is poor. During the data fusion process, these effects are transmitted to the fusion image, resulting in the same confusion in the image classification. However, the classification accuracy of the fused image surpasses that of the single optical and single SAR images. This shows the feasibility of using optical and SAR image fusion to improve classification results.

Table 8. SVM classification accuracy table of the high-resolution images. The bolded item is the highest value of classification accuracy for each feature category.

SVM	Bare Ground	Low Vegetation	Trees	Houses	Roads	OA
RGB	49.75%	60.92%	60.37%	62.36%	62.91%	59.69%
SAR	34.08%	27.88%	45.68%	22.14%	42.44%	39.29%
CVT	39.45%	53.32%	50.03%	**73.97%**	63.96%	59.85%
DTCWT	37.39%	56.28%	46.60%	62.99%	61.40%	56.68%
DWT	33.71%	57.02%	48.33%	48.98%	58.98%	52.37%
GP	40.08%	59.70%	57.16%	73.61%	65.13%	61.98%
GS	50.26%	58.50%	59.19%	68.31%	70.23%	64.08%
GTF	**50.96%**	60.45%	58.01%	70.96%	**72.25%**	**65.59%**
IHS	49.48%	**62.39%**	**61.15%**	67.61%	65.35%	63.11%
LP	37.68%	57.15%	50.77%	58.96%	66.57%	57.46%
NSCT	44.38%	55.32%	52.13%	68.98%	65.79%	60.28%
PCA	49.74%	61.59%	60.25%	68.82%	67.78%	63.96%
SR	36.08%	29.89%	45.41%	60.20%	46.13%	42.24%

Table 9. RF classification accuracy table of the high-resolution images. The bolded item is the highest value of classification accuracy for each feature category.

RF	Bare Ground	Low Vegetation	Trees	Houses	Roads	OA
RGB	46.78%	57.77%	60.74%	70.10%	61.67%	57.79%
SAR	32.41%	27.85%	44.89%	22.47%	42.44%	39.20%
CVT	32.58%	48.85%	47.21%	77.40%	62.76%	56.75%
DTCWT	31.59%	51.37%	43.11%	70.37%	60.72%	54.04%
DWT	31.37%	51.00%	45.63%	59.17%	59.72%	52.11%
GP	33.88%	55.27%	53.09%	74.74%	64.45%	58.57%
GS	45.11%	56.05%	59.23%	80.65%	69.97%	64.13%
GTF	45.87%	55.00%	55.78%	78.43%	**72.27%**	**64.45%**
IHS	**47.33%**	57.97%	**60.77%**	81.14%	64.74%	63.69%
LP	34.29%	52.22%	47.32%	70.71%	66.73%	56.68%
NSCT	37.20%	50.37%	49.08%	73.99%	64.31%	57.96%
PCA	45.95%	**58.59%**	60.22%	**81.26%**	68.12%	64.41%
SR	34.49%	29.48%	44.42%	58.35%	48.54%	42.77%

From the perspective of different types of fusion methods, the overall classification accuracy of the three CS methods (such as IHS, PCA, and GS) is nearly the same. The overall classification accuracy of PCA and GS is higher than that of others, suggesting that the two methods have better effect in classification applications though the performance of image fusion is poorer than that of MSD methods. Among the MSD methods, the overall classification accuracy of GP and NSCT is higher, indicating their superior classification effectiveness. Nonetheless, the overall classification accuracy of DTCWT, DWT and LP is lower, with the classification accuracy of fused images obtained by the three methods even lower than that of the optical image. This illustrates that not all optical–SAR fusion methods can improve land classification.

Table 10. CNN classification accuracy table of the high-resolution images. The bolded item is the highest value of classification accuracy for each feature category.

CNN	Bare Ground	Low Vegetation	Trees	Houses	Roads	OA
RGB	56.39%	71.32%	72.54%	81.01%	73.61%	70.07%
SAR	35.37%	62.39%	70.44%	57.78%	55.14%	57.86%
CVT	50.42%	66.82%	67.55%	82.74%	75.67%	71.41%
DTCWT	47.07%	67.81%	66.73%	78.46%	74.64%	69.79%
DWT	44.96%	65.50%	62.32%	73.21%	73.00%	66.76%
GP	51.06%	70.78%	68.67%	81.67%	76.21%	71.94%
GS	53.85%	72.02%	79.72%	**86.69%**	**78.24%**	75.74%
GTF	**58.43%**	72.05%	74.97%	83.75%	77.87%	**75.83%**
IHS	52.88%	74.11%	**80.35%**	85.32%	72.60%	73.89%
LP	48.83%	65.07%	65.67%	74.62%	76.44%	68.87%
NSCT	53.05%	65.89%	65.33%	82.72%	77.40%	71.18%
PCA	53.76%	**74.74%**	80.20%	85.68%	75.40%	75.12%
SR	36.17%	62.11%	70.18%	71.32%	56.79%	61.01%

Overall, GTF obtains the highest overall classification accuracy among the eleven fusion methods. Compared with the single optical image and single SAR image, the fused image has a better classification effect, with up to about 5% improvement.

5.3.2. Fusion Evaluation of Medium-Resolution Images According to Classification

In this section, similarly to the previous section, we evaluate the 11 fusion methods according to image classification for medium-resolution images. Some typical ground objects, such farmland, city, village, water, forest, and roads, are classified by SVM, RF, and CNN, respectively. Tables 11–13 show the classification accuracy of the 11 fusion methods. The overall classification accuracy for medium-resolution images is observed to be lower than that of the high-resolution images. Like the high-resolution images, the city and water have a smaller chance of being misclassified due to their distinct spectral characteristics. Among the 11 fusion methods, GTF obtains the highest overall classification accuracy. The results in Figure 18 and the classification accuracy tables indicate that, for most cases, the overall classification accuracy after fusion is better than before fusion across all of the three classification methods. This indicates that optical–SAR fusion has the potential to improve land classification. But, the visual effect is not as good as that of the high-resolution images, possibly due to the lower image resolution and a larger number of categories. Figure 18 also reveals that the overall classification result of CNN is more similar to the ground truth, indicating that this method has better classification results, whereas SVM and RF have more misclassification.

Building on the above groups of classification experiments, we can obtain the following results: (1) The classification effect of CNN is better than that of RF and SVM. (2) Features with relatively different spectral characteristics from other features have a lower probability of misclassification, while features with relatively uncertain spectral characteristics have a higher probability of misclassification. (3) Fused images obtained using fusion methods exhibit better a classification effect compared to single SAR or optical images. (4) Among the 11 fusion methods selected, GTF consistently achieves the highest overall classification accuracy for all three classification methods.

Table 11. SVM classification accuracy table of the medium-resolution images. The bolded item is the highest value of classification accuracy for each feature category.

SVM	Farmland	City	Village	Water	Forest	Road	Others	OA
RGB	39.97%	66.02%	44.46%	58.82%	61.56%	42.61%	43.75%	51.97%
SAR	21.37%	46.66%	20.39%	46.92%	27.26%	24.61%	28.45%	33.65%
CVT	34.48%	54.90%	26.92%	48.13%	45.93%	24.74%	28.60%	43.89%
DTCWT	37.15%	59.48%	23.55%	47.51%	41.30%	25.24%	25.58%	43.54%
DWT	37.41%	50.44%	25.49%	44.33%	42.41%	26.15%	26.27%	44.12%
GP	27.03%	54.97%	28.11%	45.74%	33.75%	20.73%	28.53%	45.19%
GS	**39.98%**	61.50%	**44.53%**	66.78%	51.12%	36.39%	40.98%	53.40%
GTF	35.90%	**79.87%**	40.55%	**79.14%**	**62.20%**	**45.96%**	**60.42%**	**56.95%**
IHS	30.10%	61.86%	34.67%	68.20%	47.75%	39.19%	36.38%	52.77%
LP	29.48%	45.80%	26.94%	53.76%	48.20%	30.86%	30.62%	48.53%
NSCT	31.18%	54.98%	27.11%	61.42%	45.80%	33.19%	34.65%	52.01%
PCA	30.13%	69.53%	31.01%	66.86%	49.01%	38.20%	36.23%	52.81%
SR	27.29%	34.26%	25.64%	47.45%	27.95%	29.12%	29.19%	32.61%

Table 12. RF classification accuracy table of the medium-resolution images. The bolded item is the highest value of classification accuracy for each feature category.

RF	Farmland	City	Village	Water	Forest	Road	Others	OA
RGB	37.41%	66.55%	37.54%	61.87%	58.16%	40.54%	38.30%	48.50%
SAR	20.92%	44.62%	20.14%	46.59%	27.35%	24.69%	28.44%	33.63%
CVT	31.10%	55.19%	24.59%	41.19%	41.96%	22.30%	23.44%	40.16%
DTCWT	32.37%	52.41%	22.82%	42.09%	37.09%	22.68%	22.49%	40.71%
DWT	36.78%	54.63%	25.54%	41.15%	40.63%	24.93%	24.08%	42.96%
GP	26.51%	55.10%	26.83%	45.92%	41.06%	31.84%	27.61%	43.62%
GS	**37.61%**	60.41%	**39.71%**	56.83%	56.69%	33.09%	33.79%	49.51%
GTF	34.18%	**80.71%**	38.39%	**72.33%**	**58.85%**	**44.69%**	**56.69%**	**54.82%**
IHS	28.54%	61.01%	30.42%	58.33%	43.75%	35.98%	31.34%	49.46%
LP	27.45%	48.89%	25.83%	49.26%	44.14%	28.16%	27.92%	46.36%
NSCT	29.15%	52.90%	27.16%	54.44%	40.18%	33.56%	29.81%	48.80%
PCA	29.80%	69.39%	37.81%	66.62%	49.57%	35.18%	28.63%	49.37%
SR	25.93%	44.14%	26.21%	49.20%	27.69%	27.81%	28.96%	34.47%

Table 13. CNN classification accuracy table of the medium-resolution images. The bolded item is the highest value of classification accuracy for each feature category.

CNN	Farmland	City	Village	Water	Forest	Road	Others	OA
RGB	52.83%	81.49%	61.08%	72.85%	70.89%	70.25%	67.85%	65.23%
SAR	30.49%	53.45%	35.20%	62.07%	31.84%	41.26%	29.70%	39.00%
CVT	44.48%	75.19%	56.32%	69.73%	64.91%	51.44%	57.65%	53.91%

Table 13. *Cont.*

CNN	Farmland	City	Village	Water	Forest	Road	Others	OA
DTCWT	48.35%	79.58%	54.52%	67.41%	61.56%	56.86%	55.43%	53.76%
DWT	47.81%	71.64%	56.71%	65.38%	63.62%	56.61%	57.37%	54.26%
GP	49.33%	74.68%	58.56%	66.84%	63.85%	61.64%	59.56%	55.39%
GS	51.38%	80.91%	**66.70%**	**83.02%**	66.07%	69.20%	66.87%	68.11%
GTF	**53.14%**	85.05%	65.94%	78.39%	**73.33%**	**72.45%**	**67.97%**	**69.49%**
IHS	40.62%	71.46%	54.82%	68.12%	67.78%	59.29%	56.43%	66.56%
LP	49.91%	76.82%	57.64%	63.98%	69.25%	60.55%	50.76%	60.73%
NSCT	42.73%	75.78%	58.17%	72.56%	67.18%	63.73%	55.69%	66.98%
PCA	50.28%	**86.13%**	64.26%	80.53%	68.96%	70.11%	66.43%	68.19%
SR	31.25%	61.46%	40.71%	66.54%	34.59%	45.33%	31.29%	42.91%

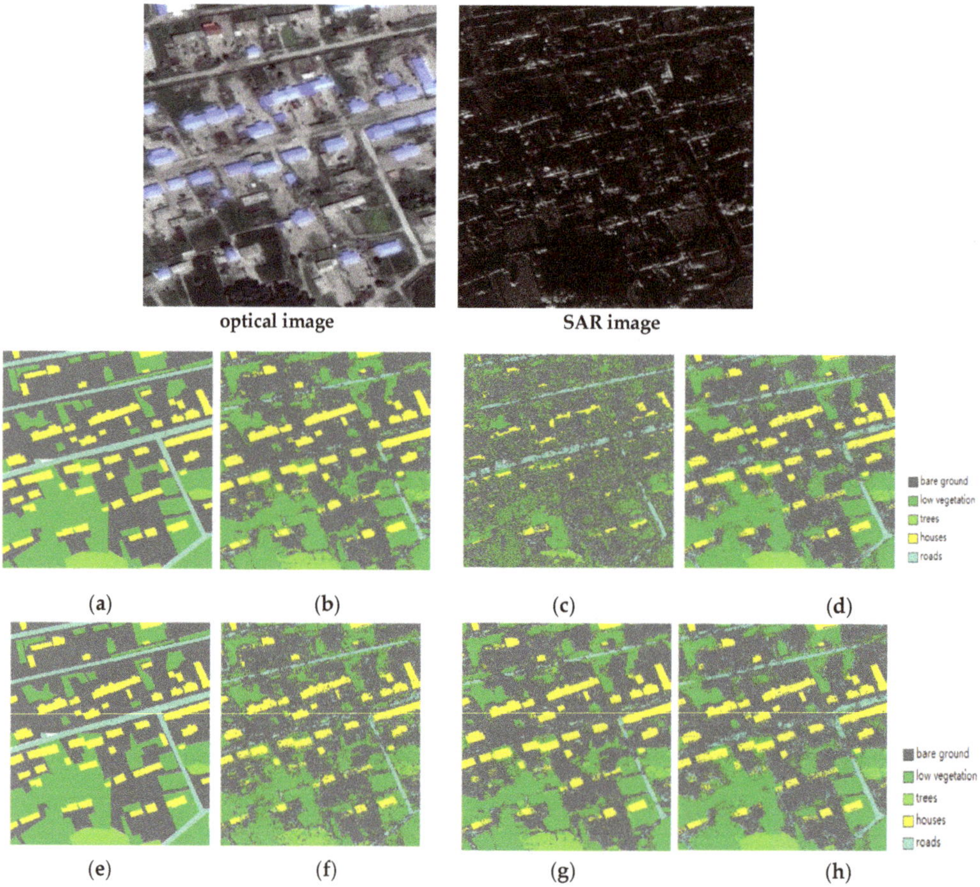

Figure 17. Classified images of (**a**) label, (**b**) RGB, (**c**) SAR, (**d**) GTF, (**e**) label, (**f**) SVM, (**g**) RF, (**h**) CNN. (YYX-OPT-SAR).

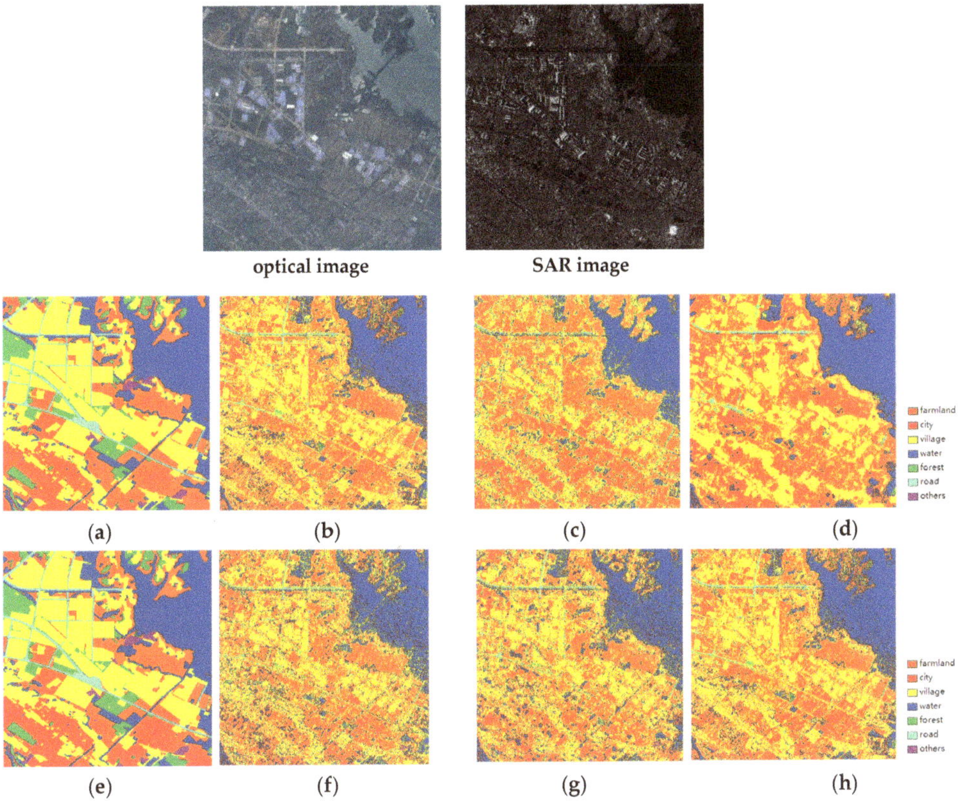

Figure 18. Classified images of (**a**) label, (**b**) RGB, (**c**) SAR, (**d**) GTF, (**e**) label, (**f**) SVM, (**g**) RF, (**h**) CNN (WHU-OPT-SAR).

For image quality metrics, among all of the fusion methods involved in comparison, NSCT has a superior visual effect in image fusion, and its quantitative metrics of fusion results are at the forefront. In the evaluation using image classification, three classic methods, including SVM, RF, and CNN, are used to perform image classification. Most experimental results show that the overall classification accuracy after fusion is better than that before fusion for all three classification methods. This demonstrates that optical–SAR fusion can improve land classification. In all of these fusion methods, GTF obtains the best classification results. Therefore, we recommend NSCT for image fusion and GTF for classification applications.

6. Conclusions

The fusion of optical and SAR images is an important research direction in remote sensing. This fusion allows for the effective integration of complementary information from SAR and optical sources, thus better meeting the requirements of remote sensing applications, such as target recognition, classification, and change detection, so as to realize the collaborative utilization of multi-modal images.

In order to select appropriate methods to achieve high-quality fusion of SAR and optical images, this paper systematically reviews the current pixel-level fusion algorithms for SAR and optical image fusion and then selects eleven representative fusion methods, including CS methods, MSD methods, and model-based methods for comparison analysis. In the experiment, we produce a high-resolution SAR and optical image dataset (named

YYX-OPT-SAR) covering three different types of scenes, including urban, suburban, and mountain scenes. Additionally, a publicly available medium-resolution dataset named WHU-OPT-SAR is utilized to evaluate these fusion methods according to three different kinds of evaluation criteria, including the visual evaluation, the objective image quality metrics, and the classification accuracy.

The evaluation based on image quality metrics reveals that MSD methods can effectively avoid the negative effects of SAR image shadows on the corresponding area of the fusion result compared with the CS methods, while the model-based methods show comparatively poorer performance. Notably, among all of the evaluated fusion methods, NSCT presents the most effective fusion result.

It is suggested that image quality metrics should not be the only option for the interpretation of fused images. Therefore, image classification should also be used as an additional metric to evaluate the quality of fused images, because some fused images with poor image quality metrics can obtain the highest classification accuracy. The experiment utilizes three classic classification methods (SVM, RF, and CNN) to perform image classification. Most experimental results show that the overall classification accuracy after fusion is better than that before fusion for all three classification methods, indicating that optical–SAR fusion can improve land classification. Notably, in all of these fusion methods, GTF obtains the best classification results. Consequently, the suggestion is to employ NSCT for image fusion and GTF for image classification based on the experimental findings.

The differences between this paper and the previous conference paper are mainly related to the following four aspects: First, we extend the original self-built dataset from the original 60 image pairs to 150 image pairs, and we add classification labels to provide data support for subsequent advanced visual tasks. While previous contributions did not expose the dataset, this paper exposes the produced dataset. Second, because the self-built dataset is a high-resolution image, in order to better evaluate the fusion effect of the fusion method at different resolutions, we added the experiment under the published medium-resolution images as a comparison, so as to prove that the excellent fusion method can obtain better results in images with different resolutions. Third, the previous contribution is a short paper, and there is no detailed introduction to optical and SAR pixel-level image fusion algorithms. This paper systematically reviews the current pixel-level fusion algorithms of optical and SAR image fusion. Fourth, we evaluate the fusion quality between different fusion methods by combining subsequent advanced visual tasks, and we verify the effectiveness of image fusion in image classification, proving that the fused image can obtain better results than the original image in image classification.

At present, most pixel-level fusion methods of optical and SAR images rely on traditional algorithms, which may lack comprehensive analysis and interpretation of these highly heterogeneous data. Consequently, these methods inevitably encounter performance bottlenecks. Therefore, this non-negligible limitation further creates a strong demand for alternative tools with powerful processing capabilities. As a cutting-edge technology, deep learning has made remarkable breakthroughs in many computer vision tasks due to its impressive capabilities in data representation and reconstruction. Naturally, it has been successfully applied to other types of multimodal image fusion, such as optical-infrared fusion [62,63]. Accordingly, we will also explore the application of deep learning methods for optical and SAR image fusion in the future.

Author Contributions: Conceptualization and methodology, J.L. and Y.Y.; writing—original draft preparation, J.L.; writing—review and editing, J.L., J.Z. and Y.Y.; supervision, C.Y., H.L. and Y.Z. All authors have read and agreed to the published version of the manuscript.

Funding: This work is supported in part by the National Natural Science Foundation of China under Grant 41971281, Grant 42271446, and in part by the Natural Science Foundation of Sichuan Province under Grant 2022NSFSC0537. We are also very grateful for the constructive comments of the anonymous reviewers and members of the editorial team.

Data Availability Statement: We produced a high-resolution SAR and optical image fusion dataset named YYX-OPT-SAR. The download link for the dataset is https://github.com/yeyuanxin110/YYX-OPT-SAR (accessed on 21 January 2023). The publicly available medium-resolution SAR and optical dataset named WHU-OPT-SAR can be downloaded from this link: https://github.com/AmberHen/WHU-OPT-SAR-dataset (accessed on 3 June 2021).

Conflicts of Interest: The authors declare no conflict of interest.

References

1. Kulkarni, S.C.; Rege, P.P. Fusion of RISAT-1 SAR Image and Resourcesat-2 Multispectral Images Using Wavelet Transform. In Proceedings of the 2019 6th International Conference on Signal Processing and Integrated Networks (SPIN), Noida, India, 7–8 March 2019.
2. Ghassemian, H. A review of remote sensing image fusion methods. *Inf. Fusion* **2016**, *32*, 75–89. [CrossRef]
3. Pohl, C.; Van Genderen, J.L. Review article Multisensor image fusion in remote sensing: Concepts, methods and applications. *Int. J. Remote Sens.* **1998**, *19*, 823–854. [CrossRef]
4. Battsengel, V.; Amarsaikhan, D.; Bat-Erdene, T.; Egshiglen, E.; Munkh-Erdene, A.; Ganzorig, M. Advanced Classification of Lands at TM and Envisat Images of Mongolia. *Adv. Remote Sens.* **2013**, *2*, 102–110. [CrossRef]
5. Sanli, F.B.; Abdikan, S.; Esetlili, M.T.; Sunar, F. Evaluation of image fusion methods using PALSAR, RADARSAT-1 and SPOT images for land use/land cover classification. *J. Indian Soc. Remote Sens.* **2016**, *45*, 591–601. [CrossRef]
6. Abdikana, S.; Sanlia, F.B.; Balcikb, F.B.; Gokselb, C. Fusion of Sar Images (Palsar and Radarsat-1) with Multispectral Spot Image: A Comparative Analysis of Resulting Images. In Proceedings of the International Archives of the Photogrammetry, Remote Sensing and Spatial Information Sciences, Beijing, China, 3–11 July 2008.
7. Klonus, S.; Ehlers, M. Pansharpening with TerraSAR-X and optical data, 3rd TerraSAR-X Science Team Meeting. In Proceedings of the 3rd TerraSAR-X Science Team Meeting, Darmstadt, Germany, 25–26 November 2008; pp. 25–26.
8. Abdikan, S.; Sanli, F.B. Comparison of different fusion algorithms in urban and agricultural areas using sar (palsar and radarsat) and optical (spot) images. *Bol. Ciênc. Geod.* **2012**, *18*, 509–531. [CrossRef]
9. Sanli, F.B.; Abdikan, S.; Esetlili, M.T.; Ustuner, M.; Sunar, F. Fusion of terrasar-x and rapideye data: A quality analysis. *ISPRS—Int. Arch. Photogramm. Remote Sens. Spat. Inf. Sci.* **2013**, *XL-7/W2*, 27–30. [CrossRef]
10. Clerici, N.; Calderón, C.A.V.; Posada, J.M. Fusion of Sentinel-1A and Sentinel-2A Data for Land Cover Mapping: A Case Study in the Lower Magdalena Region, Colombia. *J. Maps* **2017**, *13*, 718–726. [CrossRef]
11. He, W.; Yokoya, N. Multi-Temporal Sentinel-1 and -2 Data Fusion for Optical Image Simulation. *ISPRS Int. J. Geo-Inf.* **2018**, *7*, 389. [CrossRef]
12. Benedetti, P.D.; Ienco, R.; Gaetano, K.; Ose, R.G.; Pensa, S.D. M^3Fusion: A Deep Learning Architecture for Multiscale Multimodal Multitemporal Satellite Data Fusion. *IEEE J. Sel. Top. Appl. Earth Obs. Remote Sens.* **2018**, *11*, 4939–4949. [CrossRef]
13. Hughes, L.H.; Merkle, N.; Bürgmann, T.; Auer, S.; Schmitt, M. Deep Learning for SAR-Optical Image Matching. In Proceedings of the IEEE International Geoscience and Remote Sensing, Yokohama, Japan, 28 July–2 August 2019; pp. 4877–4880.
14. Abdikan, S.; Bilgin, G.; Sanli, F.B.; Uslu, E.; Ustuner, M. Enhancing land use classification with fusing dual-polarized TerraSAR-X and multispectral RapidEye data. *J. Appl. Remote Sens.* **2015**, *9*, 096054. [CrossRef]
15. Gaetano, R.; Cozzolino, D.; D'Amiano, L.; Verdoliva, L.; Poggi, G. Fusion of sar-optical data for land cover monitoring. In Proceedings of the 2017 IEEE International Geoscience and Remote Sensing Symposium (IGARSS), Fort Worth, TX, USA, 23–28 July 2017; pp. 5470–5473.
16. Gibril, M.B.A.; Bakar, S.A.; Yao, K.; Idrees, M.O.; Pradhan, B. Fusion of RADARSAT-2 and multispectral optical remote sensing data for LULC extraction in a tropical agricultural area. *Geocarto Int.* **2017**, *32*, 735–748. [CrossRef]
17. Hu, J.; Ghamisi, P.; Schmitt, A.; Zhu, X.X. Object based fusion of polarimetric SAR and hyperspectral imaging for land use classification. In Proceedings of the 2016 8th Workshop on Hyperspectral Image and Signal Processing: Evolution in Remote Sensing (WHISPERS), Los Angeles, CA, USA, 21–24 August 2016; pp. 1–5.
18. Kulkarni, S.C.; Rege, P.P.; Parishwad, O. Hybrid fusion approach for synthetic aperture radar and multispectral imagery for improvement in land use land cover classification. *J. Appl. Remote Sens.* **2019**, *13*, 034516. [CrossRef]
19. Dabbiru, L.; Samiappan, S.; Nobrega, R.A.A.; Aanstoos, J.A.; Younan, N.H.; Moorhead, R.J. Fusion of synthetic aperture radar and hyperspectral imagery to detect impacts of oil spill in Gulf of Mexico. In Proceedings of the 2015 IEEE International Geoscience and Remote Sensing Symposium (IGARSS), Milan, Italy, 26–31 July 2015; pp. 1901–1904.
20. Gao, J.; Yuan, Q.; Li, J.; Zhang, H.; Su, X. Cloud Removal with Fusion of High Resolution Optical and SAR Images Using Generative Adversarial Networks. *Remote Sens.* **2020**, *12*, 191. [CrossRef]
21. Kang, W.; Xiang, Y.; Wang, F.; You, H. CFNet: A Cross Fusion Network for Joint Land Cover Classification Using Optical and SAR Images. *IEEE J. Sel. Top. Appl. Earth Obs. Remote Sens.* **2022**, *15*, 1562–1574. [CrossRef]
22. Ye, Y.; Liu, W.; Zhou, L.; Peng, T.; Xu, Q. An Unsupervised SAR and Optical Image Fusion Network Based on Structure-Texture Decomposition. *IEEE Geosci. Remote Sens. Lett.* **2022**, *19*, 1–5. [CrossRef]

23. Liu, H.; Ye, Y.; Zhang, J.; Yang, C.; Zhao, Y. Comparative Analysis of Pixel Level Fusion Algorithms in High Resolution SAR and Optical Image Fusion. In Proceedings of the IGARSS 2022—2022 IEEE International Geoscience and Remote Sensing Symposium, Kuala Lumpur, Malaysia, 17–22 July 2022; pp. 2829–2832.
24. Chen, C.-M.; Hepner, G.; Forster, R. Fusion of hyperspectral and radar data using the IHS transformation to enhance urban surface features. *ISPRS J. Photogramm. Remote Sens.* **2003**, *58*, 19–30. [CrossRef]
25. Yin, N.; Jiang, Q.-G. Feasibility of multispectral and synthetic aperture radar image fusion. In Proceedings of the 6th International Congress on Image and Signal Processing (CISP), Hangzhou, China, 16–18 December 2013; pp. 835–839.
26. Yang, J.; Ren, G.; Ma, Y.; Fan, Y. Coastal wetland classification based on high resolution SAR and optical image fusion. In Proceedings of the 2016 IEEE International Geoscience and Remote Sensing Symposium (IGARSS), Beijing, China, 10–15 July 2016.
27. Liu, Y.; Liu, S.; Wang, Z. A general framework for image fusion based on multi-scale transform and sparse representation. *Inf. Fusion* **2015**, *24*, 147–164. [CrossRef]
28. Eltaweel, G.S.; Helmy, A.K. Fusion of Multispectral and Full Polarimetric SAR Images in NSST Domain. *Comput. Sci. J.* **2014**, *8*, 497–513.
29. Pajares, G.; de la Cruz, J.M. A wavelet-based image fusion tutorial. *Pattern Recognit.* **2004**, *37*, 1855–1872. [CrossRef]
30. Yang, B.; Li, S. Multifocus image fusion and restoration with sparse representation. *IEEE Trans. Instrum. Meas.* **2009**, *59*, 884–892. [CrossRef]
31. Zhang, W.; Yu, L. SAR and Landsat ETM+ image fusion using variational model. In Proceedings of the 2010 International Conference on Computer and Communication Technologies in Agriculture Engineering (CCTAE), Chengdu, China, 12–13 June 2010; pp. 205–207.
32. Li, H.; Manjunath, B.S.; Mitra, S.K. Multisensor Image Fusion Using the Wavelet Transform. *Graph. Models Image Process.* **1995**, *57*, 235–245. [CrossRef]
33. Li, S.; Yang, B.; Hu, J. Performance comparison of different multi-resolution transforms for image fusion. *Inf. Fusion* **2011**, *12*, 74–84. [CrossRef]
34. Aharon, M.; Elad, M.; Bruckstein, A. K-SVD: An Algorithm for Designing Overcomplete Dictionaries for Sparse Representation. *IEEE Trans. Signal Process.* **2006**, *54*, 4311–4322. [CrossRef]
35. Bruckstein, A.M.; Donoho, D.L.; Elad, M. From Sparse Solutions of Systems of Equations to Sparse Modeling of Signals and Images. *SIAM Rev.* **2009**, *51*, 34–81. [CrossRef]
36. Tu, T.-M.; Huang, P.S.; Hung, C.-L.; Chang, C.-P. A Fast Intensity—Hue—Saturation Fusion Technique with Spectral Adjustment for IKONOS Imagery. *IEEE Geosci. Remote Sens. Lett.* **2004**, *1*, 309–312. [CrossRef]
37. Wang, Z.; Ziou, D.; Armenakis, C.; Li, D.; Li, Q. A comparative analysis of image fusion methods. *IEEE Trans. Geosci. Remote Sens.* **2005**, *43*, 1391–1402. [CrossRef]
38. Aiazzi, B.; Baronti, S.; Selva, M. Improving Component Substitution Pansharpening through Multivariate Regression of MS +Pan Data. *IEEE Trans. Geosci. Remote Sens.* **2007**, *45*, 3230–3239. [CrossRef]
39. Burt, P.J.; Adelson, E.H. *The Laplacian Pyramid as a Compact Image Code*; Fischler, M.A., Firschein, O., Eds.; Readings in Computer Vision; Morgan Kaufmann: San Francisco, CA, USA, 1987; pp. 671–679.
40. Burt, P.J. A Gradient Pyramid Basis for Pattern-Selective Image Fusion. *Proc. SID* **1992**, 467–470.
41. Mallat, S.G. A theory for multiresolution signal decomposition: The wavelet representation. *IEEE Trans. Pattern Anal. Mach. Intell.* **1989**, *11*, 674–693. [CrossRef]
42. Selesnick, I.; Baraniuk, R.; Kingsbury, N. The dual-tree complex wavelet transform. *IEEE Signal Process. Mag.* **2005**, *22*, 123–151. [CrossRef]
43. Candès, E.; Demanet, L.; Donoho, D.; Ying, L. Fast Discrete Curvelet Transforms. *Multiscale Model. Simul.* **2006**, *5*, 861–899. [CrossRef]
44. Da Cunha, A.; Zhou, J.; Do, M. The Nonsubsampled Contourlet Transform: Theory, Design, and Applications. *IEEE Trans. Image Process.* **2006**, *15*, 3089–3101. [CrossRef] [PubMed]
45. Ma, J.; Chen, C.; Li, C.; Huang, J. Infrared and visible image fusion via gradient transfer and total variation minimization. *Inf. Fusion* **2016**, *31*, 100–109. [CrossRef]
46. Piella, G.; Heijmans, H. A new quality metric for image fusion. In Proceedings of the 2003 International Conference on Image Processing (Cat. No.03CH37429), Barcelona, Spain, 14–17 September 2003; p. III-173-6.
47. Yang, C.; Zhang, J.-Q.; Wang, X.-R.; Liu, X. A novel similarity based quality metric for image fusion. *Inf. Fusion* **2008**, *9*, 156–160. [CrossRef]
48. Jagalingam, P.; Hegde, A.V. A Review of Quality Metrics for Fused Image. *Aquat. Procedia* **2015**, *4*, 133–142. [CrossRef]
49. Xydeas, C.; Petrović, V. Objective image fusion performance measure. *Electron. Lett.* **2000**, *36*, 308–309. [CrossRef]
50. Wang, Z.; Bovik, A.C. A universal image quality index. *IEEE Signal Process. Lett.* **2002**, *9*, 81–84. [CrossRef]
51. Chen, Y.; Blum, R.S. A new automated quality assessment algorithm for image fusion. *Image Vis. Comput.* **2007**, *27*, 1421–1432. [CrossRef]
52. Liu, Z.; Blasch, E.; Xue, Z.; Zhao, J.; Laganiere, R.; Wu, W. Objective Assessment of Multiresolution Image Fusion Algorithms for Context Enhancement in Night Vision: A Comparative Study. *IEEE Trans. Pattern Anal. Mach. Intell.* **2011**, *34*, 94–109. [CrossRef]
53. Zhang, X.; Ye, P.; Xiao, G. VIFB. A Visible and Infrared Image Fusion Benchmark. In Proceedings of the 2020 IEEE/CVF Conference on Computer Vision and Pattern Recognition Workshops (CVPRW), Seattle, WA, USA, 14–19 June 2020.

54. Ye, Y.; Ren, X.; Zhu, B.; Tang, T.; Tan, X.; Gui, Y.; Yao, Q. An adaptive attention fusion mechanism convolutional network for object detection in remote sensing images. *Remote Sens.* **2022**, *14*, 516. [CrossRef]
55. Melgani, F.; Bruzzone, L. Support vector machines for classification of hyperspectral remote-sensing images. In Proceedings of the IEEE International Geoscience and Remote Sensing Symposium, Toronto, ON, Canada, 4–28 June 2002; Volume 1, pp. 506–508.
56. Feng, T.; Ma, H.; Cheng, X. Greenhouse Extraction from High-Resolution Remote Sensing Imagery with Improved Random Forest. In Proceedings of the IGARSS 2020—2020 IEEE International Geoscience and Remote Sensing Symposium, Waikoloa, HI, USA, 26 September–2 October 2020; pp. 553–556.
57. Krizhevsky, A.; Sutskever, I.; Hinton, G.E. Imagenet Classification with Deep Convolutional Neural Networks. In Proceedings of the Conference on Neural Information Processing Systems, Lake Tahoe, NV, USA, 3–6 December 2012; pp. 1097–1105.
58. Ye, Y.; Zhu, B.; Tang, T.; Yang, C.; Xu, Q.; Zhang, G. A robust multimodal remote sensing image registration method and system using steerable filters with first-and second-order gradients. *ISPRS J. Photogramm. Remote Sens.* **2022**, *188*, 331–350. [CrossRef]
59. Ye, Y.; Tang, T.; Zhu, B.; Yang, C.; Li, B.; Hao, S. A multiscale framework with unsupervised learning for remote sensing image registration. *IEEE Trans. Geosci. Remote Sens.* **2022**, *60*, 1–15. [CrossRef]
60. Ye, Y.; Bruzzone, L.; Shan, J.; Bovolo, F.; Zhu, Q. Fast and Robust Matching for Multimodal Remote Sensing Image Registration. *IEEE Trans. Geosci. Remote Sens.* **2019**, *57*, 9059–9070. [CrossRef]
61. Li, X.; Zhang, G.; Cui, H.; Hou, S.; Wang, S.; Li, X.; Chen, Y.; Li, Z.; Zhang, L. MCANet: A joint semantic segmentation framework of optical and SAR images for land use classification. *Int. J. Appl. Earth Obs. Geoinf.* **2022**, *106*, 102638. [CrossRef]
62. Ma, J.; Yu, W.; Liang, P.; Li, C.; Jiang, J. FusionGAN: A generative adversarial network for infrared and visible image fusion. *Inf. Fusion* **2018**, *48*, 11–26. [CrossRef]
63. Tang, W.; He, F.; Liu, Y. TCCFusion: An infrared and visible image fusion method based on transformer and cross correlation. *Pattern Recognit.* **2023**, *137*, 109295. [CrossRef]

remote sensing

MDPI

Article

Fusion of Identification Information from ESM Sensors and Radars Using Dezert–Smarandache Theory Rules

Tadeusz Pietkiewicz

Institute of Radioelectronics, Faculty of Electronics, Military University of Technology, 00-908 Warsaw, Poland; tadeusz.pietkiewicz@wat.edu.pl

Abstract: This paper presents a method of fusion of identification (attribute) information provided by two types of sensors: combined primary and secondary (IFF) surveillance radars and ESMs (electronic support measures). In the first section, the basic taxonomy of attribute identification is adopted in accordance with the standards of STANAG 1241 ed. 5 and STANAG 1241 ed. 6 (draft). These standards provide the following basic values of the attribute identifications: FRIEND; HOSTILE; NEUTRAL; UNKNOWN; and additional values, namely ASSUMED FRIEND and SUSPECT. The basis of theoretical considerations is Dezert–Smarandache theory (DSmT) of inference. This paper presents and uses in practice six information-fusion rules proposed by DSmT, i.e., the proportional conflict redistribution rules (PCR1, PCR2, PCR3, PCR4, PCR5, and PCR6), for combining identification information from different ESM sensors and radars. This paper demonstrates the rules of determining attribute information by an ESM sensor equipped with the database of radar emitters. It is proposed that each signal vector sent by the ESM sensor contains an extension specifying a randomized identification declaration (hypothesis)—a basic belief assignment (BBA). This paper also presents a model for determining the basic belief assignment for a combined primary and secondary radar. Results of the PCR rules of sensor information combining for different scenarios of a radio electronic situation (deterministic and Monte Carlo) are presented in the final part of this paper. They confirm the legitimacy of the use of Dezert–Smarandache theory in information fusion for primary radars, secondary radars, and ESM sensors.

Keywords: information fusion; Dezert–Smarandache theory (DSmT) of inference; conflict redistribution rules; radar emitters recognition; electronic support measures (ESMs); primary and secondary radars

1. Introduction

Designing systems for creating a recognized air picture in the air defense system requires, among other factors, the development of algorithms for combining identification information about detected targets from various types of sensors. The basic elements of the air-situation-recognition system are two types of sensors: ESM sensors and combined primary and secondary radars (IFF: identification friend or foe). They provide identification information to information-processing centers that develop a recognized air picture. Each air-situation information-processing system should have an attribute information set, specifying acceptable values for the identification information of the detected targets transmitted by sensors to information-processing centers and information produced by those centers. This article uses a certain interpretation of attribute identification in accordance with the NATO STANAG 1241 standard [1,2]. It should be noted that this is one of the possible interpretations of the adopted definitions. It is assumed that five identification classes are used: three primary and two secondary ones. Sensors can transmit identification information in the form of a hard decision, sometimes determined as non-randomized, or a soft decision, sometimes determined as a randomized decision. In this paper, it is assumed that the sensors send identification information to the system in a randomized form, i.e., in the form of a basic belief assignment on the set of identification classes. This assignment determines the sensor's belief that the detected emitter belongs to separate identification classes.

Another problem that should be solved by the designers of air-situation-recognition systems is the choice of the method of combining information from sensors. One possible solution is the STANAG 4162 standard proposed by NATO. It is a standard based on Bayesian decision functions. Another solution is to use Dempster–Shafer reasoning [3,4]. Works [5,6] show that such solutions have their drawbacks. The disadvantages of Dempster–Shafer reasoning are also confirmed in this work for situations of high conflict. These disadvantages can be avoided by applying Dezert–Smarandache theory (DSmT).

DSmT in its basic version contains five rules of proportional conflict redistribution, namely PCR1-PCR5. Later, the authors presented a new PCR6 rule [7], which was tested in this paper, alongside the earlier PCR1-PCR5 rules.

At this point, it should be noted that DSmT is not the only development of Dempster–Shafer theory. DSmT provides different rules for the proportional redistribution of the conflict mass between the resulting BBA masses in the process of information fusion. Examples of this development of Dempster–Shafer theory can be found in [8–10]. In [8,9], it was proposed to use the negation evidence (BBAs from sensors) and Deng's [11] entropy to determine BBAs after information fusion. The new BBA distribution is calculated as a weighted sum of the original BBA, with the weights being a function of Deng's entropy. The paper [10] shows the application of the negation evidence method in fault diagnosis.

In [12], a new risk priority model based on the belief Jensen–Shannon divergence measure and Deng's entropy is proposed. In the new method, Deng's entropy and the belief Jensen–Shannon divergence measure are used to model the uncertainty of risk assessments in the "Failure mode and effects analysis" procedure and to deal with potential conflict information. This allows one to calculate the appropriate weighted average probabilities (WAPs) value. Classic Dempster's combination rule is used to fuse data to generate integrated values of the risk factors. Unfortunately, the latest publications do not compare the DSmT method with the proposed new solutions. The examples provided there concern other applications than those presented in this work.

In [13], a generalized evidential Jensen–Shannon (GEJS) divergence to measure the conflict and disparity among multiple sources of evidence. This generalization was used to determine the weights of the information sources. Subsequently, it was used to determine the results of information fusion using Dempster's combination rule.

In [14], an extension of Dempster–Shafer theory was presented, which received the name complex evidence theory. It implements complex weighted discounting multisource information fusion. The complex evidence theory defines complex basic belief assignments and a complex evidential correlation coefficient. A weighted discounting multisource information-fusion algorithm with complex evidential correlation coefficient improves the performance of expert systems based on complex evidence theory.

Another development of Dempster–Shafer theory is the generalized quantum evidence theory [15,16]. In these papers, multisource quantum information fusion was presented. The papers are complemented by an example of a pattern classifier from the motor-rotor-fault-diagnosis domain, which confirmed the efficiency of the multisource quantum information-fusion algorithm.

This paper is a continuation of another work [17] and contains new research results obtained for new scenarios of the electronic situation, new DSmT rules, and new information-fusion schemes.

Below, the substantive content of individual chapters is subsequently discussed. The first part of the paper presents the applied interpretation of attribute identification in accordance with the NATO STANAG 1241 standard. It should be noted that this is one of the possible interpretations of the adopted definitions. It leads to the Bayesian model of the basic belief assignment.

The next part of the paper presents the mathematical form of the DSmT rules PCR1, PCR2, PCR3, PCR4, PCR5, and PCR6 [5,18] for two sensor inputs and PCR5 and PCR6 for three sensor inputs, assuming the Bayesian model of the basic belief assignment of the hypothesis.

The next two sections, namely Sections 4 and 5, show how to determine the basic belief assignment for a combined primary and secondary (IFF) radar and ESM sensors. These assignments are the input information in the PCR1-PCR6 information-fusion algorithms. Section 4 presents a method for determining the basic belief assignments (BBAs) of airborne targets moving in the observation space of a combined primary and secondary (IFF) radar sensor. This method uses the primary radar model, taken from [17]. The result of executing the algorithm of this method are scenarios containing reports from BBAs. Section 5 presents a method for determining BBAs for airborne targets that emit electromagnetic radiation (airborne radars and other emitters). It requires databases of reference signals of various airborne emitters, equipment of airborne platforms, and the nationality of the platforms.

Each sensor report sent to the information-fusion center contains a vector of belief mass for all attribute identification values. The results of the proportional conflict redistribution sensor information, combining rules for selected deterministic and Monte Carlo scenarios, are presented in Sections 6 and 7 of the paper. Section 6 presents the results of research on the fusion of information sent only from ESM sensors. This corresponds to a situation when the ESM sensors operate in a system: one master station and one or two slave stations. Section 7 presents the results of research on the fusion of information sent from ESM sensors and combined primary and secondary radars. This corresponds to a situation where one ESM sensor (master station) and radars cooperate with the information-processing center (the producer of the recognized air picture).

Conclusions are provided at the end of the paper. They confirm the legitimacy of the use of DSmT in information fusion for primary radars, secondary radars, and ESM sensors.

2. Interpretation of Attribute Identification according to STANAG 1241

The set of possible values of attribute identifications used by sensors can be adopted based on standardization documents of organizations that exploit these sensors [1,2,19–21].

This paper assumes a basic taxonomy of identification in accordance with the draft of STANAG 1241 ed. 6 [2]. Other similar documents may include the following standards: STANAG 4420 and STANAG 1241 ed. 5, which provide the following basic values of the attribute identifications:

- FRIEND (F);
- HOSTILE (H);
- NEUTRAL (N);
- UNKNOWN (U).

Each of these documents contain their own definitions of the declarations.

The following definitions of these basic values of the attribute identification are used in this paper (in accordance with [2]):

- FRIEND: an allied/coalition military track, object, or entity; a track, object, or entity, supporting friendly forces and belonging to an allied/coalition nation or a declared or recognized friendly faction or group;
- HOSTILE: a track, object, or entity whose characteristics, behavior, or origin indicate that it belongs to opposing forces, or that it poses a threat to friendly forces or their mission;
- NEUTRAL: a military or civilian track, object, or entity, neither belonging to allied/coalition military forces nor to opposing military forces, whose characteristics, behavior, origin, or nationality indicate that it is neither supporting nor opposing friendly forces or their mission;
- UNKNOWN: an evaluated track, object, or entity, which do not meet the criteria for any other standard identity.

These standards bring additional values of the attribute identification:

- ASSUMED FRIEND (AF);
- SUSPECT (S).

Attention should be paid to these two last identities contained in [1], as well as their definitions [2]:

- ASSUMED FRIEND: a track, object, or entity, which is assumed to be friend or neutral because of its characteristics, behavior, or origin;
- SUSPECT: a track, object, or entity whose characteristics, behavior, or origin indicate that it potentially belongs to opposing forces or potentially poses a threat to friendly forces or their mission.

The identification definitions in [1,2] can lead to different interpretations. This paper adopts the interpretation that is shown by the graphical form of in Figure 1.

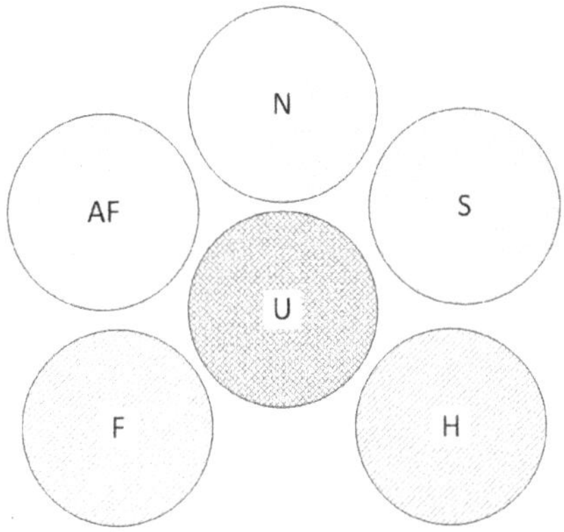

Figure 1. The interpretation of STANAG 1241 using the Venn diagram.

3. Fusion of Information from ESM Sensors and Radars in the Information-Fusion Center (IFC)

3.1. Diagram of the Process of Information Fusion for Two Sensors in the Information-Fusion Center

In this work, it is assumed that ESM sensors send messages asynchronously to the information-fusion center. These reports contain sensor decisions regarding the identification of objects emitting the detected signals. The set of possible identifications is as follows:

$$\Theta = \{\theta_i, = 1, \ldots, 6\}, \tag{1}$$

where in the following interpretation is used:

θ_1: FRIEND (F);
θ_2: HOSTILE (H);
θ_3: NEUTRAL (N);
θ_4: ASSUMED FRIEND(AF);
θ_5: SUSPECT(S);
θ_6: UNKNOWN (U).

According to Figure 1, the hypotheses are mutually exclusive, i.e.,

$$\theta_i \cap \theta_j = \begin{cases} \theta_i, & \text{if } i = j, \\ \varnothing, & \text{if } i \neq j. \end{cases} \tag{2}$$

Each sensor with the number i ($i \in \mathbb{N}$) sends its decisions as so-called soft decisions, i.e., as BBA measure vectors (BBA: basic belief assignment).

$$\boldsymbol{m}_i = [m_i(\theta_1), \ldots, m_i(\theta_6)]. \tag{3}$$

A vector of generalized BBA measures for the information-fusion center should also be introduced as follows:

$$\boldsymbol{m}_F = [m_F(\theta_1), \ldots, m_F(\theta_6)]. \quad (4)$$

This paper adopts the Bayesian BBA model as it has been adopted as valid in the STANAG 4162 standard [20]. This means that Equation (5) applies in addition to (1) and (2).

$$\sum_{i=1}^{6} m_F(\theta_i) = \sum_{i=1}^{6} m_i(\theta_i) = 1. \quad (5)$$

In the first case that is considered, two sensors send, asynchronously in one cycle, one report each, containing decisions regarding the BBAs related to the target. The IFC system, after receiving the report from the sensor, fuses the information contained in the two vectors: in the current generalized BBA vector $\boldsymbol{m}_F = [m_F(\theta_1), \ldots, m_F(\theta_6)]$ and in the BBA vector $\boldsymbol{m}_1$ from sensor 1 or in the BBA vector $\boldsymbol{m}_2$ from sensor 2.

The information-fusion procedure performed in the IFC is carried out in accordance with the following formula:

$$\boldsymbol{m}_F^{'} = R_F(\boldsymbol{m}_F, \boldsymbol{m}_i) \ (i = 1 \text{ or } 2), \quad (6)$$

wherein $\boldsymbol{m}_F^{'}$ is a vector of the generalized BBA measure determined by the R_F rule based on the previous generalized BBA measure vector $\boldsymbol{m}_F$ and the new BBA measure vector $\boldsymbol{m}_i$ sent by the i-th sensor. The diagram of identification information fusion from the ESM sensors is shown in Figure 2.

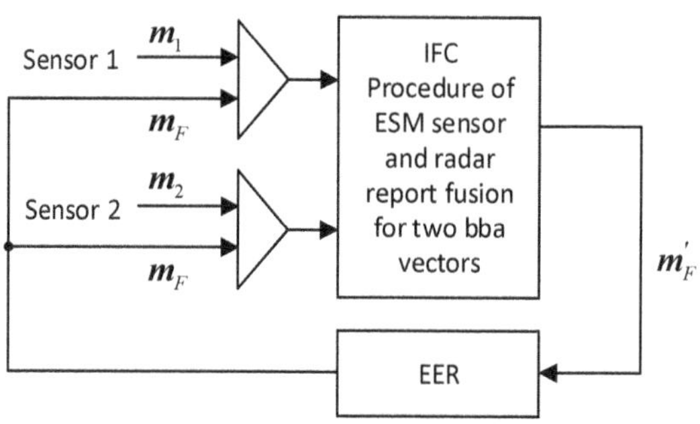

Figure 2. The diagram of the information-fusion process in the information-fusion center (IFC) for two sensors. Explanations: m_i a BBA measure vector of i-th sensor; m_F, a generalized BBA measure vector that is a part of the electronic entity record in the IFC; EER, an electronic entity record in the IFC database.

In the second case that is considered, two sensors send, asynchronously in one cycle, one report each, containing decisions regarding the BBAs related to the target. The IFC system waits for reports from both sensors in one cycle, using registers. Only when both registers are full does the IFC system perform a fusion of the information contained in three vectors: BBA vector $\boldsymbol{m}_F = [m_F(\theta_1), \ldots, m_F(\theta_6)]$, BBA vector $\boldsymbol{m}_1$ from sensor 1, and BBA vector $\boldsymbol{m}_2$ from sensor 2. It should be noted that this method has a drawback: the information stored in the registers lose credibility.

In this case, the information-fusion procedure performed in the IFC is carried out in accordance with the following formula:

$$\boldsymbol{m}_F^{'} = R_F(\boldsymbol{m}_F, \boldsymbol{m}_1, \boldsymbol{m}_2), \quad (7)$$

wherein $\boldsymbol{m}_F^{'}$ is a vector of the generalized BBA measure determined by the R_F rule based on the previous generalized BBA measure vector $\boldsymbol{m}_F$ and the new BBA measure vectors $\boldsymbol{m}_1$ and $\boldsymbol{m}_2$ sent by both sensors. The diagram of identification information fusion from the ESM sensors is shown in Figure 3.

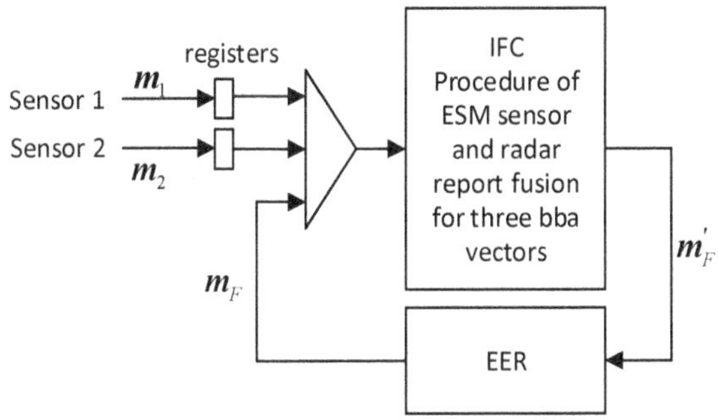

Figure 3. The diagram of the information-fusion process in the information-fusion center (IFC) for two sensors and electronic entity record from the IFC database. Explanations: $\boldsymbol{m}_i$, a BBA measure vector of i-th sensor; $\boldsymbol{m}_F$, a generalized BBA measure vector that is a part of the electronic entity record in the IFC; EER, an electronic entity record in the IFC database.

Further in this paper, the combination rules of the BBA vector from the i-th sensor and the generalized BBA vector in the CFI are described.

3.2. The Rules of Combination of BBA Measures Vectors

This section presents formulas defining various combination rules for calculating basic belief assignments for the system shown in Figures 2 and 3. The general forms are described in detail in [6,18,22]. The information-fusion rules of the DSm theory are presented below with the following constraints:

- the properties of a set of hypotheses are described by Formulas (1) and (2);
- for the first scheme (Figure 2), the information-fusion procedure handles two information inputs: on one input, reports from two ESM sensors appear alternately, while on the second input, electronic entity records from the IFC database appear;
- for the second scheme (Figure 3), the information-fusion procedure handles three information inputs: on the first input, reports from a combined primary and secondary surveillance radar appear; on the second input, reports from an ESM sensor appear; and on the third input, electronic entity records from the IFC database appear.

3.2.1. Dempster's Rule

Dempster's rule [3,4] of the BBA measure vector $\boldsymbol{m}_i$ sent by the i-th sensor and the generalized BBA measure vector $\boldsymbol{m}_F$ s in the IFC is described for each $\theta_j \in \boldsymbol{\Theta}$ by the following formula:

$$\begin{gathered} m'_F(\theta_j) = m_D(\theta_j) = \frac{\sum_{\substack{k=1,\ldots,6 \\ l=1,\ldots,6 \\ \theta_k \cap \theta_l = \theta_j}} m_F(\theta_k) m_i(\theta_l)}{1 - \sum_{\substack{k=1,\ldots,6 \\ l=1,\ldots,6 \\ \theta_j \cap \theta_k = \varnothing}} m_F(\theta_k) m_i(\theta_l)} = \\ = \frac{m_F(\theta_j) m_i(\theta_j)}{1 - \sum_{k=1}^{6} \sum_{\substack{l=1 \\ l \neq k}}^{6} m_F(\theta_k) m_i(\theta_l)} = \frac{m_{Fi}(\theta_j)}{1 - k_{Fi}}, \end{gathered} \tag{8}$$

where in the k_{Fi} degree of conflict is defined by the formula:

$$k_{Fi} = \sum_{\substack{k=1,\ldots,6 \\ l=1,\ldots,6 \\ \theta_j \cap \theta_k = \varnothing}} m_F(\theta_k) m_i(\theta_l) = \sum_{k=1}^{6} \sum_{\substack{l=1 \\ l \neq k}}^{6} m_F(\theta_k) m_i(\theta_l), \tag{9}$$

while

$$m_{Fi}(\theta_j) = m_F(\theta_j)m_i(\theta_j). \tag{10}$$

It can be noticed that

$$\sum_{\substack{k=1,\ldots,6 \\ l=1,\ldots,6}} m_F(\theta_k)m_i(\theta_l) = 1\,, \tag{11}$$

$$\begin{aligned}
&\sum_{\substack{k=1,\ldots,6 \\ l=1,\ldots,6 \\ \theta_j\cap\theta_k=\varnothing}} m_F(\theta_k)m_i(\theta_l) + \sum_{\substack{k=1,\ldots,6 \\ l=1,\ldots,6 \\ \theta_j\cap\theta_k\neq\varnothing}} m_F(\theta_k)m_i(\theta_l) = \\
&= \sum_{k=1}^{6}\sum_{\substack{l=1 \\ l\neq k}}^{6} m_F(\theta_k)m_i(\theta_l) + \sum_{k=1}^{6}\sum_{\substack{l=1 \\ l=k}}^{6} m_F(\theta_k)m_i(\theta_l) = \\
&= \sum_{k=1}^{6}\sum_{\substack{l=1 \\ l\neq k}}^{6} m_F(\theta_k)m_i(\theta_l) + \sum_{k=1}^{6} m_F(\theta_k)m_i(\theta_k) = 1\,.
\end{aligned} \tag{12}$$

From (12), it follows that if

$$\sum_{k=1}^{6} m_F(\theta_k)m_i(\theta_k) = 1\,, \text{i.e.,} \sum_{\substack{k=1,\ldots,6 \\ l=1,\ldots,6}} m_F(\theta_k)m_i(\theta_l) = 0\,, \tag{13}$$

then the degree of conflict is full.

If

$$\sum_{k=1}^{6} m_F(\theta_k)m_i(\theta_k) = 0\,, \text{i.e.,} \sum_{\substack{k=1,\ldots,6 \\ l=1,\ldots,6}} m_F(\theta_k)m_i(\theta_l) = 1\,, \tag{14}$$

then there is no conflict.

$m_D(.)$ is the Dempster–Shafer fusion result if, and only if, the denominator of expression (8) is non-zero, i.e., if the degree of conflict k_{Fi} is less than 1.

3.2.2. The Proportional Conflict Redistribution Rule PCR1

The PCR1 rule is the simplest and the easiest version of the proportional conflict redistribution rule. The concept of the PCR1 rule assumes the calculation of the total conflicting mass (not worrying about the partial conflicting masses). The total conflicting mass is redistributed to all non-empty sets of hypotheses proportionally, with respect to their corresponding non-empty column sum of the associated mass matrix. The PCR1 rule is defined for every non-empty hypothesis in the following way:

$$\begin{aligned}
m'_F(\theta_j) = m_{PCR1}(\theta_j) &= [\sum_{\substack{k=1,\ldots,6 \\ l=1,\ldots,6 \\ \theta_k\cap\theta_l=\theta_j}} m_F(\theta_k)m_i(\theta_l)] + \frac{c_{Fi}(\theta_j)}{d_{Fi}}\cdot k_{Fi} = \\
&= m_F(\theta_j)m_i(\theta_j) + \frac{c_{Fi}(\theta_j)}{d_{Fi}}\cdot k_{Fi} = m_{Fi}(\theta_j) + \frac{c_{Fi}(\theta_j)}{d_{Fi}}\cdot k_{Fi}
\end{aligned} \tag{15}$$

where $c_{Fi}(\theta_j)$ is the non-zero sum of the column corresponding to the hypotheses θ_j in the mass matrix

$$\boldsymbol{M} = \begin{bmatrix} \boldsymbol{m}_F \\ \boldsymbol{m}_i \end{bmatrix} \tag{16}$$

specified by the following formula:

$$c_{Fi}(\theta_j) = m_F(\theta_j) + m_i(\theta_j) \tag{17}$$

where:

- $\boldsymbol{m}_i$ (i = 1,2) is a row vector of the basic belief assignment masses of the *i*-th sensor's hypotheses;
- $\boldsymbol{m}_F$ is a row vector of the basic belief assignments masses of the IFC system's hypotheses;
- k_{Fi} is the degree of mass conflict specified by the following formula:

$$k_{Fi} = \sum_{\substack{k=1,\ldots,6 \\ l=1,\ldots,6 \\ \theta_k \cap \theta_l = \varnothing}} m_F(\theta_k) m_i(\theta_l) = \sum_{k=1}^{6} \sum_{\substack{l=1 \\ l \neq k}}^{6} m_F(\theta_k) m_i(\theta_l), \tag{18}$$

- d_{Fi} is the sum of all non-zero column sums of all non-empty sets, as follows:

$$d_{Fi} = \sum_{j=1}^{6} \left[m_F(\theta_j) + m_i(\theta_j) \right] = \sum_{j=1}^{6} c_{Fi}(\theta_j)\,. \tag{19}$$

In the case from this paper, $d_{Fi} = 2$ because

$$\sum_{j=1}^{6} m_F(\theta_j) \; = \sum_{j=1}^{6} m_i(\theta_j) = 1 \tag{20}$$

In addition,

$$m_{Fi}(\theta_j) = m_F(\theta_j) m_i(\theta_j) \tag{21}$$

3.2.3. The Proportional Conflict Redistribution Rule PCR2

In the PCR2 rule, the total conflicting mass k_{Fi} is distributed only to the non-empty sets involved in the conflict (not to all non-empty sets) and taken proportionally with respect to their corresponding non-empty column sum.

A non-empty set $\theta_k \in \Theta$ is considered to be involved in the conflict if there exists another set $\theta_l \in \Theta$ that is neither included in θ_k nor includes a θ_k such value that $\theta_k \cap \theta_l = \varnothing$ and $m_{Fi}(\theta_k \cap \theta_l) > 0$. The PCR2 rule is defined for every non-empty hypothesis $\theta_j \in \Theta$ in the following way:

$$\begin{aligned} m'_F(\theta_j) = m_{PCR2}(\theta_j) &= \Big[\sum_{\substack{k=1,\ldots,6 \\ l=1,\ldots,6 \\ \theta_k \cap \theta_l = \theta_j}} m_F(\theta_k) m_i(\theta_l) \Big] + C(\theta_j) \frac{c_{Fi}(\theta_j)}{e_{Fi}} \cdot k_{Fi} = \\ &= m_F(\theta_j) m_i(\theta_j) + C(\theta_j) \frac{c_{Fi}(\theta_j)}{e_{Fi}} \cdot k_{Fi} = m_{Fi}(\theta_j) + C(\theta_j) \frac{c_{Fi}(\theta_j)}{e_{Fi}} \cdot k_{Fi} \end{aligned} \tag{22}$$

where

$$C(\theta_j) = \begin{cases} 1, & \text{if } \theta_j \text{ is involved in the conflict,} \\ 0, & \text{otherwise.} \end{cases} \tag{23}$$

Formula (23) can be written differently in form (25), taking into account the definition of the involvement in a conflict and Formula (24) [6]:

$$m_{Fi}(\theta_j \cap \theta_k) = m_F(\theta_j) \cdot m_i(\theta_k) + \; m_F(\theta_k) \cdot m_i(\theta_j) \tag{24}$$

$$C(\theta_j) = \begin{cases} 1, & \text{if } \exists\, \theta_k \in \Theta, k \neq j \; : \; m_F(\theta_j) \cdot m_i(\theta_k) + \; m_F(\theta_k) \cdot m_i(\theta_j) > 0 \\ 0, & \text{otherwise.} \end{cases} \tag{25}$$

$c_{Fi}(\theta_j)$ is the non-zero sum of the column corresponding to the hypotheses in the mass matrix M (16) specified by the following formula:

$$c_{Fi}(\theta_j) = m_F(\theta_j) + m_i(\theta_j) \tag{26}$$

where:

- $\boldsymbol{m}_i$ (i = 1,2) is a row vector of the basic belief assignment; masses of the *i*-th sensor's hypotheses;
- $\boldsymbol{m}_F$ is a row vector of the basic belief assignment masses of the IFC system's hypotheses;
- k_{Fi} is the degree of mass conflict specified by Formula (18);
- e_{Fi} is the sum of all non-zero column sums of all non-empty sets involved in the conflict, as follows:

$$e_{Fi} = \sum_{j \in CF} [m_F(\theta_j) + m_i(\theta_j)] = \sum_{j \in CF} c_{Fi}(\theta_j) = \sum_{j=1}^{6} C(\theta_j)[m_F(\theta_j) + m_i(\theta_j)] = \sum_{j=1}^{6} C(\theta_j) \cdot c_{Fi}(\theta_j) \quad (27)$$

where

$$CF = \{j = 1, \ldots, 6 \; : \; \forall \theta_k \in \Theta \;\; m_{Fi}(\theta_j \cap \theta_k) > 0\} \quad (28)$$

and $m_{Fi}(\theta_j \cap \theta_k)$ is defined by (24).

In addition

$$m_{Fi}(\theta_j) = m_F(\theta_j) m_i(\theta_j) \quad (29)$$

It is shown below that in the case of data used in numerical experiments (Section 6), $e_{Fi} = 2$, which means that the PCR2 rule is equivalent to the PCR1 rule. The BBA vectors used there contain values less than 1, which means the following:

$$\forall j = 1, \ldots, 6 \; : \; m_F(\theta_j) < 1 \wedge m_i(\theta_j) < 1 \quad (30)$$

It follows that each BBA vector contains at least two non-zero components, that is $\exists j = 1, \ldots, 6, \;\; \exists k = 1, \ldots, 6$ with $k \neq j$, such that

$$0 < m_F(\theta_j) < 1 \wedge 0 < m_F(\theta_k) < 1 \quad (31)$$

$$0 < m_i(\theta_j) < 1 \wedge 0 < m_i(\theta_k) < 1 \quad (32)$$

From (31) and (32), it follows that if $m_F(\theta_j) > 0$, then there exists at least one value $k \neq j$, such that $m_i(\theta_k) > 0$, which can be written in the following form:

$$\forall j = 1, \ldots, 6 \; : \; 0 < m_F(\theta_j) < 1 \;\Rightarrow\; \exists k \ldots\ldots, 6, k \neq j \; : \; 0 < m_i(\theta_k) < 1 \quad (33)$$

From (33), it follows that

$$\forall j = 1, \ldots, 6 \; : \; 0 < m_F(\theta_j) < 1 \;\Rightarrow\; \exists k = 1, \ldots, 6, k \neq j \; : \; m_F(\theta_j) \cdot m_i(\theta_k) > 0 \quad (34)$$

The same applies to the following:

$$\forall j = 1, \ldots, 6 \; : \; 0 < m_i(\theta_j) < 1 \;\Rightarrow\; \exists k = 1, \ldots, 6, k \neq j \; : \; m_i(\theta_j) \cdot m_F(\theta_k) > 0 \quad (35)$$

Taking into account (34), (35) and (25) can obtain the following:

$$\begin{gathered}\forall j = 1, \ldots, 6 \; : \; 0 < m_F(\theta_j) < 1 \;\Rightarrow\; \exists k = 1, \ldots, 6, k \neq j \text{ such that} \\ m_F(\theta_j) \cdot m_i(\theta_k) + m_i(\theta_j) \cdot m_F(\theta_k) > 0\end{gathered} \quad (36)$$

$$\begin{gathered}\forall j = 1, \ldots, 6 \; : \; 0 < m_i(\theta_j) < 1 \;\Rightarrow\; \exists k = 1, \ldots, 6, k \neq j \text{ such that} \\ m_i(\theta_j) \cdot m_F(\theta_k) + m_F(\theta_j) \cdot m_i(\theta_k) > 0\end{gathered} \quad (37)$$

From (36) and (37), it follows that

$$\forall j = 1, \ldots, 6 \; : \; 0 < m_F(\theta_j) < 1 \;\Rightarrow\; C(\theta_j) = 1 \quad (38)$$

$$\forall j = 1, \ldots, 6 \; : \; 0 < m_i(\theta_j) < 1 \;\Rightarrow\; C(\theta_j) = 1 \quad (39)$$

This means that any hypothesis with a non-zero BBA value for any of the two sensors is involved in a conflict.

From (27), it follows that

$$e_{Fi} = \sum_{j=1}^{6} C(\theta_j)\left[m_F(\theta_j) + m_i(\theta_j)\right] = \sum_{j=1}^{6} C(\theta_j)m_F(\theta_j) + \sum_{j=1}^{6} C(\theta_j)m_i(\theta_j) \tag{40}$$

Using (36), (37) and (40), the value e_{Fi} is determined.
Because

$$\sum_{j=1}^{6} C(\theta_j)m_F(\theta_j) = \sum_{\substack{j=1 \\ m_F(\theta_j)>0}}^{6} C(\theta_j)m_F(\theta_j) + \sum_{\substack{j=1 \\ m_F(\theta_j)=0}}^{6} C(\theta_j)m_F(\theta_j) = \sum_{\substack{j=1 \\ m_F(\theta_j)>0}}^{6} m_F(\theta_j) = 1 \tag{41}$$

$$\sum_{j=1}^{6} C(\theta_j)m_i(\theta_j) = \sum_{\substack{j=1 \\ m_i(\theta_j)>0}}^{6} C(\theta_j)m_i(\theta_j) + \sum_{\substack{j=1 \\ m_i(\theta_j)=0}}^{6} C(\theta_j)m_i(\theta_j) = \sum_{\substack{j=1 \\ m_i(\theta_j)>0}}^{6} m_i(\theta_j) = 1 \tag{42}$$

we obtain

$$e_{Fi} = \sum_{j=1}^{6} C(\theta_j)m_F(\theta_j) + \sum_{j=1}^{6} C(\theta_j)m_i(\theta_j) = 2 \tag{43}$$

Considering (43), it can be said that in this case, the PCR2 rule is equivalent to the PCR1 rule. For this reason, the results of the PCR2 rule are not presented in Section 6, as they would be identical to the results of the PCR1 rule as only the Bayesian BBAs are used in this application.

3.2.4. The Proportional Conflict Redistribution Rule PCR3

In the PCR3 rule, the partial conflicting masses are distributed instead of the total conflicting mass, k_{Fi}, to the non-empty sets involved in the partial conflict. If an intersection is empty, for instance $\theta_k \cap \theta_l = \varnothing$, then the mass $m(\theta_k \cap \theta_l)$ of the partial conflict is transferred to the non-empty sets θ_k and θ_l proportionally, with respect to the non-zero sum of masses assigned to θ_k and, respectively, to θ_l by BBAs $m_F(.)$ and $m_i(.)$. The PCR3 rule works if at least one set between θ_k and θ_l is non-empty and its column sum is non-zero.

The PCR3 rule is defined for every non-empty hypothesis $\theta_j \in \Theta$ in the following way:

$$\begin{gathered} m'_F(\theta_j) = m_{PCR3}(\theta_j) = \left[\sum_{\substack{k=1,\dots,6 \\ l=1,\dots,6 \\ \theta_k\cap\theta_l=\theta_j}} m_F(\theta_k)m_i(\theta_l)\right] + \left[c_{Fi}(\theta_j)\sum_{\substack{k=1,\dots,6 \\ \theta_k\cap\theta_j=\varnothing}} S_{Fi}^{PCR3}(\theta_j,\theta_k)\right] = \\ = m_{Fi}(\theta_j) + \left[c_{Fi}(\theta_j)\sum_{\substack{k=1,\dots,6 \\ k\neq j}} S_{Fi}^{PCR3}(\theta_j,\theta_k)\right] \end{gathered} \tag{44}$$

where

$$S_{Fi}^{PCR3}(\theta_j,\theta_k) = \begin{cases} \frac{m_F(\theta_k)m_i(\theta_j)+m_F(\theta_j)m_i(\theta_k)}{c_{Fi}(\theta_j)+c_{Fi}(\theta_k)} & \text{for } c_{Fi}(\theta_j)+c_{Fi}(\theta_k) \neq 0 \\ 0 \text{ for } c_{Fi}(\theta_j)+c_{Fi}(\theta_k) = 0 \end{cases} \tag{45}$$

$c_{Fi}(\theta_j)$ is the non-zero sum of the column corresponding to the hypotheses θ_j in the mass matrix $\boldsymbol{M}$ (16), specified by the following formula:

$$c_{Fi}(\theta_j) = m_F(\theta_j) + m_i(\theta_j) \tag{46}$$

3.2.5. The Proportional Conflict Redistribution Rule PCR4

The PCR4 rule redistributes the partial conflicting masses only to the sets involved in the partial conflict in proportion to the non-zero mass sum assigned to θ_k and θ_l by the conjunction rule according to the following formula:

$$\begin{aligned} m'_F(\theta_j) = m_{PCR4}(\theta_j) = m_{Fi}(\theta_j) + m_{Fi}(\theta_j) \sum_{\substack{k=1,\ldots,6 \\ \theta_k \cap \theta_j = \varnothing}} S_{Fi}^{PCR4}(\theta_j, \theta_k) = \\ = m_{Fi}(\theta_j) + m_{Fi}(\theta_j) \sum_{\substack{k=1,\ldots,6 \\ k \neq j}} S_{Fi}^{PCR4}(\theta_j, \theta_k) \end{aligned} \tag{47}$$

where

$$S_{Fi}^{PCR4}(\theta_j, \theta_k) = \begin{cases} \frac{m_{Fi}(\theta_j \cap \theta_k)}{m_{Fi}(\theta_j) + m_{Fi}(\theta_k)} \text{ for } c_{Fi}(\theta_j) + c_{Fi}(\theta_k) \neq 0 \text{ and } m_{Fi}(\theta_j) \cdot m_{Fi}(\theta_k) \neq 0 \\ \frac{m_{Fi}(\theta_j \cap \theta_k)}{c_{Fi}(\theta_j) + c_{Fi}(\theta_k)} \text{ for } c_{Fi}(\theta_j) + c_{Fi}(\theta_k) \neq 0 \text{ and } m_{Fi}(\theta_j) \cdot m_{Fi}(\theta_k) = 0 \\ 0 \text{ for } c_{Fi}(\theta_j) + c_{Fi}(\theta_k) = 0 \end{cases} \tag{48}$$

wherein

$$m_{Fi}(\theta_j \cap \theta_k) = m_F(\theta_k) m_i(\theta_j) + m_F(\theta_j) m_i(\theta_k) \tag{49}$$

$$m_{Fi}(\theta_j) = m_F(\theta_j) \cdot m_i(\theta_j) \tag{50}$$

$$c_{Fi}(\theta_j) = m_F(\theta_j) + m_i(\theta_j) \tag{51}$$

If at least one of the BBAs, $m_F(.)$ or $m_i(.)$, is zero, the fraction is discarded and the mass $m_{Fi}(\theta_j \cap \theta_k)$ is transferred to θ_j and θ_k proportionally, with respect to their non-zero column sum of masses $c_{Fi}(\theta_j)$.

3.2.6. The Proportional Conflict Redistribution Rule PCR5 for Two BBAs (Two Sources)

Similar to the PCR2-PCR4 rules, PCR5 redistributes the partial conflicting mass to the hypothesis involved in the partial conflict. PCR5 provides the most mathematically precise [6,18,22] redistribution of conflicting mass to non-empty sets in accordance with the logic of the conjunctive rule. However, it is more difficult to implement. The PCR5 rule is defined for every non-empty hypothesis $\theta_j \in \Theta$ in the following way:

$$m'_F(\theta_j) = m_{PCR5}(\theta_j) = m_{Fi}(\theta_j) + \sum_{\substack{k=1,\ldots,6 \\ \theta_k \cap \theta_j = \varnothing}} S_{Fi}^{PCR5}(\theta_j, \theta_k) = m_{Fi}(\theta_j) + \sum_{\substack{k=1,\ldots,6 \\ k \neq j}} S_{Fi}^{PCR5}(\theta_j, \theta_k) \tag{52}$$

where

$$S_{Fi}^{PCR5}(\theta_j, \theta_k) = \begin{cases} \frac{m_F(\theta_j)^2 \cdot m_i(\theta_k)}{m_F(\theta_j) + m_i(\theta_k)} + \frac{m_i(\theta_j)^2 \cdot m_F(\theta_k)}{m_i(\theta_j) + m_F(\theta_k)} \text{for} \\ m_F(\theta_j) + m_i(\theta_k) \neq 0 \text{ and } m_i(\theta_j) + m_F(\theta_k) \neq 0 \\ 0 \text{ for } m_F(\theta_j) + m_i(\theta_k) = 0 \text{ or } m_i(\theta_j) + m_F(\theta_k) = 0 \end{cases} \tag{53}$$

wherein

$$m_{Fi}(\theta_j) = m_F(\theta_j) \cdot m_i(\theta_j) \tag{54}$$

In Formula (52), the component S_{Fi}^{PCR5} is equal to zero if both denominators are equal to zero. In Formula (53), if a denominator is zero, then the component is discarded.

3.2.7. The Proportional Conflict Redistribution Rules PCR5 and PCR6 for Three BBAs (Three Sources)

In [6,22], improved proportional conflict redistribution rules of the combination of basic belief assignments PCR6, PCR5+, and PCR6+ are presented. The authors point out that these rules should be applied if, and only if, we are to combine more than two BBAs. If we only have two BBAs to combine (s = 2), we always obtain $m_{\text{PCR5}} = m_{\text{PCR5+}} = m_{\text{PCR6}} = m_{\text{PCR6+}}$, because in this case, the PCR5, PCR5+, PCR6, and PCR6+ rules coincide. Below are the formulas that define the PCR5 and PCR6 rules for three BBAs.

The PCR5 rule for three BBAs (three sources) is defined for every non-empty hypothesis in the following way:

$$m'_F(\theta_j) = m_{PCR5}(\theta_j) = \frac{m''(\theta_j)}{\sum_{i=1}^{6} m''(\theta_i)} \tag{55}$$

wherein

$$\begin{aligned} m''(\theta_j) &= m_{F12}(\theta_j) + \sum_{\substack{k=1,\ldots,6 \\ l=1,\ldots,6 \\ \theta_k \cap \theta_l \cap \theta_j = \varnothing}} S_{F12}^{PCR5}(\theta_j,\theta_k,\theta_l) + \sum_{\substack{k=1,\ldots,6 \\ \theta_k \cap \theta_j = \varnothing}} S1_{F12}^{PCR5}(\theta_j,\theta_k) + + \sum_{\substack{k=1,\ldots,6 \\ \theta_k \cap \theta_j = \varnothing}} S2_{F12}^{PCR5}(\theta_j,\theta_k) = \\ &= m_{F12}(\theta_j) + \sum_{\substack{k=1,\ldots,6 \\ k \neq j}} \sum_{\substack{l=1,\ldots,6 \\ l \neq j \wedge l \neq k}} S_{F12}^{PCR5}(\theta_j,\theta_k,\theta_l) + \sum_{\substack{k=1,\ldots,6 \\ j \neq k}} S1_{F12}^{PCR5}(\theta_j,\theta_k) + + \sum_{\substack{k=1,\ldots,6 \\ j \neq k}} S2_{F12}^{PCR5}(\theta_j,\theta_k) = \\ &= m_{F12}(\theta_j) + \sum_{\substack{k=1,\ldots,6 \\ k \neq j}} \left[\sum_{\substack{l=1,\ldots,6 \\ l \neq j \wedge l \neq k}} S_{F12}^{PCR5}(\theta_j,\theta_k,\theta_l) + S1_{F12}^{PCR5}(\theta_j,\theta_k) + S2_{F12}^{PCR5}(\theta_j,\theta_k) \right] \end{aligned} \tag{56}$$

$$S_{F12}^{PCR5}(\theta_j,\theta_k,\theta_l) = \frac{m_F(\theta_j)^2 \cdot m_1(\theta_k) \cdot m_2(\theta_l)}{m_F(\theta_j) + m_1(\theta_k) + m_2(\theta_l)} + \frac{m_F(\theta_l) \cdot m_1(\theta_j)^2 \cdot m_2(\theta_k)}{m_F(\theta_l) + m_1(\theta_j) + m_2(\theta_k)} + \frac{m_F(\theta_k) \cdot m_1(\theta_l) \cdot m_2(\theta_j)^2}{m_F(\theta_k) + m_1(\theta_l) + m_2(\theta_j)} \tag{57}$$

$$S1_{F12}^{PCR5}(\theta_j,\theta_k) = \frac{m_F(\theta_j)^2 \cdot m_1(\theta_k) \cdot m_2(\theta_k)}{m_F(\theta_j) + m_1(\theta_k) + m_2(\theta_k)} + \frac{m_F(\theta_k) \cdot m_1(\theta_j)^2 \cdot m_2(\theta_k)}{m_F(\theta_k) + m_1(\theta_j) + m_2(\theta_k)} + \frac{m_F(\theta_k) \cdot m_1(\theta_k) \cdot m_2(\theta_j)^2}{m_F(\theta_k) + m_1(\theta_k) + m_2(\theta_j)} \tag{58}$$

$$S2_{F12}^{PCR5}(\theta_j,\theta_k) = \frac{m_F(\theta_j)^2 \cdot m_1(\theta_j)^2 \cdot m_2(\theta_k)}{m_F(\theta_j) + m_1(\theta_j) + m_2(\theta_k)} + \frac{m_F(\theta_k) \cdot m_1(\theta_j)^2 \cdot m_2(\theta_j)^2}{m_F(\theta_k) + m_1(\theta_j) + m_2(\theta_j)} + \frac{m_F(\theta_j)^2 \cdot m_1(\theta_k) \cdot m_2(\theta_j)^2}{m_F(\theta_j) + m_1(\theta_k) + m_2(\theta_j)} \tag{59}$$

$$m'_F(\theta_j) = m_{PCR5}(\theta_j) = \frac{m''(\theta_j)}{\sum_{i=1}^{6} m''(\theta_i)} \tag{60}$$

In Formulas (57)–(59), if a denominator is zero, then the component is discarded.

The quotient in Formula (55) ensures the normalization of the BBA vector m'_F, which ensures the following:

$$\sum_{i=1}^{6} m'_F(\theta_i) = \sum_{i=1}^{6} m_{PCR5}(\theta_i) = 1$$

The PCR6 rule for three BBAs (three sources) is defined for every non-empty hypothesis $\theta_j \in \Theta$ in the following way:

$$m'_F(\theta_j) = m_{PCR5}(\theta_j) = \frac{m''(\theta_j)}{\sum_{i=1}^{6} m''(\theta_i)} \tag{61}$$

wherein

$$\begin{aligned} m''(\theta_j) &= m_{F12}(\theta_j) + \sum_{\substack{k=1,\ldots,6 \\ l=1,\ldots,6 \\ \theta_k\cap\theta_l\cap\theta_j=\varnothing}} S_{F12}^{PCR6}(\theta_j,\theta_k,\theta_l) + \sum_{\substack{k=1,\ldots,6 \\ \theta_k\cap\theta_j=\varnothing}} S1_{F12}^{PCR6}(\theta_j,\theta_k) + \\ &+ \sum_{\substack{k=1,\ldots,6 \\ \theta_k\cap\theta_j=\varnothing}} S2_{F12}^{PCR6}(\theta_j,\theta_k) = \\ &= m_{F12}(\theta_j) + \sum_{\substack{k=1,\ldots,6 \\ k\neq j}} \sum_{\substack{l=1,\ldots,6 \\ l\neq j\wedge l\neq k}} S_{F12}^{PCR6}(\theta_j,\theta_k,\theta_l) + \sum_{\substack{k=1,\ldots,6 \\ j\neq k}} S1_{F12}^{PCR6}(\theta_j,\theta_k) + \\ &+ \sum_{\substack{k=1,\ldots,6 \\ j\neq k}} S2_{F12}^{PCR6}(\theta_j,\theta_k) = \\ &= m_{F12}(\theta_j) + \sum_{\substack{k=1,\ldots,6 \\ k\neq j}} \left[\sum_{\substack{l=1,\ldots,6 \\ l\neq j\wedge l\neq k}} S_{F12}^{PCR6}(\theta_j,\theta_k,\theta_l) + S1_{F12}^{PCR6}(\theta_j,\theta_k) + S2_{F12}^{PCR6}(\theta_j,\theta_k) \right] \end{aligned} \tag{62}$$

with

$$\begin{aligned} S_{F12}^{PCR5}(\theta_j,\theta_k,\theta_l) = & \frac{m_F(\theta_j)^2\cdot m_1(\theta_k)\cdot m_2(\theta_l)}{m_F(\theta_j)+m_1(\theta_k)+m_2(\theta_l)} + \frac{m_F(\theta_l)\cdot m_1(\theta_j)^2\cdot m_2(\theta_k)}{m_F(\theta_l)+m_1(\theta_j)+m_2(\theta_k)} + \\ & + \frac{m_F(\theta_k)\cdot m_1(\theta_l)\cdot m_2(\theta_j)^2}{m_F(\theta_k)+m_1(\theta_l)+m_2(\theta_j)} \end{aligned} \tag{63}$$

$$\begin{aligned} S1_{F12}^{PCR5}(\theta_j,\theta_k) = & \frac{m_F(\theta_j)^2\cdot m_1(\theta_k)\cdot m_2(\theta_k)}{m_F(\theta_j)+m_1(\theta_k)+m_2(\theta_k)} + \frac{m_F(\theta_k)\cdot m_1(\theta_j)^2\cdot m_2(\theta_k)}{m_F(\theta_k)+m_1(\theta_j)+m_2(\theta_k)} + \\ & + \frac{m_F(\theta_k)\cdot m_1(\theta_k)\cdot m_2(\theta_j)^2}{m_F(\theta_k)+m_1(\theta_k)+m_2(\theta_j)} \end{aligned} \tag{64}$$

$$\begin{aligned} S2_{F12}^{PCR5}(\theta_j,\theta_k) = & \frac{m_F(\theta_j)^2\cdot m_1(\theta_j)\cdot m_2(\theta_k) + m_F(\theta_j)\cdot m_1(\theta_j)^2\cdot m_2(\theta_k)}{m_F(\theta_j)+m_1(\theta_j)+m_2(\theta_k)} + \\ & + \frac{m_F(\theta_k)\cdot m_1(\theta_j)^2\cdot m_2(\theta_j) + m_F(\theta_k)\cdot m_1(\theta_j)\cdot m_2(\theta_j)^2}{m_F(\theta_k)+m_1(\theta_j)+m_2(\theta_j)} + \\ & + \frac{m_F(\theta_j)^2\cdot m_1(\theta_k)\cdot m_2(\theta_j) + m_F(\theta_j)\cdot m_1(\theta_k)\cdot m_2(\theta_j)^2}{m_F(\theta_j)+m_1(\theta_k)+m_2(\theta_j)} \end{aligned} \tag{65}$$

and

$$m_{F12}(\theta_j) = m_F(\theta_j)\ \cdot m_1(\theta_j)\cdot m_2(\theta_j) \tag{66}$$

In Formulas (63)–(65), if a denominator is zero, then the component is discarded.

The quotient in Formula (61) ensures the normalization of the BBA vector m'_F, which ensures the following:

$$\sum_{i=1}^{6} m'_F(\theta_i) = \sum_{i=1}^{6} m_{PCR5}(\theta_i) = 1$$

Comparing the two fusion schemes (Figures 2 and 3), it should be noted that sequential and global information fusion generally produces different results [18], i.e.,

$$PCR5(m_F, m_1, m_2) \neq PCR5(PCR5(m_F, m_1), m_2) \neq PCR5(PCR5(m_F, m_2), m_1) \tag{67}$$

In addition, the article experimentally verified the theorem on the inequality of the results of both PCR5 and PCR6 rules for three BBAs (three sources) presented in [18]:

$$PCR5(m_F, m_1, m_2) \neq PCR6(m_F, m_1, m_2) \tag{68}$$

4. Basic Belief Assignment for Combined Primary and Secondary Surveillance Radars

Combined primary and secondary (IFF) radars are the main source of identification information regarding air and maritime objects. A primary radar only yields the detection of an object in a supervised area. The detection of the object is the precondition for sending a request to the object by the secondary radar (interrogator). Interpretation of the object response is dependent on the type of request. The so-called civilian modes only yield a determination of whether the detected object replies to an interrogation or not.

This paper assumes that the analyzed radar sensor consists of two radars: primary and secondary. Therefore, the probability of the correct detection and the correct identification of a target is expressed by the following formula:

$$P_{pi} = P_d \cdot P_{IFF} \tag{69}$$

where P_d is the probability of correct detection of the target by a primary radar, and P_{IFF} is the probability of a correct reply to an interrogation. If a target is detected by the primary radar and there is a lack of proper identification by the secondary radar, it can be assumed that the target has a value of attribute identification of UNKNOWN—U. Thus, the following relation can be written:

$$m(U) = P_d(1 - P_{IFF}), \tag{70}$$

where $m(U)$ is the mass of probability for a value of UNKNOWN identification attribute.

A method for calculating the probabilities P_d and P_{IFF} is presented in [7,17,23].

This section explains the way the remaining mass of probability is calculated $(1 - m(U))$. It is assumed there that every simulated target should have a base value of attribute identification from the set as follows:

$$\mathbf{Z}_{BI} = \{N_B, F_B, H_B\} \tag{71}$$

where:

- N_B: base NEUTRAL identity;
- F_B: base FRIEND identity;
- H_B: base HOSTILE identity.

STANAG 1241 introduces, in addition to the basic set of attribute identification values, secondary (additional) attribute identification values: SUSPECT (S) and ASSUMED FRIEND (AF). According to Figure 1, a table of possible attribute value transitions between set (10) and the set of secondary attribute identification values can be introduced:

$$\mathbf{Z}_{SI} = \{N_S, F_S, H_S, AF, S\} \tag{72}$$

The belief mass values contained in Table 1 determine how the mass of the base belief assignment is transformed into the mass of the secondary belief assignment. They can be estimated as empirical frequencies based on recorded archive events.

Table 1. Transformation of the base belief assignment mass into the secondary belief assignment mass.

Base Identification →	F_B	N_B	H_B
F_S	$m(F_S \vert F_B)$	0	0
N_S	0	$m(N_S \vert N_B)$	0
H_S	0	0	$m(H_S \vert H_B)$
AF	$m(AF \vert F_B)$	$m(AF \vert N_B)$	0
S	0	$m(S \vert F_B)$	$m(S \vert H_B)$

Of course, the normalization conditions are satisfied: $\sum_{x \in Z_{SI}} m(x|F_B) = 1, \sum_{x \in Z_{SI}} m(x|N_B) = 1$, and $\sum_{x \in Z_{SI}} m(x|H_B) = 1$.

The final values of the belief mass of secondary attribute identification values are calculated according to the formulas as follows:

1. For a target with the FRIEND base value of an attribute identification,

$$\begin{aligned}
&m(U) = P_d(1 - P_{IFF});\\
&m(AF) = m(AF|F_B)(1 - m(U));\\
&m(F_S) = m(F_S|F_B)(1 - m(U)).
\end{aligned}$$

2. For a target with the NEUTRAL base value of an attribute identification,

$$\begin{aligned}
&m(U) = P_d(1 - P_{IFF});\\
&m(AF) = m(AF|N_B)(1 - m(U));\\
&m(S) = m(S|N_B)(1 - m(U));\\
&m(N_S) = (1 - m(AF|N_B) - m(S|N_B))(1 - m(U)).
\end{aligned}$$

3. For a target with the HOSTILE base value of an attribute identification,

$$\begin{aligned}
&m(U) = P_d(1 - P_{IFF});\\
&m(H_S) = m(H_S|H_B)(1 - m(U));\\
&m(S) = m(S|H_B)(1 - m(U)).
\end{aligned}$$

Other final values of the belief mass of secondary attribute identification values are equal to zero.

5. Basic Belief Assignment for ESM Sensors

ESM sensors consist of passive receivers and direction finders, which allow them to capture emitter signals coming from certain directions. In this way, the electronic recognition system can receive, among other data, information on radar emitters mounted on air or maritime platforms. Reports sent from the ESM sensors include, among others, the characteristics of the intercepted signal, the emitter's azimuth, and the so-called identification information.

This paper also assumes that sensors are equipped with specialized databases called the databases of emitter signal patterns, in which information about previously captured, processed, analyzed, recognized, and described radar emitter signals is stored, along with additional information about the type and mode of the emitter work, the platform on which these emitters can be installed, and the national or organizational affiliation of these platforms. The detected signals are the subject of an analysis procedure, which yields the determination of the so-called distinctive features of the signal, and then assigns this information to a specific electronic entity (already existing or created ad hoc) [24]. The basis for assigning distinctive information to an electronic entity is the azimuth angle of the incoming signal.

In the case of a high density of targets, identification information may fluctuate due to incorrect assignment of signal information to the electronic entity [25]. The impact of this negative phenomenon can be significantly reduced by an efficient estimation of the emitter positions [24]. Assuming that the sensors send all reports on the tracked electronic entities to the superior operation center in the electronic recognition system, such a center (in this paper, called the information-fusion center (IFC)) can perform the fusion function of the identification information. The fusion of identification information ensures greater stability of this information, i.e., resistance to accidental changes in sensor decisions.

An ESM sensor is a passive sensor that captures incoming electromagnetic signals generated firstly by radar emitters mounted on air or maritime platforms. This sensor recognizes radar signals determining the values of their distinctive features. In this paper, we do not handle methods of radar signals recognition in detail. We do, however, use the information about these methods to identify platforms generating the signals according to STANAG 1241—NATO Standardization Agreement and DSmT. As previously stated, we are interested in three basic values of identification: friend, hostile, and neutral, as well as two secondary values: suspicious and assumed friendly. In addition, we assume that in some situations, it is not possible to determine the identity of the emitter-carrier platform. To clarify this issue, we briefly describe the method used to determine the identification of the emitter-carrier platform that generated the captured signal. The sensor-recognition system is equipped with a database that can be divided into three components: a platform database, an emitter list, and a geopolitical list [21]. The platform database (**PDB**) contains information about platforms that can be met in the area of interest, along with their equipment with emitters. The emitter name list (**ENL**) includes all emitters corresponding to each platform of the **PDB** and contains the values of the signal distinctive features for each emitter. The values of distinctive features are the basis for the procedure of recognizing a captured signal. The geopolitical list (**GPL**) provides the allegiance of various countries and platforms and yields the identification of them in accordance with STANAG 1241.

The algorithm of signal recognition is realized in two stages:

1. Verification at the level of signal quality features. The second stage is executed after a positive assessment of the conformity of quality features;

2. The signal-recognition procedure determines the distances between the distinctive features of the recognized signal and the distinctive features of all pattern signals stored within the emitter list.

Let us introduce the following notation:

x_s : vector of distinctive features of the recognized signal;

x_i: vector of distinctive features of the i-th pattern signal (i: the number of the pattern signal, $i \in \{1, \ldots, M\}$);

$d_{s,i} = d(x_s, x_i)$: the distance between the distinctive features vector of the recognized signal and the distinctive features vector of the i-th pattern signal; the distance $d_{s,i}$ is the Mahalanobis distance, taking into account the correlations of the distinctive features.

The signal-recognition classifier compares the distance $d(x_s, x_i)$ with the acceptable positive distance of the classification δ. The distance δ is the limit that we interpret as a boundary of emitter pattern recognition. We divide the set of pattern signals into two subsets: the patterns satisfying the positive classification condition in relation to the recognized signal s (D_s^+) and the patterns that do not satisfy the positive classification condition (D_s^-). The formal definition is as follows:

$$D_s^+ = \{\, i \in \{1, \ldots, M\} | d_{s,i} \leq \delta\} \tag{73}$$

$$D_s^- = \{\, i \in \{1, \ldots, M\} | d_{s,i} > \delta\} \tag{74}$$

In this paper, we propose the following method of determining the basic belief assignment on a set of pattern signals, which is related to the distance between a signal and a pattern in the distinctive features space:

$$m_s(i) = e^{-d(\mathbf{x}_s, \mathbf{x}_i)} \tag{75}$$

As can be seen from Formula (75), if $d(x_s, x_i) = 0$, then $m_s(i) = 1$, whereas if $d(x_s, x_i) > 0$, then $0 < m_s(i) < 1$. The above measure is not normalized; hence, we normalize it as follows:

$$\tilde{m}_s(i) = \frac{m_s(i)}{\sum_{i=1}^{M} m_s(i)} \tag{76}$$

The sum of the measures assigned to all the emitters, the distinctive features of which lie outside the limit δ, are treated as a measure assigned to the base hypothesis "unknown" (U):

$$\tilde{m}_s(U) = \sum_{i \in D_s^-} \tilde{m}_s(i) \tag{77}$$

Another way of recognizing emitters based on their signals is presented in [26]. For this, the authors use a convolutional neural network with a softmax layer.

To determine the belief measure of other base hypotheses (H, F, N) and secondary hypotheses (AF and S), we introduce formal definitions of sets contained in the sensor database and used for the recognition of captured signals. As mentioned above, the set of all the necessary data for platform identification can be divided into three sets: **PDB**, a platform database; **ENL**, an emitter name list; and **GPL**, a geopolitical list:

PDB: the platform database contains information about all platforms observed in the area of interest, including information on all emitters mounted on each platform; we assume that one platform can have many emitters and the same type of emitters can be installed on many platforms; the **PDB** also contains information on the national affiliation of each platform;

ENL: the emitter name list is a set of information about all recognized emitters in the area of interest; this set contains the mean values of the distinctive features of emitter signals (the so-called signal patterns) and their standard deviations;

GPL: the geopolitical list contains base values of identification attributes (H, F, N) assigned to the various countries.

We also introduce additional notations used in this paper:

- **PDBL**: the list of platform numbers that are stored in the **PDB**;
- **PL**(i): the set of numbers of platforms that have an emitter with number "i";

IPL(j): the base identification attribute of the platform with number "j" determined on the basis of the information contained in the **PDB** and **ENL** ($IPL(j) \in \{ F, H, N\}$).

The set of signal patterns satisfying the positive classification condition in relation to the recognized signal s, denoted as $\mathbf{D}_s^+$, can be divided into disjunctive subsets according to the values of the carrier platform identification features:

$$\mathbf{D}_s^+ = \mathbf{D}_s^{+F} \cup \mathbf{D}_s^{+H} \cup \mathbf{D}_s^{+N} \cup \mathbf{D}_s^{+AF} \cup \mathbf{D}_s^{+S}, \tag{78}$$

$$\mathbf{D}_s^{+k} \cap \mathbf{D}_s^{+l} = \varnothing,\ k \neq l,\ k,l \in \{F,H,N,AF,S\} \tag{79}$$

Each subset of the set $\mathbf{D}_s^+$ for the base identification is defined as follows:

$$\mathbf{D}_s^{+F} = \{\ i \in \mathbf{D}_s^+ | \forall j \in \mathbf{PL}(i)\ \ IPL(j) = F\}, \tag{80}$$

$$\mathbf{D}_s^{+H} = \{\ i \in \mathbf{D}_s^+ | \forall j \in \mathbf{PL}(i)\ \ IPL(j) = H\} \tag{81}$$

$$\mathbf{D}_s^{+N} = \{\ i \in \mathbf{D}_s^+ | \forall j \in \mathbf{PL}(i)\ \ IPL(j) = N\} \tag{82}$$

In a similar way, subsets of the set $\mathbf{D}_s^+$ for the secondary identification (AF, S) can be defined as follows:

$$\mathbf{D}_s^{+AF} = \{\ i \in \mathbf{D}_s^+ | \exists\, j \in \mathbf{PL}(i)\ \ IPL(j) = F\ \wedge\ \exists\, j \in \mathbf{PL}(i)\ \ IPL(j) = N\} \tag{83}$$

$$\mathbf{D}_s^{+S} = \{\ i \in \mathbf{D}_s^+ : \exists\, j \in \mathbf{PL}(i)\ \ IPL(j) = H\ \wedge\ \exists\, j \in \mathbf{PL}(i)\ \ IPL(j) = N\} \tag{84}$$

It can be noticed that we assume in this paper that no emitter type can be installed simultaneously on platforms with identifications F and H:

$$\{\ i \in \mathbf{D}_s^+ | \exists\, j \in \mathbf{PL}(i)\ \ IPL(j) = F\ \ \wedge\ \ \exists\, j \in \mathbf{PL}(i)\ \ IPL(j) = H\} = \varnothing \tag{85}$$

Introducing the definition of subsets of the set determines the belief masses for all identification features:

$$\widetilde{m}_s(F) = \sum_{i \in \mathbf{D}_s^{+F}} \widetilde{m}_s(i),\ \widetilde{m}_s(H) = \sum_{i \in \mathbf{D}_s^{+H}} \widetilde{m}_s(i), \widetilde{m}_s(N) = \sum_{i \in \mathbf{D}_s^{+N}} \widetilde{m}_s(i), \tag{86}$$

$$\widetilde{m}_s(\mathrm{AF}) = \sum_{i \in D_s^{+AF}} \widetilde{m}_s(i), \widetilde{m}_s(S) = \sum_{i \in D_s^{+S}} \widetilde{m}_s(i) \quad (87)$$

It should be emphasized that the method presented here is different than that presented in [25,27]. These papers assume that ESM sensors can only generate basic declarations with attribute values FRIEND, HOSTILE, and NEUTRAL, but in this paper, we assume that ESM sensors can generate declarations from an extended set of attribute values (including ASSUME FRIEND, SUSPECT, and UNKNOWN).

6. Numerical Experiments of Fusion of Identification Information from ESM Sensors

6.1. General Research Scheme of Fusion of Identification Information from ESM Sensors

Figure 4 shows a general scheme of simulation experiments, which indicates the places of description of individual models.

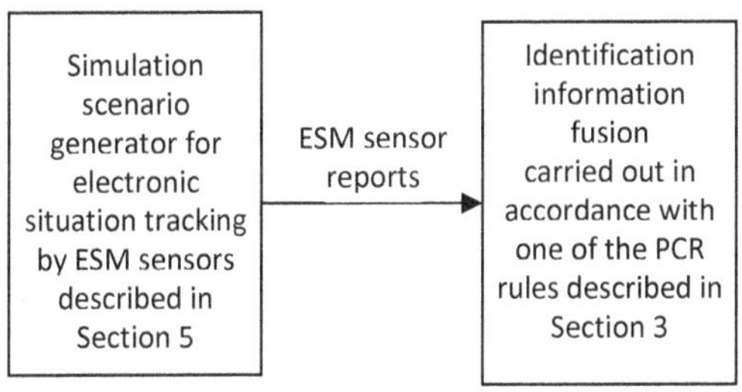

Figure 4. The general diagram of simulation experiments of fusion of identification information from ESM sensors.

6.2. Simulation Scenarios

Paper [25] presents a typical simulation scenario for testing identification information fusion. The authors formulated several requirements that should be met by such a scenario. It should meet the following requirements:

(1) adequately represent the known ground truth of the emitter identification;
(2) include sufficient numbers of incorrect associations to be realistic and to test the robustness of the rules in temporary incorrect sensor decisions;
(3) provide only partial knowledge about the ESM sensor declarations and thus contain uncertainty;
(4) be able to show stability in the case of countermeasures;
(5) be able to switch identification when the ground truth changes.

The authors of [25] propose the following parameters of the scenario:

(1) ground truth of identification is FRIEND (*F*) for the first 50 iterations of the scenario and HOSTILE (*H*) for the last 50 iterations;
(2) the percentage of correct associations is 80% of all iterations, and the percentage of incorrect associations caused by countermeasures is 20% of all iterations in randomly selected moments of time;
(3) ESM sensor declarations have a mass of 0.7 for the most credible identification and 0.3 for the identification of UNKNOWN (*U*).

Assumption (5) is not considered in this paper, assuming that the real object does not change its real identity while performing the mission. Therefore, assumption (1) regarding the scenario parameters becomes obsolete.

The following assumptions concerning the parameters of the scenario have been made in this paper:

(1) the real value of identification is constant in each scenario and is equal to FRIEND (*F*) in scenarios 1, 2, and 5 and HOSTILE (*H*) in scenarios 3, 4, and 6;

(2) the above declarations are transmitted by sensor number 1 with the real identification mass equal to 0.7 and the mass of complementary identification (UNKNOWN) equal to 0.3;

(3) the second sensor transmits its declarations in accordance with Tables 2 and 3 for scenarios 1 and 2 and in accordance with Tables 4 and 5 for scenarios 3 and 4.

Table 2. Belief mass values for the second sensor for scenarios 1 and 5.

Type of Identification	*F*	*N*	*H*	*AF*	*S*	*u*
Correct identification (80% of events)	0.6	0.1	0	0.2	0	0.1
Incorrect identification (20% of events)	0	0.1	0.6	0	0.2	0.1

Table 3. Belief mass values for the second sensor for scenario 2.

Type of Identification	*F*	*N*	*H*	*AF*	*S*	*u*
Correct identification (80% of events)	0.7	0.1	0	0.1	0	0.1
Incorrect identification (20% of events)	0	0.1	0.7	0	0.7	0.1

Table 4. Belief mass values for the second sensor for scenario 3.

Type of Identification	*F*	*N*	*H*	*AF*	*S*	*u*
Correct identification (80% of events)	0	0.1	0.6	0	0.2	0.1
Incorrect identification (20% of events)	0.6	0.1	0	0.2	0	0.1

Table 5. Belief mass values for the second sensor for scenarios 4 and 6.

Type of Identification	*F*	*N*	*H*	*AF*	*S*	*u*
Correct identification (80% of events)	0	0.1	0.7	0	0.1	0.1
Incorrect identification (20% of events)	0.7	0.1	0	0.1	0	0.1

It should be noted that scenario 2 differs from scenario 1 by a greater belief mass assigned to an incorrect identification of the recognized emitter. Scenarios 3 and 4 are similarly different.

Scenarios 1–6 for sensor 1 are presented in Figures 5 and 6. All deterministic scenarios for sensor 2 are presented in Figures 7–10.

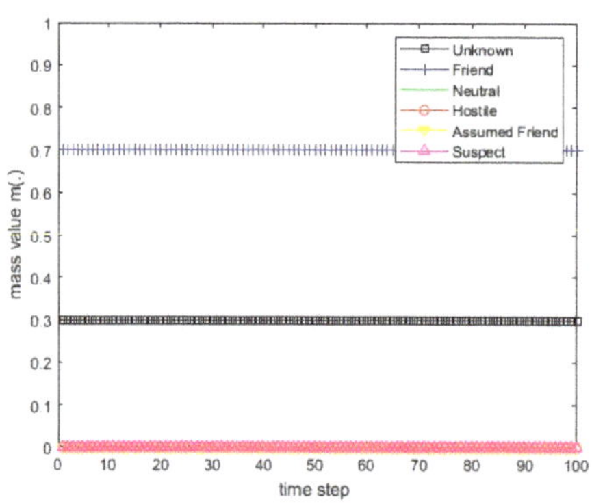

Figure 5. The course of scenarios number 1, 2, and 5 for sensor 1.

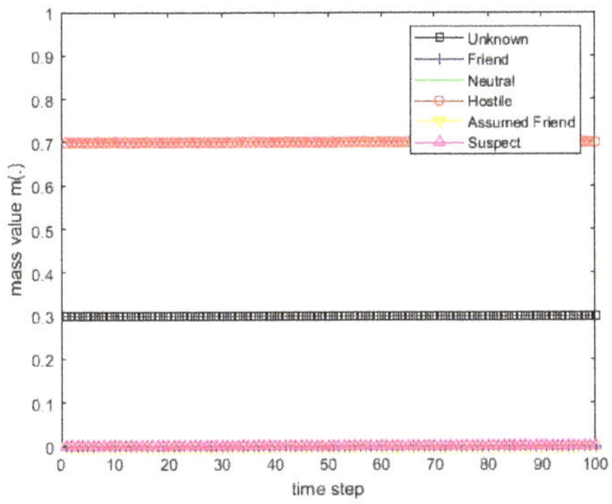

Figure 6. The course of scenarios number 3, 4, and 6 for sensor 1.

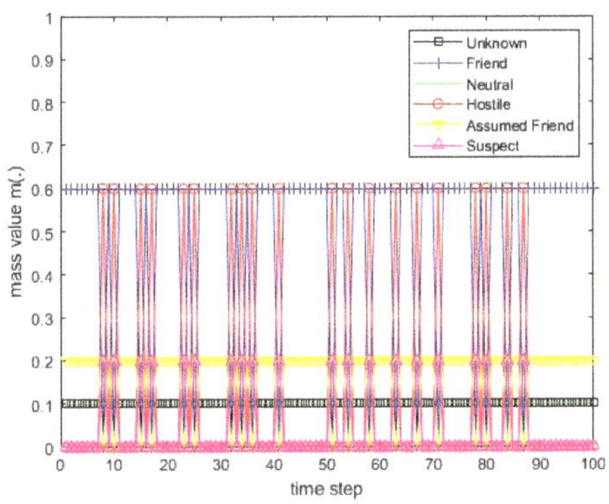

Figure 7. The course of scenario number 1 for sensor 2.

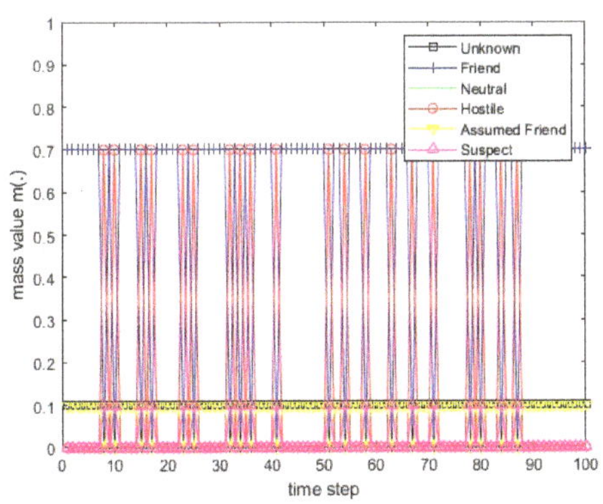

Figure 8. The course of scenario number 2 for sensor 2.

This section is divided by subheadings. It provides a concise and precise description of the experimental results, their interpretation, and the experimental conclusions that can be drawn.

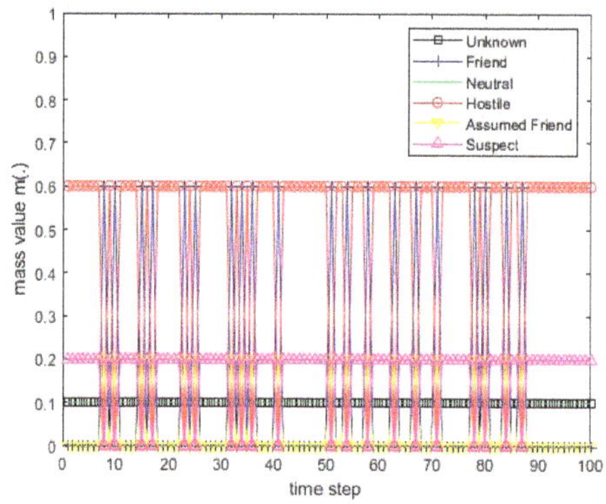

Figure 9. The course of scenario number 3 for sensor 2.

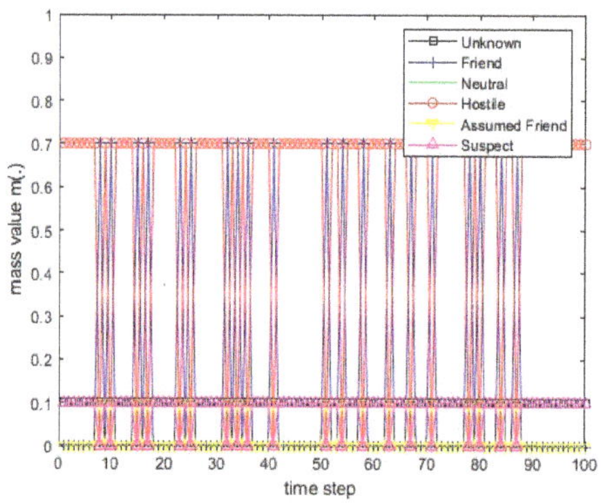

Figure 10. The course of scenario number 4 for sensor 2.

The paper also uses the Monte Carlo method for generating the scenario for sensor 2. Moments in which incorrect identifications occurred are generated by the pseudorandom integer number generator from the range [0, 100]. Examples of scenarios are shown in Figures 11 and 12.

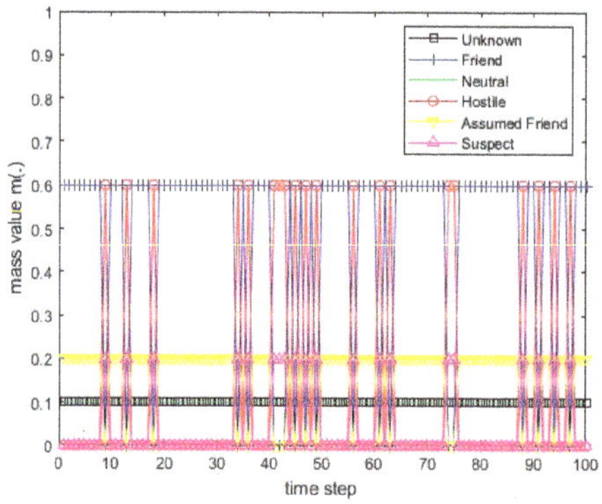

Figure 11. The course of Monte Carlo scenario number 5 for sensor 2.

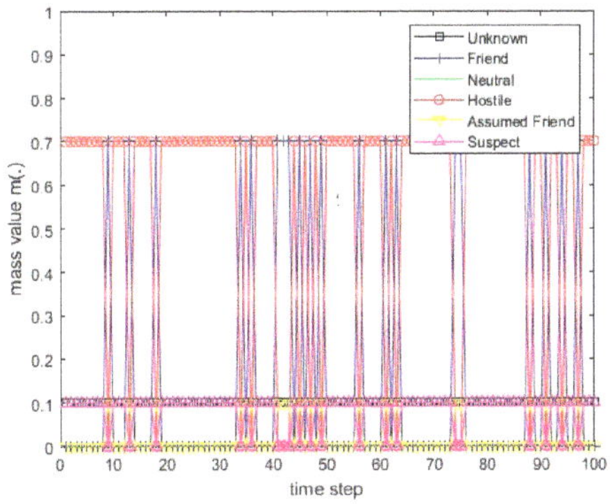

Figure 12. The course of Monte Carlo scenario number 6 for sensor 2.

6.3. Calculation Results for Deterministic Scenarios

In all figures presenting the values of the resulting belief mass, the decision threshold is marked with a horizontal line. An identification whose belief mass at a given moment is above the decision threshold is the so-called hard decision.

6.3.1. Dempster's Rule

Dempster's rule is not resistant to a situation where the degree of conflict $k_{Fi} = 1$. This means the total conflict between the mass vector sent by the sensor and the mass vector of the information-fusion center, which occurs when each non-zero belief mass value sent by the sensor corresponds to the zero belief mass value of the vector determined by the information-fusion center and vice versa.

The simulation results of identification information fusion using Dempster's rule are presented for deterministic scenarios 1 and 3 in Figures 13 and 14, respectively. When the degree of conflict $k_{Fi} = 1$, according to Equation (8), it is impossible to perform sensor information fusion, i.e., it is impossible to determine the resulting BBA.

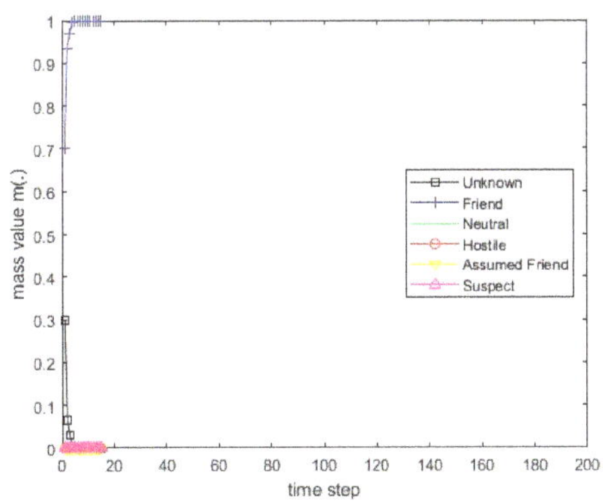

Figure 13. The values of the resulting belief mass for scenario 1 and Dempster's rule.

6.3.2. The PCR1 Rule

The simulation results of identification information fusion using the PCR1 rule for deterministic scenarios 1, 2, 3, and 4 are presented in Figures 15–18, respectively.

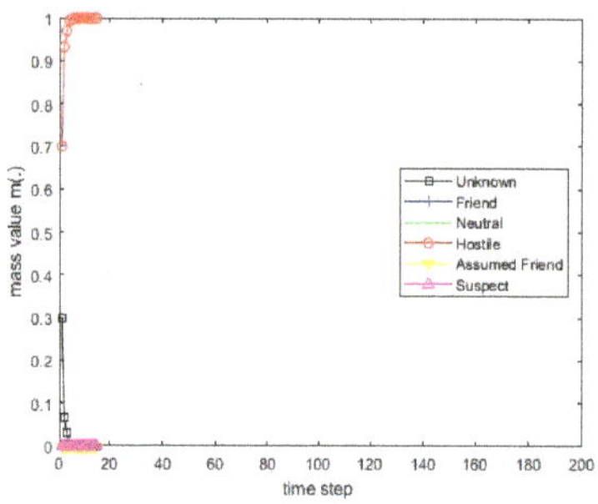

Figure 14. The values of the resulting belief mass for scenario 3 and Dempster's rule.

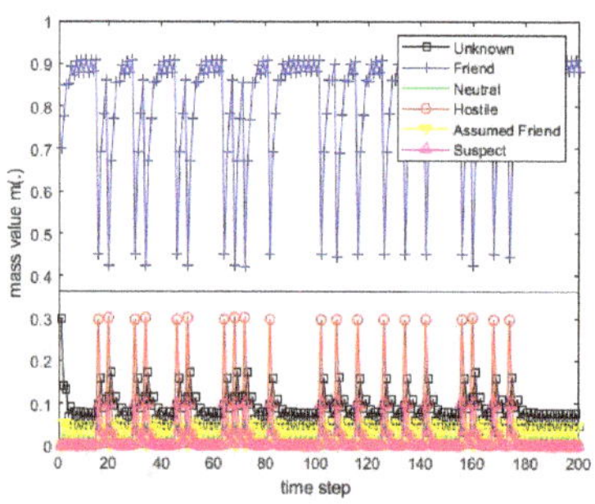

Figure 15. The values of the resulting belief mass for scenario 1 and the PCR1 rule.

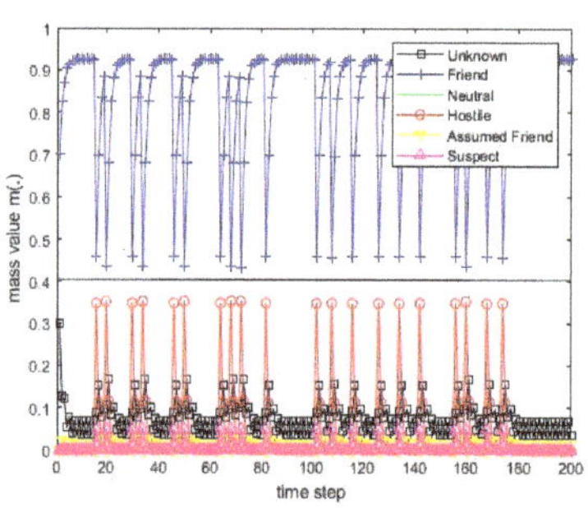

Figure 16. The values of the resulting belief mass for scenario 2 and the PCR1 rule.

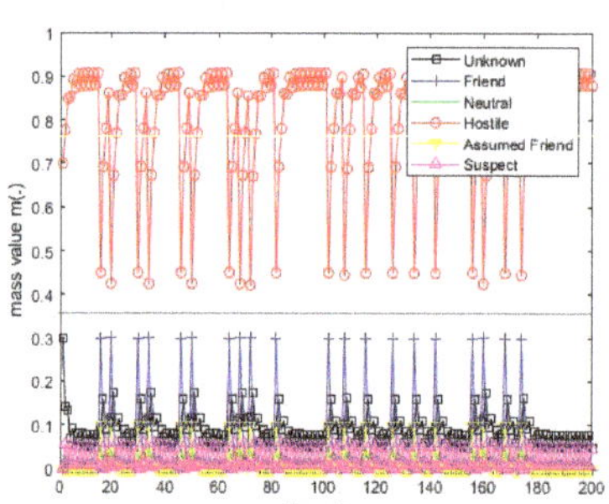

Figure 17. The values of the resulting belief mass for scenario 3 and the PCR1 rule.

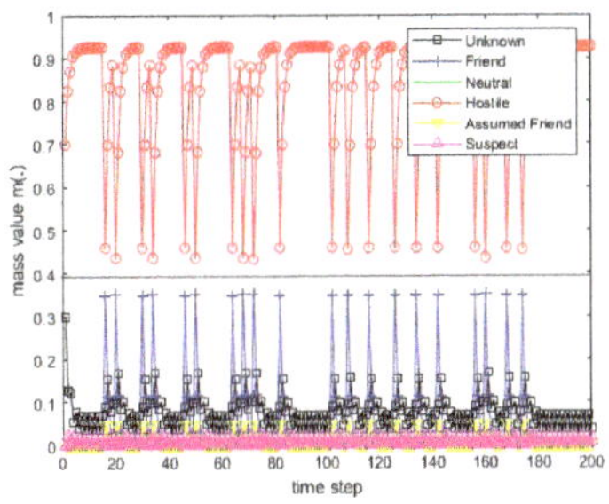

Figure 18. The values of the resulting belief mass for scenario 4 and the PCR1 rule.

6.3.3. The PCR3 Rule

The simulation results of identification information fusion using the PCR3 rule for deterministic scenarios 1, 2, 3, and 4 are presented in Figures 19–22, respectively.

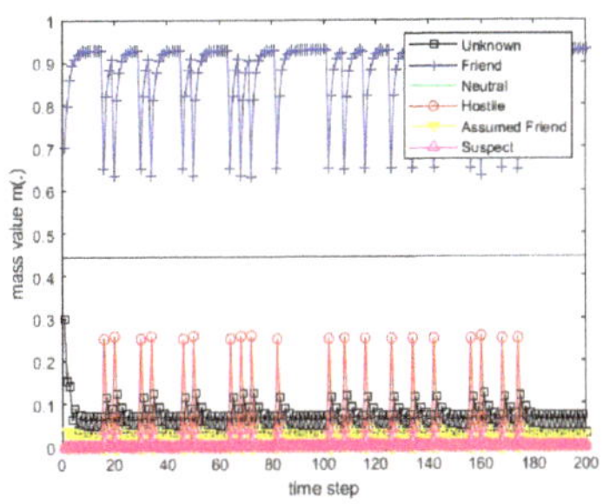

Figure 19. The values of the resulting belief mass for scenario 1 and the PCR3 rule.

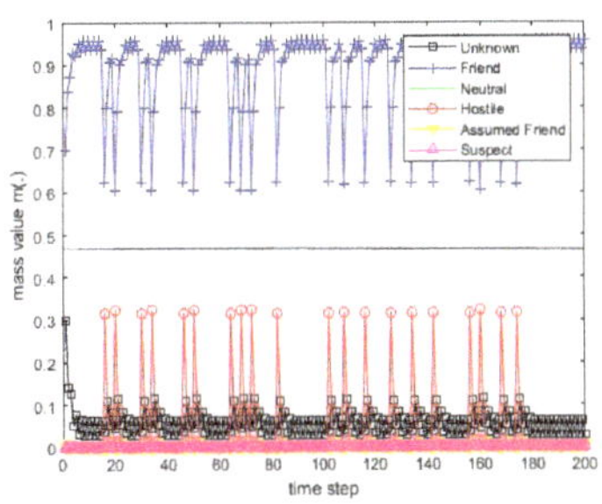

Figure 20. The values of the resulting belief mass for scenario 2 and the PCR3 rule.

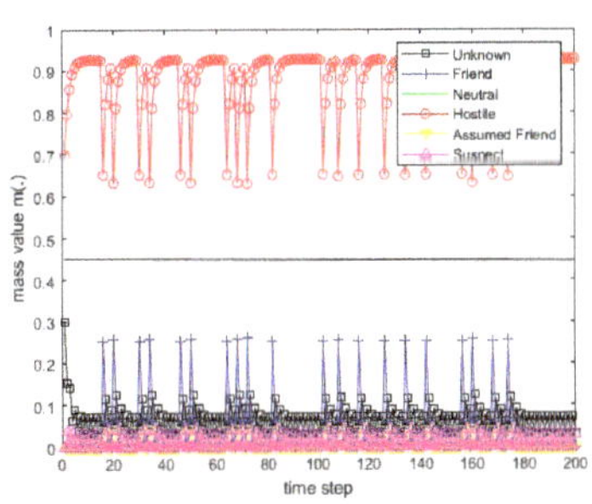

Figure 21. The values of the resulting belief mass for scenario 3 and the PCR3 rule.

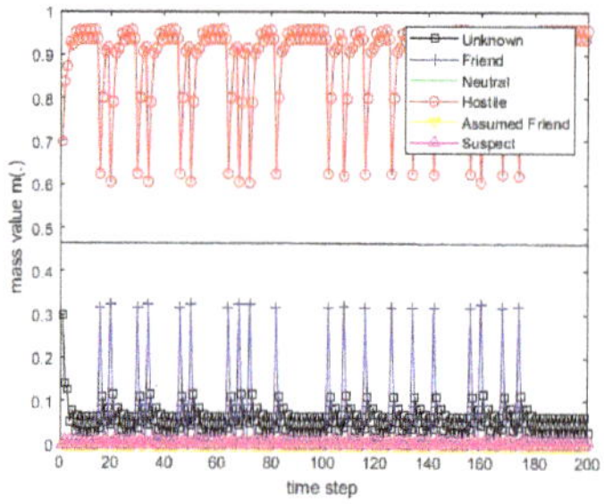

Figure 22. The values of the resulting belief mass for scenario 4 and the PCR3 rule.

6.3.4. The PCR4 Rule

The simulation results of identification information fusion using the PCR4 rule for deterministic scenarios 1, 2, 3, and 4 are presented in Figures 23–26, respectively.

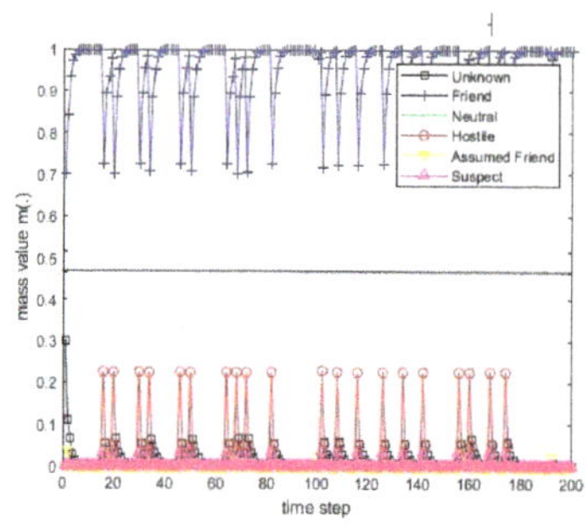

Figure 23. The values of the resulting belief mass for scenario 1 and the PCR4 rule.

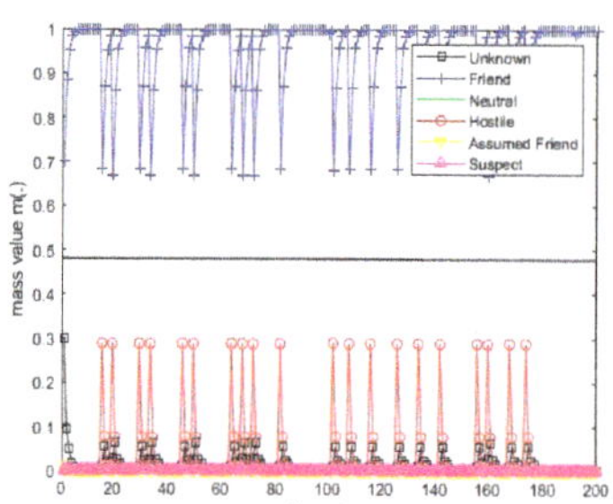

Figure 24. The values of the resulting belief mass for scenario 2 and the PCR4 rule.

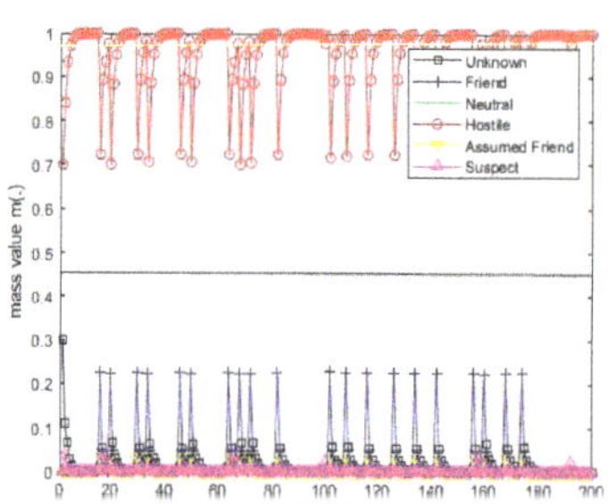

Figure 25. The values of the resulting belief mass for scenario 3 and the PCR4 rule.

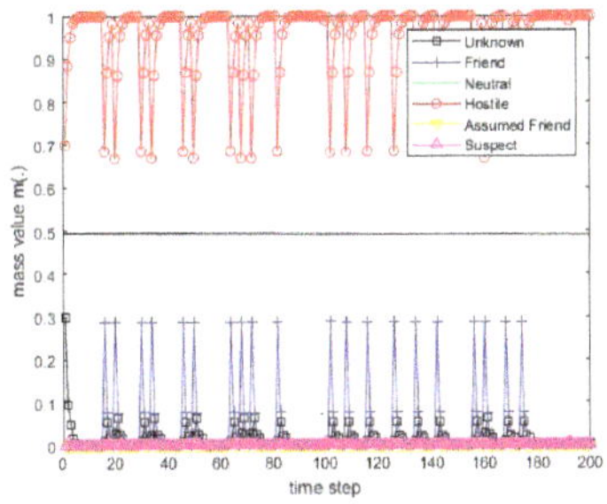

Figure 26. The values of the resulting belief mass for scenario 4 and the PCR4 rule.

6.3.5. The PCR5 Rule for 2 BBAs

The simulation results of identification information fusion using the PCR5 rule for two BBAs for deterministic scenarios 1, 2, 3, and 4 are presented in Figures 27–30, respectively.

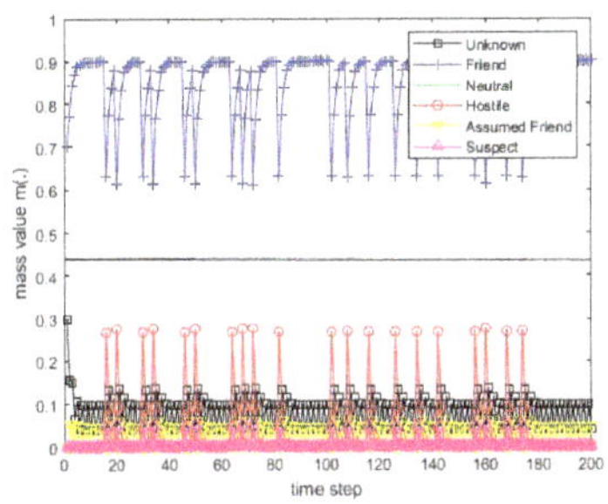

Figure 27. The values of the resulting belief mass for scenario 1 and the PCR5 rule for 2 BBAs.

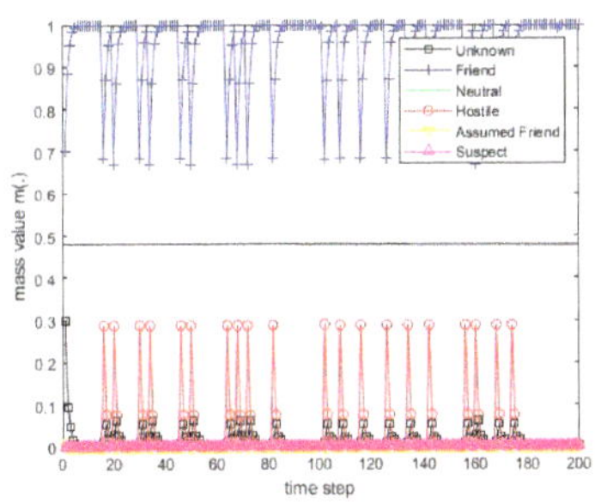

Figure 28. The values of the resulting belief mass for scenario 2 and the PCR5 rule for 2 BBAs.

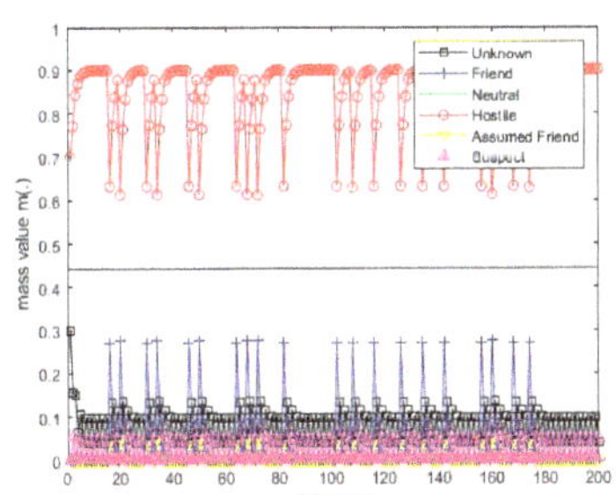

Figure 29. The values of the resulting belief mass for scenario 3 and the PCR5 rule for 2 BBAs.

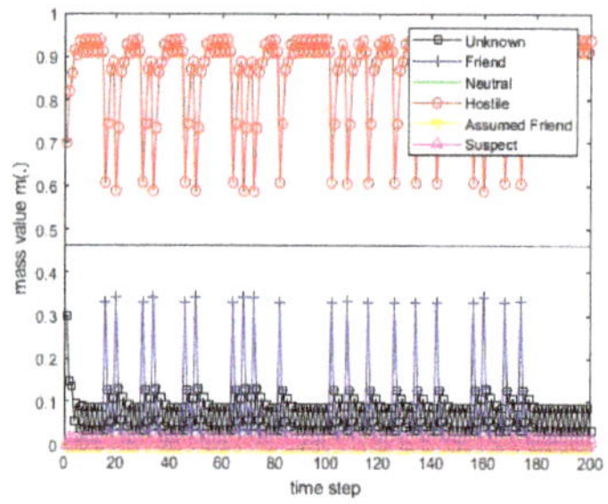

Figure 30. The values of the resulting belief mass for scenario 4 and the PCR5 rule for 2 BBAs.

6.3.6. The PCR5 Rule for 3 BBAs

The simulation results of identification information fusion using the PCR5 rule for three BBAs for deterministic scenarios 1, 2, 3, and 4 are presented in Figures 31–34, respectively.

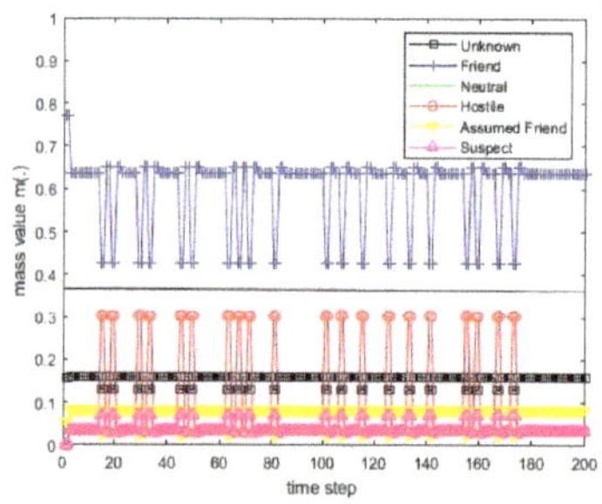

Figure 31. The values of the resulting belief mass for scenario 1 and the PCR5 rule for 3 BBAs.

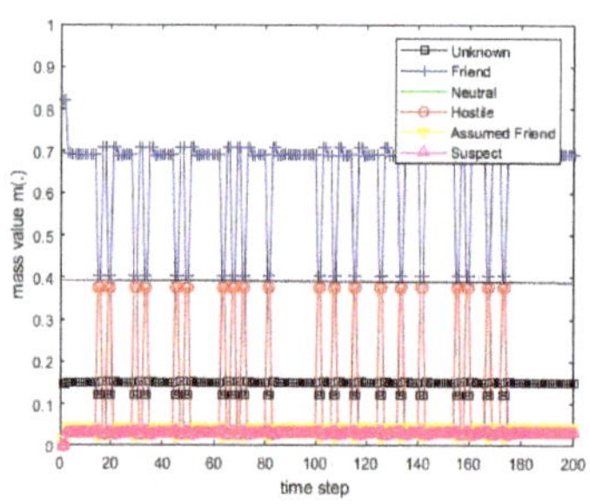

Figure 32. The values of the resulting belief mass for scenario 2 and the PCR5 rule for 3 BBAs.

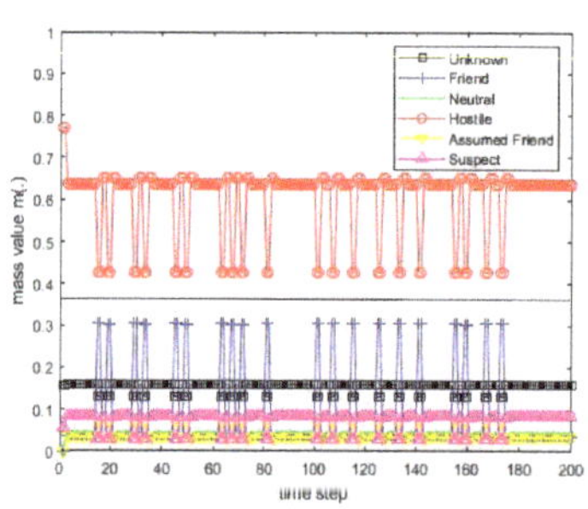

Figure 33. The values of the resulting belief mass for scenario 3 and the PCR5 rule for 3 BBAs.

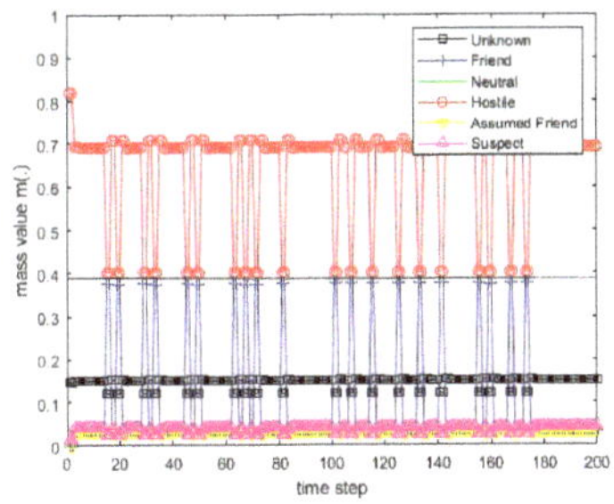

Figure 34. The values of the resulting belief mass for scenario 4 and the PCR5 rule for 3 BBAs.

6.3.7. The PCR6 Rule for 3 BBAs

The simulation results of identification information fusion using the PCR6 rule for three BBAs for deterministic scenarios 1, 2, 3, and 4 are presented in Figures 35–38, respectively.

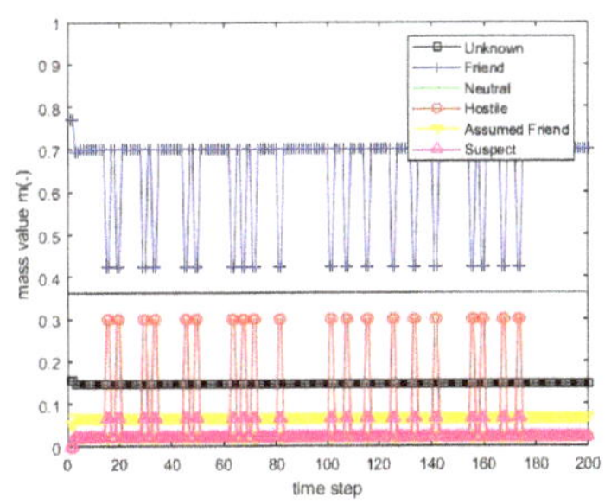

Figure 35. The values of the resulting belief mass for scenario 1 and the PCR6 rule for 3 BBAs.

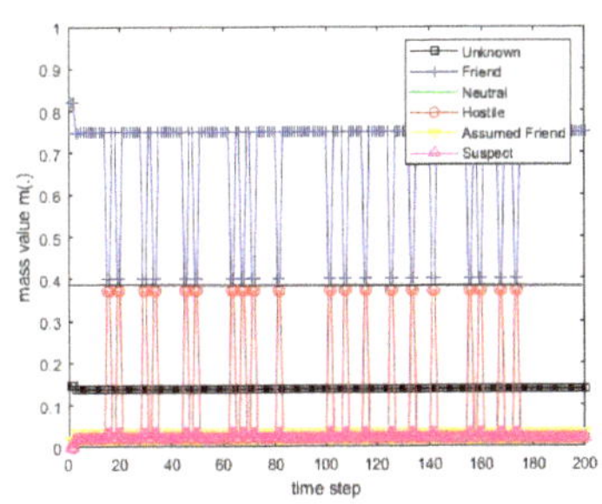

Figure 36. The values of the resulting belief mass for scenario 2 and the PCR6 rule for 3 BBAs.

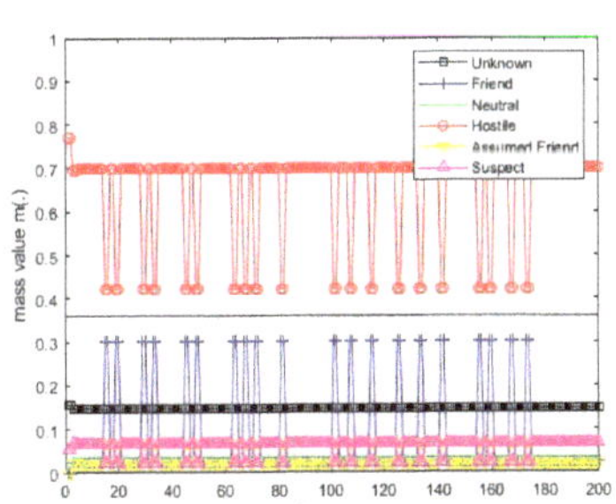

Figure 37. The values of the resulting belief mass for scenario 3 and the PCR6 rule for 3 BBAs.

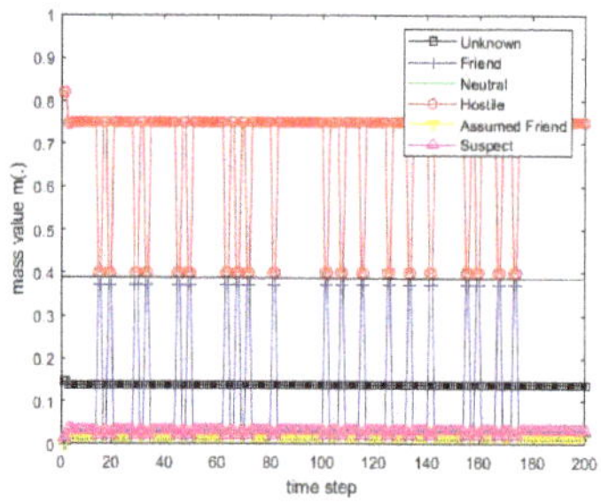

Figure 38. The values of the resulting belief mass for scenario 4 and the PCR6 rule for 3 BBAs.

The presented results (Figures 13–38) yield the conclusion that the applied methods of managing conflicts in information fusion enables correct conclusions to be drawn about the real identification of the recognized object.

The application of the decision threshold for the belief mass at the level $m_\alpha = 0.37$ for the PCR1 rule (Figures 15 and 17) and $m_\alpha = 0.45$ for PCR3, PCR4 (Figures 19, 21, 23 and 25), and PCR5 for two BBAs (Figures 27 and 29) for scenarios 1 and 3 allows for a proper evaluation of the identification of the recognized object: scenario 1, FRIEND; scenario 3, HOSTILE. For scenarios 2 and 4, the optimal thresholds are $m_\alpha = 0.4$ for the PCR1 rule (Figures 16 and 18) and $m_\alpha = 0.48$ for the PCR3, PCR4 (Figures 20, 22, 24 and 26), and PCR5 rules for two BBAs (Figures 28 and 30). When assessing the interval between the minimum resultant mass for correct identification and the maximum resultant mass for misidentification, the worst results are reached by the PCR1 rule, and the rules of PCR3, PCR4, and PCR5 behave similarly and are better than rule PCR1.

The research carried out for the deterministic scenarios shows that the PCR5 rule for three BBAs and the PCR6 rule for three BBAs behave very similarly (Figures 31, 33, 35 and 37 for scenarios 1 and 3, $m_\alpha = 0.37$ and Figures 32, 34, 36 and 37 for scenarios 2 and 4, $m_\alpha = 0.39$). They restore the correct identification after the occurrence of temporary misidentification much faster than the rules PCR1–PCR5 for two BBAs.

6.4. Calculation Results for the Monte Carlo Scenarios

6.4.1. Dempster's Rule

In the Monte Carlo scenario, Dempster's rule behaves similarly to a deterministic scenario. It is not resistant to a situation where the degree of conflict $k_{Fi} = 1$. This means that the total conflict between the mass vector sent by the sensor and the mass vector of the information-fusion center, which occurs when each non-zero belief mass value sent by the sensor corresponds to the zero belief mass value of the vector determined by the information-fusion center and vice versa.

The simulation results of identification information fusion using Dempster's rule are presented for Monte Carlo scenarios 5 and 6 in Figures 39 and 40, respectively.

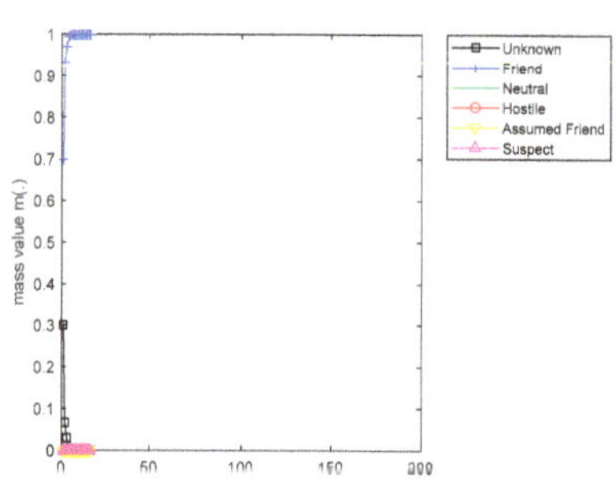

Figure 39. The values of the resulting belief mass for Monte Carlo scenario 5 and Dempster's rule.

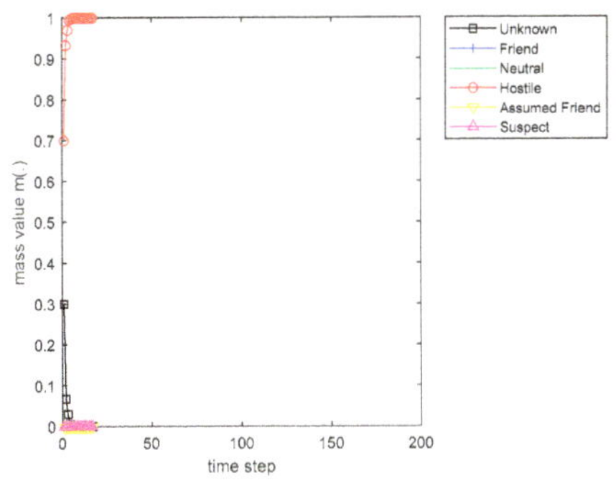

Figure 40. The values of the resulting belief mass for Monte Carlo scenario 6 and Dempster's rule.

6.4.2. The PCR1 Rule

The simulation results of identification information fusion using the PCR1 rule for Monte Carlo scenarios 5 and 6 are presented in Figures 41 and 42, respectively. The application of the decision threshold for the belief mass at the level $m_\alpha = 0.34$ for the PCR1 rule (Figures 41 and 42) for scenarios 5 and 6 allows for a proper evaluation of the identification of the recognized object. There are only three time points when the rule misidentifies due to increased misidentification intensity.

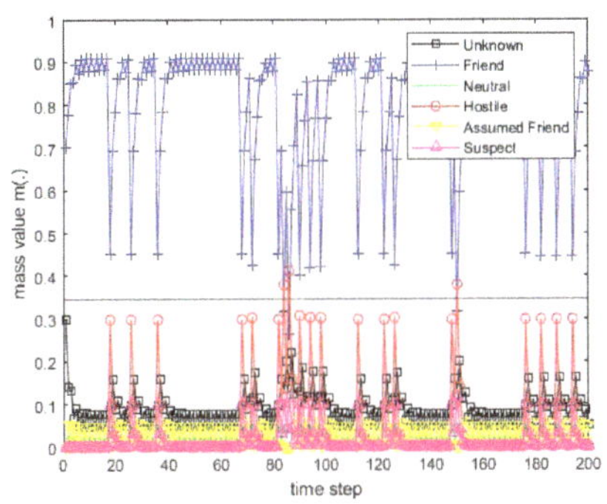

Figure 41. The values of the resulting belief mass for Monte Carlo scenario 5 and the PCR1 rule.

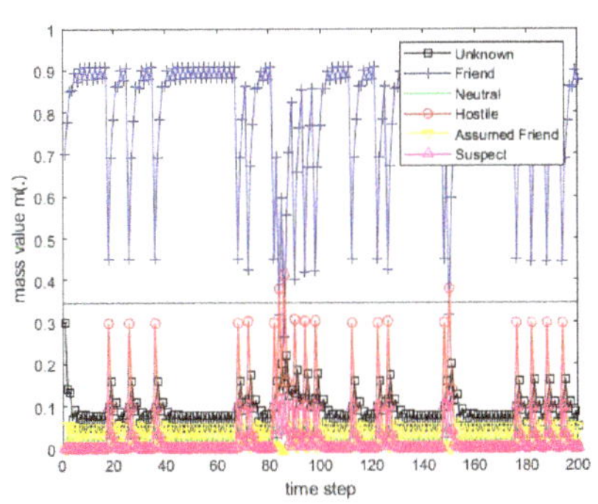

Figure 42. The values of the resulting belief mass for Monte Carlo scenario 6 and the PCR1 rule.

6.4.3. The PCR3 Rule

The simulation results of identification information fusion using the PCR3 rule for Monte Carlo scenarios 5 and 6 are presented in Figures 43 and 44, respectively. The application of the decision threshold for the belief mass at the level $m_\alpha = 0.42$ for the PCR3 rule (Figures 43 and 44) for scenarios 5 and 6 allows for a proper evaluation of the identification of the recognized object. There is only one time point for scenario 6 where the rule misidentifies due to increased misidentification intensity.

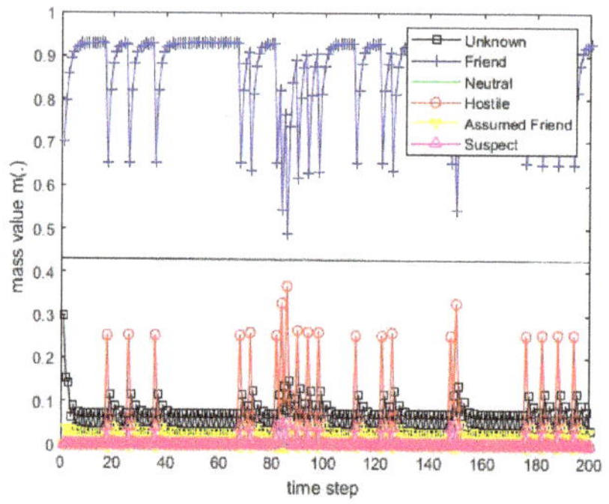

Figure 43. The values of the resulting belief mass for Monte Carlo scenario 5 and the PCR3 rule.

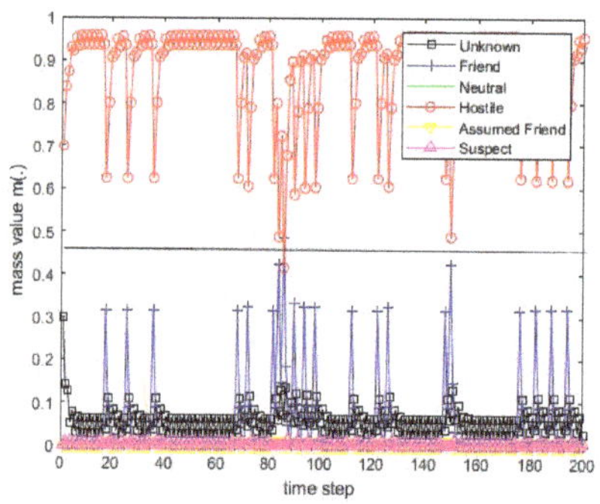

Figure 44. The values of the resulting belief mass for Monte Carlo scenario 6 and the PCR3 rule.

6.4.4. The PCR4 Rule

The simulation results of identification information fusion using the PCR4 rule for Monte Carlo scenarios 5 and 6 are presented in Figures 45 and 46, respectively. The application of the decision threshold for the belief mass at the level $m_\alpha = 0.47$ for the PCR4 rule (Figures 45 and 46) for scenarios 5 and 6 allows for a proper evaluation of the identification of the recognized object.

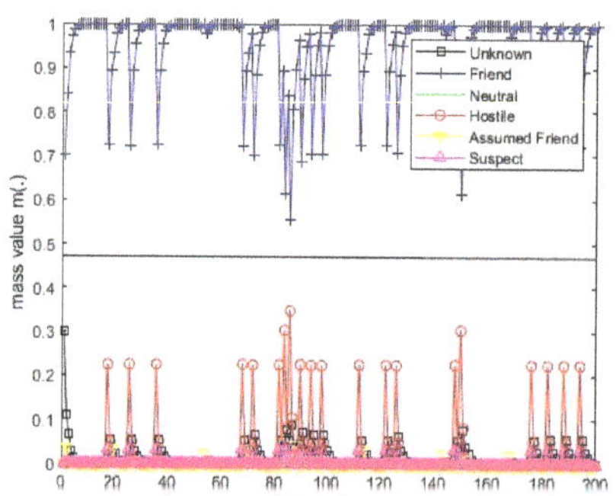

Figure 45. The values of the resulting belief mass for Monte Carlo scenario 5 and the PCR4 rule.

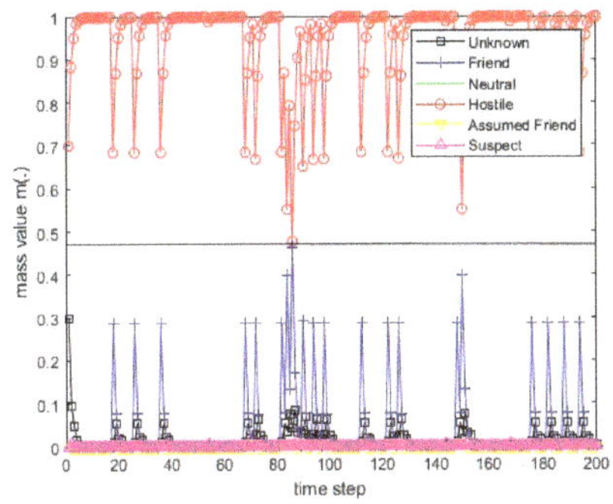

Figure 46. The values of the resulting belief mass for Monte Carlo scenario 6 and the PCR4 rule.

6.4.5. The PCR5 Rule for 2 BBAs

The simulation results of identification information fusion using the PCR5 rule for two BBAs for Monte Carlo scenarios 5 and 6 are presented in Figures 47 and 48, respectively. The application of the decision threshold for the belief mass at the level $m_\alpha = 0.47$ for the PCR5 rule for two BBAs (Figures 47 and 48) for scenarios 5 and 6 allows for a proper evaluation of the identification of the recognized object. There is only one time point for scenario 6 where the rule misidentifies due to increased misidentification intensity.

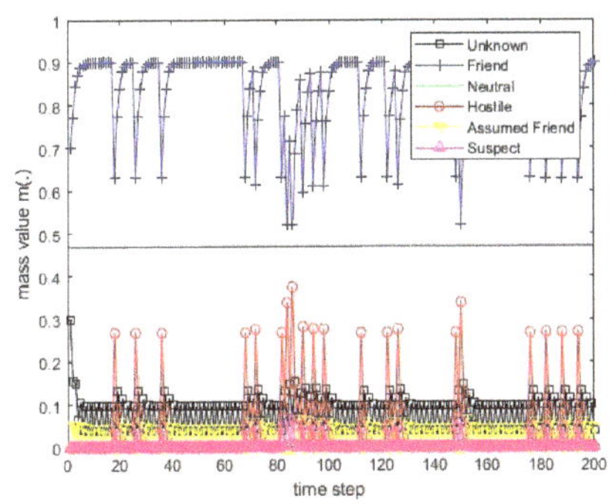

Figure 47. The values of the resulting belief mass for Monte Carlo scenario 5 and the PCR5 rule for 2 BBAs.

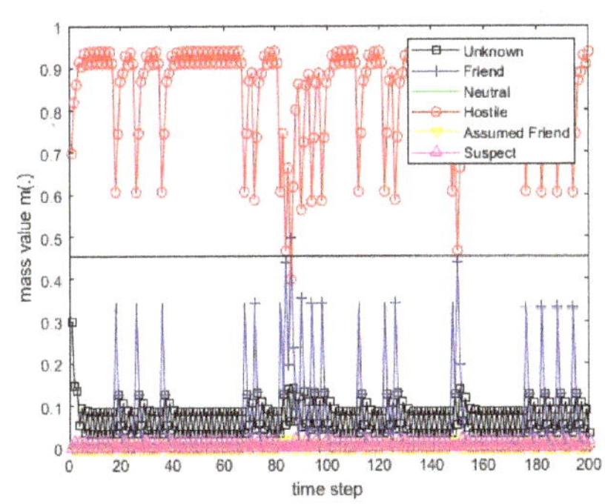

Figure 48. The values of the resulting belief mass for Monte Carlo scenario 6 and the PCR5 rule for 2 BBAs.

6.4.6. The PCR5 Rule for 3 BBAs

The simulation results of identification information fusion using the PCR5 rule for three BBAs for Monte Carlo scenarios 5 and 6 are presented in Figures 49 and 50, respectively. The application of the decision threshold for the belief mass at the level $m_{\alpha} = 0.39$ for the PCR5 rule for three BBAs (Figures 49 and 50) for scenarios 5 and 6 allows for a proper evaluation of the identification of the recognized object.

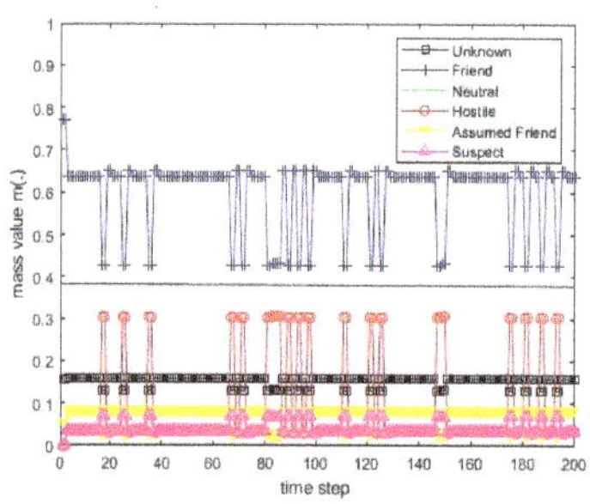

Figure 49. The values of the resulting belief mass for Monte Carlo scenario 5 and the PCR5 rule for 3 BBAs.

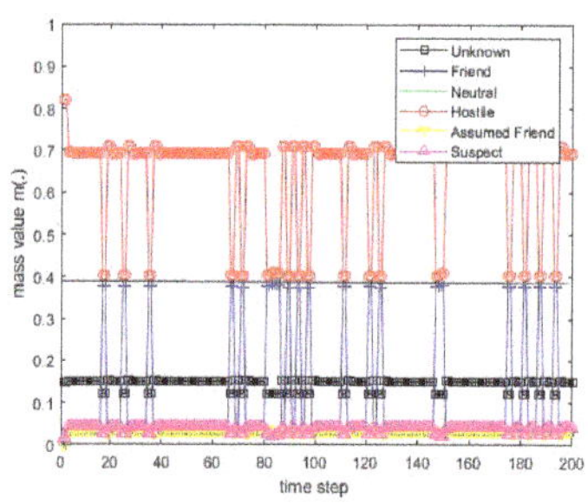

Figure 50. The values of the resulting belief mass for Monte Carlo scenario 6 and the PCR5 rule for 3 BBAs.

6.4.7. The PCR6 Rule for 3 BBAs

The simulation results of identification information fusion using the PCR6 rule for three BBAs for Monte Carlo scenarios 5 and 6 are presented in Figures 51 and 52. The application of the decision threshold for the belief mass at the level $m_{\alpha} = 0.39$ for the PCR6 rule for three BBAs (Figures 51 and 52) for scenarios 5 and 6 allows for a proper evaluation of the identification of the recognized object.

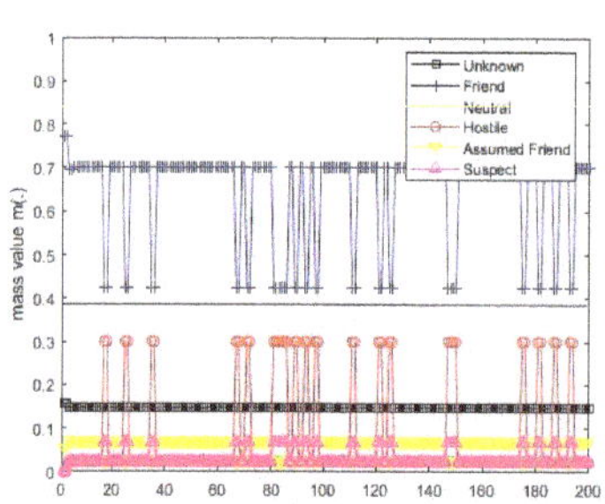

Figure 51. The values of the resulting belief mass for Monte Carlo scenario 5 and the PCR6 rule for 3 BBAs.

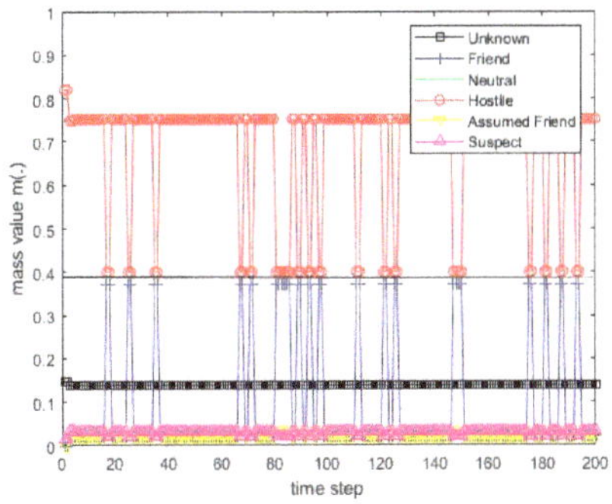

Figure 52. The values of the resulting belief mass for Monte Carlo scenario 6 and the PCR6 rule for 3 BBAs.

The presented results show that due to the high intensity of sending reports with incorrect identifications in the middle part of the scenarios, the information-fusion rules (apart from the PCR4, PCR5, and PCR6 rules) determine the maximum resulting mass for incorrect identification. The PCR5 for three BBAs and PCR6 for three BBAs rules are the fastest to restore the correct identification after receiving several incorrect reports.

7. Numerical Experiments of Fusion of Identification Information from ESM Sensors and Radars

7.1. General Research Scheme of Fusion of Identification Information from Radars and ESM Sensors

Figure 53 shows a general scheme of simulation experiments, which indicates the places of description of individual models.

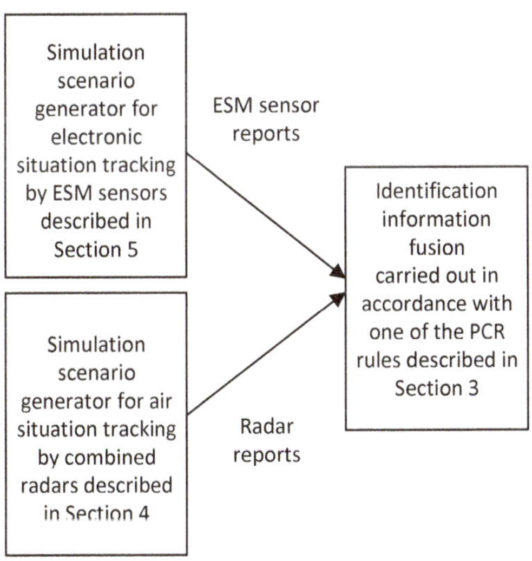

Figure 53. The general diagram of simulation experiments of fusion of identification information from radars and ESM sensors.

7.2. Numerical Experiments Scenarios

We assume that we will combine attribute information from two sensors: a combined primary and secondary surveillance radar and ESM sensor. These sensors work asynchronously. Upon receipt of the sensor's declaration in the form of a vector of masses, we fuse this vector with the vector of the actual values of the declaration masses for the fuser's frame of discernment. The frequency of transmission of the sensor declarations depends on the rules of the data exchange network and on the technical characteristics of the sensors. Various combination methods are presented in [3,4]. This paper used two of the methods of proportional redistribution conflict (PRC5 and PCR6 [6,22]). Information fusion has been simulated for two processing schemes (Figures 2 and 3). The numerical model of combined primary and secondary surveillance radars was taken from [17,23]. It allows for the determination of the probability P_d during the simulation of the object's movement, i.e., the change in the object's position relative to the radar. Detailed rules for determining BBAs' vectors, assuming the knowledge of probabilities P_d and P_{IFF}, are presented in Section 4.

Numerical experiments have been performed for the following data:

- for combined primary and secondary surveillance radars sensor:

$$P_{fa}= 10^{-6},\ R^*_{max}= 100\ [\mathrm{km}],\ P^*_d= 0.7,\ \sigma^*_c= 2\ \left[\mathrm{m}^2\right],\ P_{IFF} = 0.962$$

and the following table of masses (compare Table 6):

Table 6. Transformation of the base belief assignment mass into the secondary belief assignment mass for combined primary and secondary surveillance radar.

(Scenario Nr, Base Identification) →	(1,F_B)	(2,N_B)	(3,H_B)
F_S	0.8	0	0
N_S	0	0.5	0
H_S	0	0	0.7
AF	0.2	0.3	0
S	0	0.2	0.3

The flight path of the air object was 30 km away from the sensor (in the horizontal plane), the flight altitude was 1 km, and the radar cross-section was 1 m^2.

The following assumptions concerning the parameters of the scenario for the ESM sensor were made in this paper:

(1) The real value of identification is constant in each scenario and is equal to FRIEND (F) in the first scenario and HOSTILE (H) in the second scenario;

(2) The above declarations are transmitted by sensor number 1 with the real identification mass equal to 0.7 and the mass of complementary identification (UNKNOWN) equal to 0.3;

(3) The second sensor shall transmit its declarations in accordance with Tables 1 and 2 for scenarios 1 and 2, respectively, and with Tables 3 and 4 for scenarios 3 and 4, respectively.

Tables 7–12 present the mass values for all possible declarations for the six scenarios for the ESM sensor.

Table 7. Belief mass values for the second sensor (ESM) for scenario 1.

Type of Identification	F	N	H	AF	S	U
Correct identification (80% of events)	0.6	0.1	0	0.2	0	0.1
Incorrect identification (20% of events)	0	0.1	0.6	0	0.2	0.1

Table 8. Belief mass values for the second sensor (ESM) for scenario 2.

Type of Identification	F	N	H	AF	S	U
Correct identification (80% of events)	0	0.5	0.3	0	0.2	0
Incorrect identification (20% of events)	0	0.4	0.2	0	0.3	0.1

Table 9. Belief mass values for the second sensor (ESM) for scenario 3.

Type of Identification	*F*	*N*	*H*	*AF*	*S*	*U*
Correct identification (80% of events)	0	0.1	0.7	0	0.1	0.1
Incorrect identification (20% of events)	0	0.1	0.6	0	0.2	0.1

Table 10. Belief mass values for the second sensor (ESM) for scenario 4.

Type of Identification	*F*	*N*	*H*	*AF*	*S*	*U*
Correct identification (80% of events)	0.1	0.7	0.1	0	0	0.1
Incorrect identification (20% of events)	0	0.1	0.6	0	0.2	0.1

Table 11. Belief mass values for the second sensor (ESM) for scenario 5.

Type of Identification	*F*	*N*	*H*	*AF*	*S*	*U*
Correct identification (80% of events)	0.6	0.1	0	0.2	0	0.1
Incorrect identification (20% of events)	0	0.1	0.6	0	0.2	0.1

Table 12. Belief mass values for the second sensor (ESM) for scenario 6.

Type of Identification	*F*	*N*	*H*	*AF*	*S*	*U*
Correct identification (80% of events)	0.1	0.7	0.1	0	0	0.1
Incorrect identification (20% of events)	0.6	0.1	0	0.2	0	0.1

Scenarios 1–6 for sensor 1 are presented in Figures 54–56. All deterministic scenarios for sensor 2 are presented in Figures 57–62.

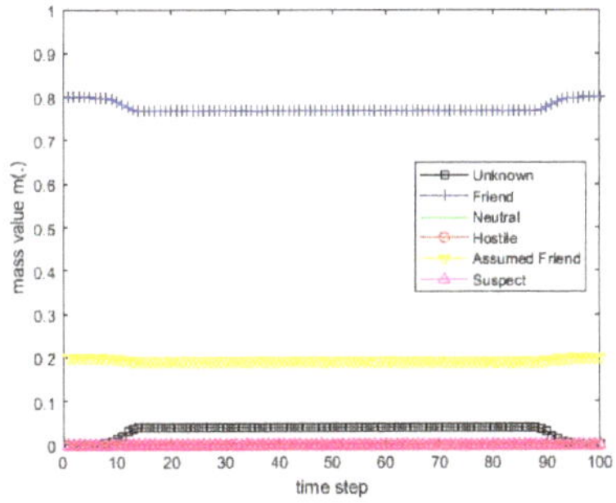

Figure 54. The course of scenarios number 1 and 4 for sensor 1.

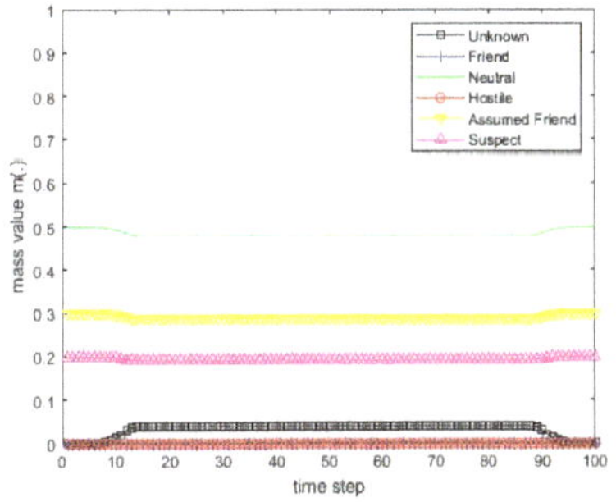

Figure 55. The course of scenarios number 2 and 5 for sensor 1.

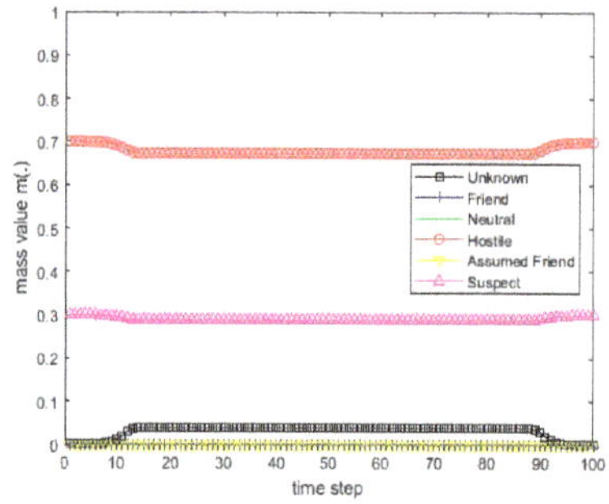

Figure 56. The course of scenarios number 3 and 6 for sensor 1.

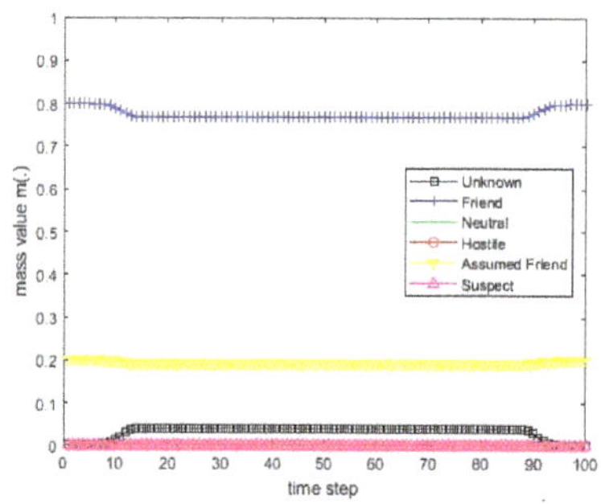

Figure 57. The course of scenario number 1 for sensor 2.

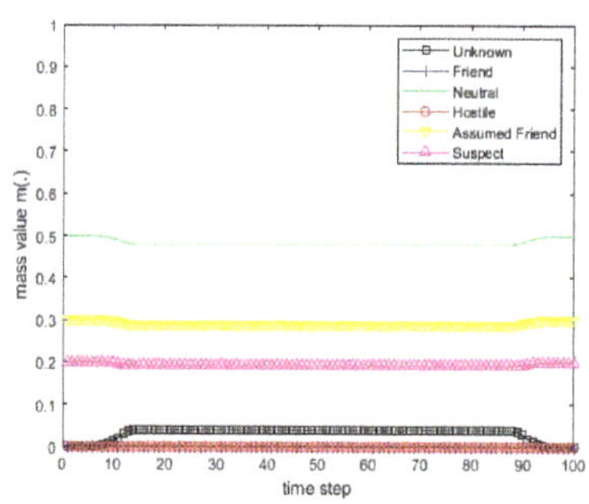

Figure 58. The course of scenario number 2 for sensor 2.

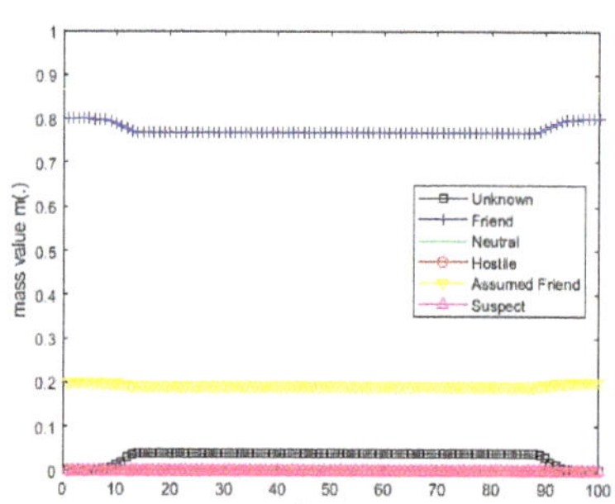

Figure 59. The course of scenario number 3 for sensor 2.

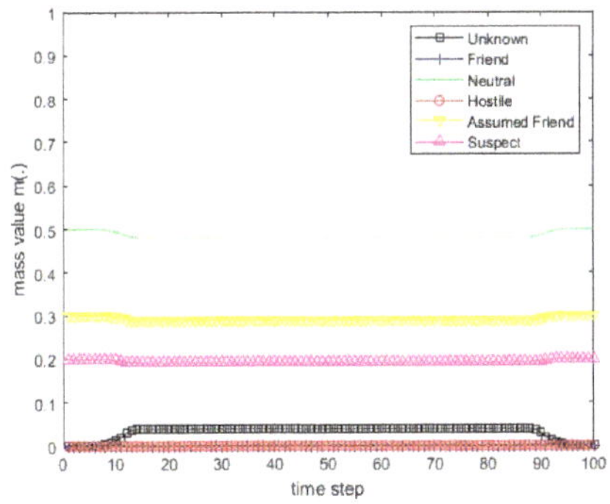

Figure 60. The course of scenario number 4 for sensor 2.

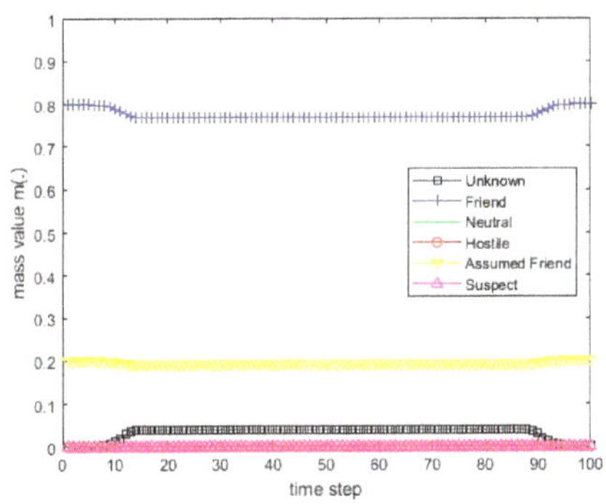

Figure 61. The course of scenario number 5 for sensor 2.

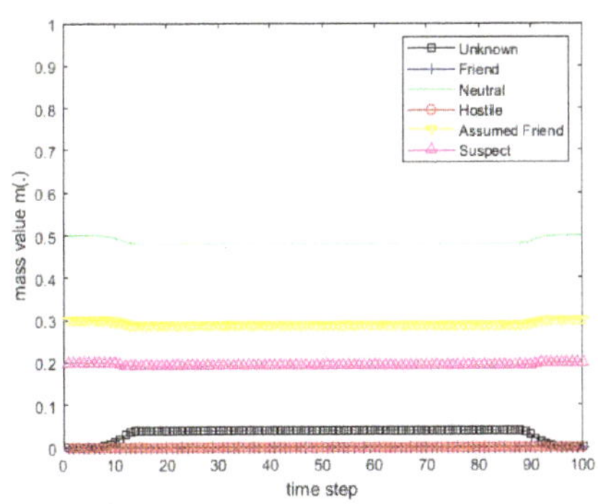

Figure 62. The course of scenario number 6 for sensor 2.

Scenarios 1–3 assume relatively small changes in the mass of all declarations. Scenarios 1–3 assume significant changes in the credibility mass of all declarations (small errors). Scenarios 4–6 assume significant changes in the mass of all declarations (large errors).

7.3. Calculation Results for Four Proportional Conflict Redistribution Rules

7.3.1. The PCR5 Rule for 2 BBAs

The simulation results of identification information fusion using the PCR5 rule for two BBAs for deterministic scenarios 1–6 are presented in Figures 63–68.

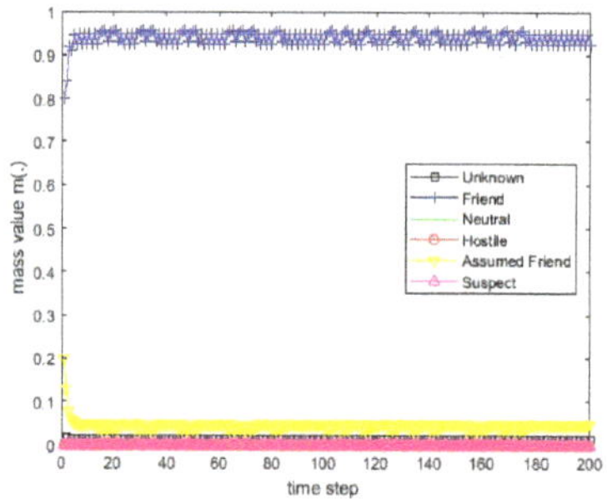

Figure 63. The values of the resulting belief mass for scenario 1 and the PCR5 rule for 2 BBAs.

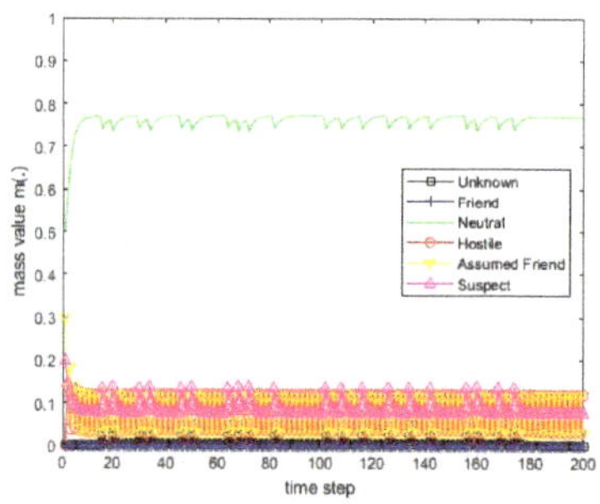

Figure 64. The values of the resulting belief mass for scenario 2 and the PCR5 rule for 2 BBAs.

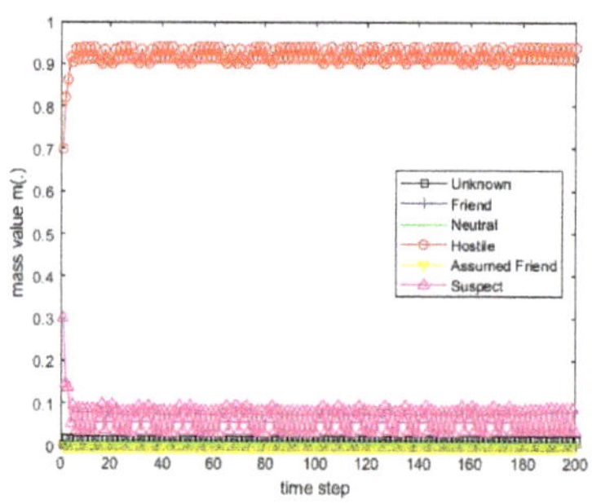

Figure 65. The values of the resulting belief mass for scenario 3 and the PCR5 rule for 2 BBAs.

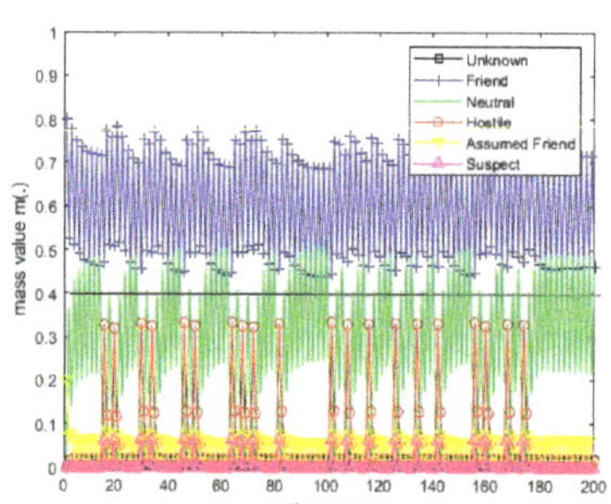

Figure 66. The values of the resulting belief mass for scenario 4 and the PCR5 rule for 2 BBAs.

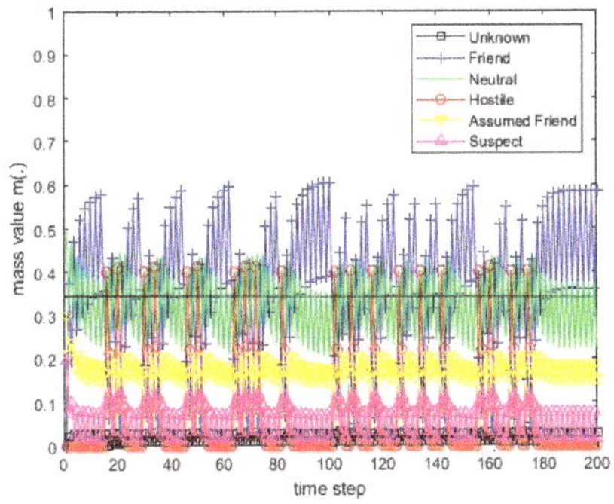

Figure 67. The values of the resulting belief mass for scenario 5 and the PCR5 rule for 2 BBAs.

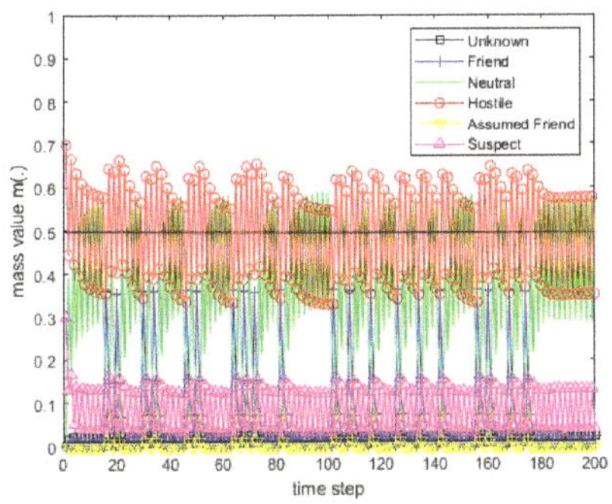

Figure 68. The values of the resulting belief mass for scenario 6 and the PCR5 rule for 2 BBAs.

For the PCR5 rule for two BBAs, the application of the decision thresholds at the belief mass level m_α = 0.40 for scenario 4, m_α = 0.36 for scenario 5, and m_α = 0.5 for scenario 6 allows for a proper evaluation of the identification of the recognized object for most time moments.

7.3.2. The PCR5 Rule for 3 BBAs

The simulation results of identification information fusion using the PCR5 rule for three BBAs for deterministic scenarios 1–6 are presented in Figures 69–74.

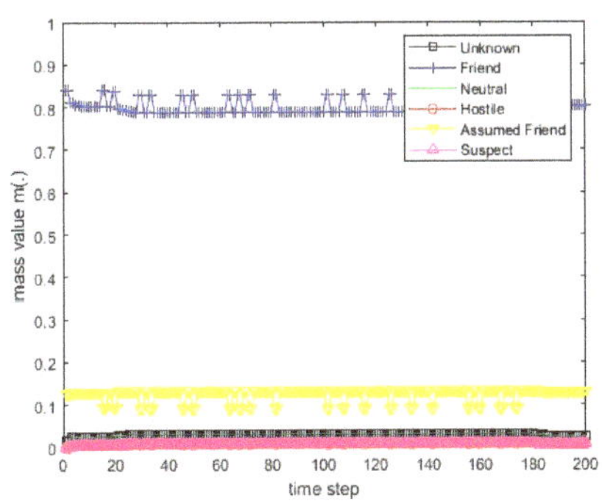

Figure 69. The values of the resulting belief mass for scenario 1 and the PCR5 rule for 3 BBAs.

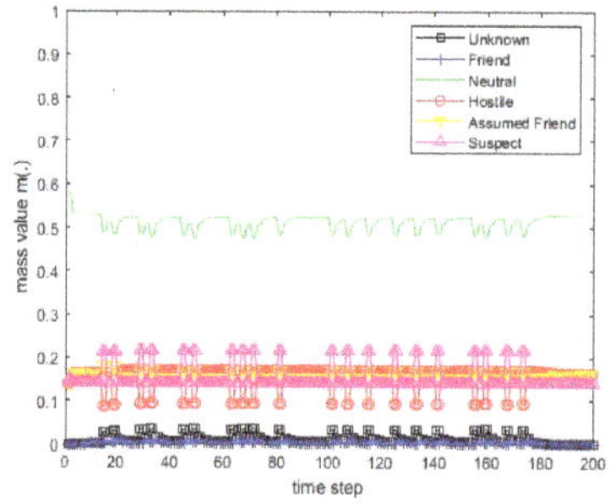

Figure 70. The values of the resulting belief mass for scenario 2 and the PCR5 rule for 3 BBAs.

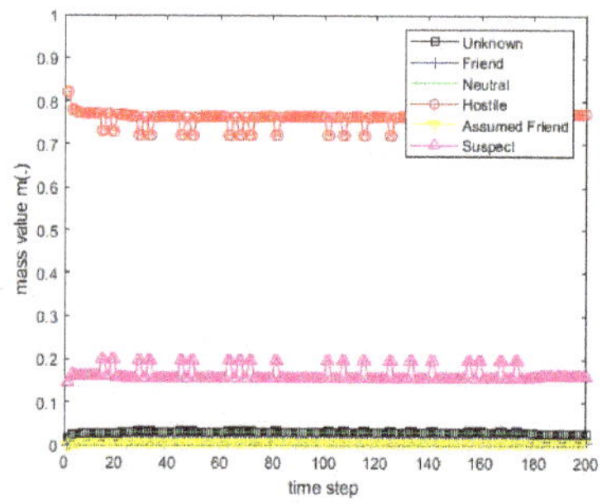

Figure 71. The values of the resulting belief mass for scenario 3 and the PCR5 rule for 3 BBAs.

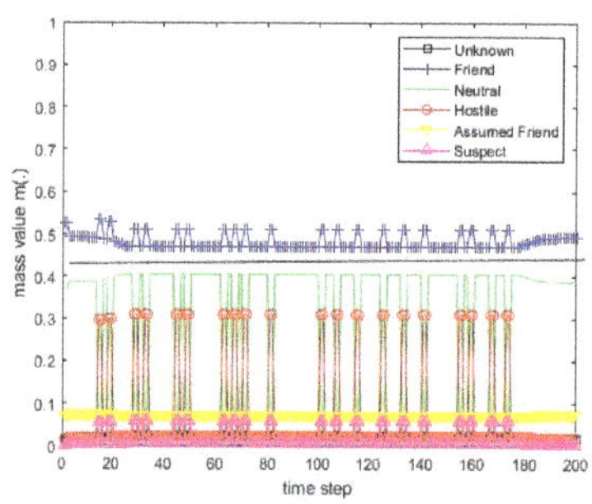

Figure 72. The values of the resulting belief mass for scenario 4 and the PCR5 rule for 3 BBAs.

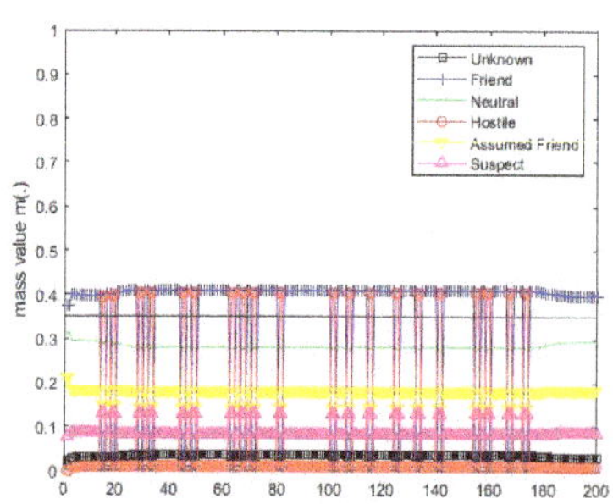

Figure 73. The values of the resulting belief mass for scenario 5 and the PCR5 rule for 3 BBAs.

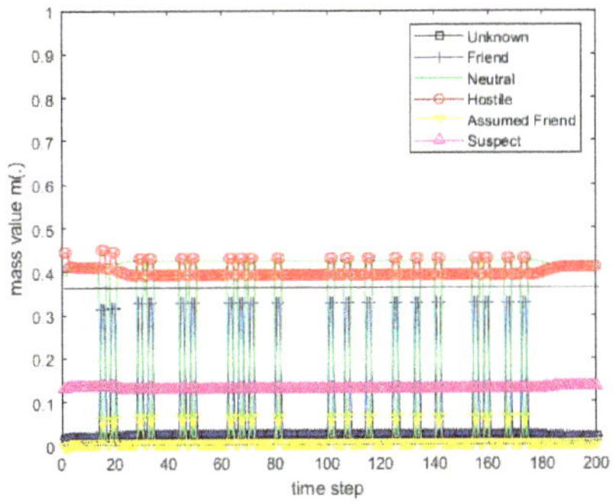

Figure 74. The values of the resulting belief mass for scenario 6 and the PCR5 rule for 3 BBAs.

For the PCR5 rule for three BBAs, the application of the decision thresholds at the belief mass level $m_\alpha = 0.42$ for scenario 4 allows for a proper evaluation of the identification of the recognized object.

For the PCR5 rule for three BBAs, the application of the decision thresholds at the belief mass level $m_\alpha = 0.37$ for scenarios 5 and 6 allows for a proper evaluation of the identification of the recognized object for most time moments.

7.3.3. The PCR6 Rule for 3 BBAs

The simulation results of identification information fusion using the PCR6 rule for three BBAs for deterministic scenarios 1–6 are presented in Figures 75–80.

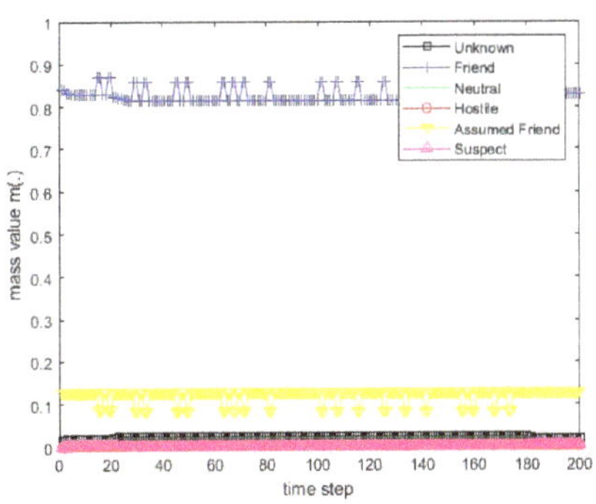

Figure 75. The values of the resulting belief mass for scenario 1 and the PCR6 rule for 3 BBAs.

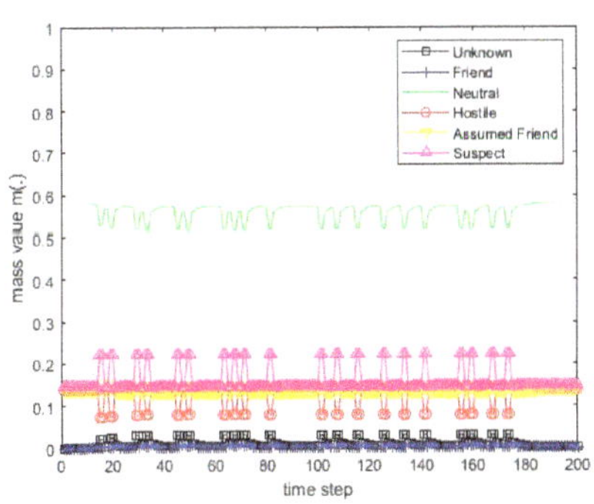

Figure 76. The values of the resulting belief mass for scenario 2 and the PCR6 rule for 3 BBAs.

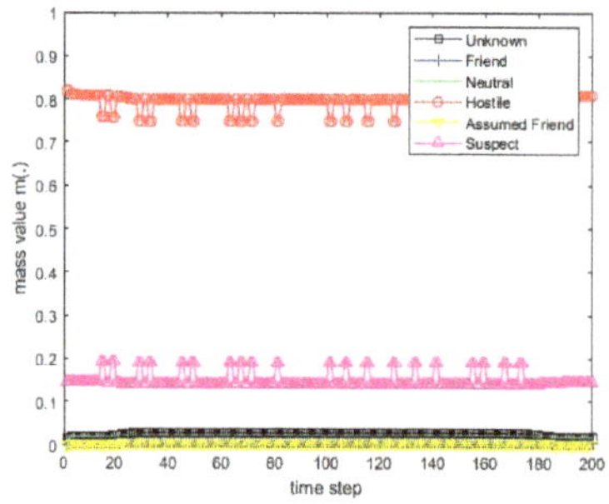

Figure 77. The values of the resulting belief mass for scenario 3 and the PCR6 rule for 3 BBAs.

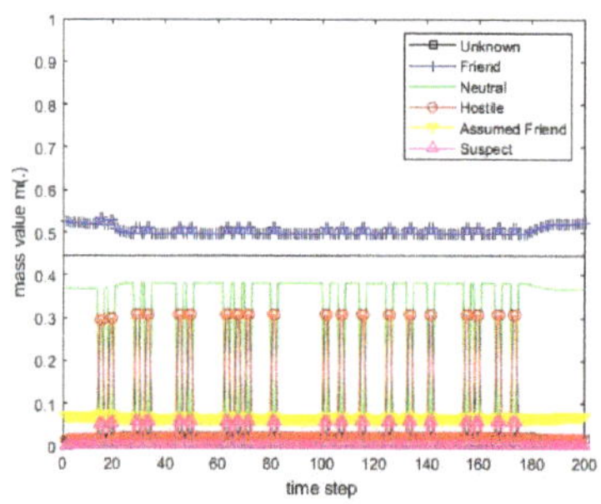

Figure 78. The values of the resulting belief mass for scenario 4 and the PCR6 rule for 3 BBAs.

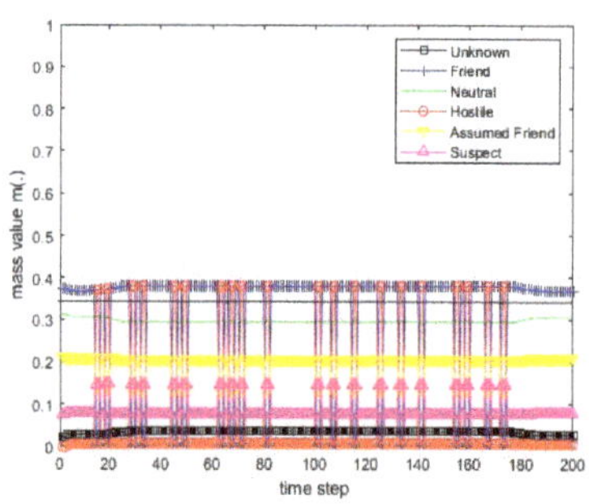

Figure 79. The values of the resulting belief mass for scenario 5 and the PCR6 rule for 3 BBAs.

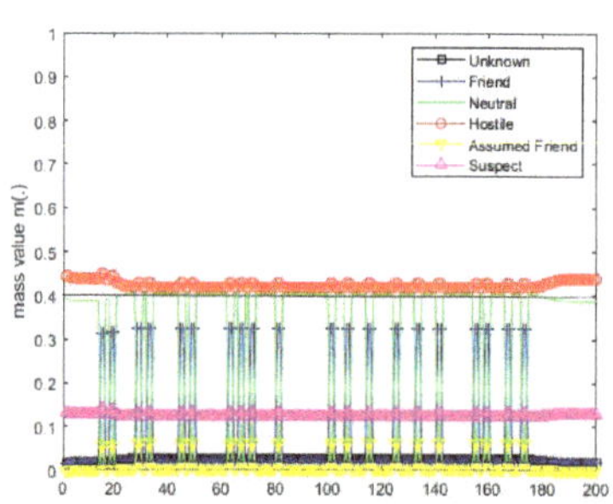

Figure 80. The values of the resulting belief mass for scenario 6 and the PCR6 rule for 3 BBAs.

For the PCR6 rule for three BBAs, the application of the decision thresholds at the belief mass level $m_\alpha = 0.45$ for scenario 4 allows for a proper evaluation of the identification of the recognized object.

For the PCR6 rule for three BBAs, the application of the decision thresholds at the belief mass level $m_\alpha = 0.35$ for scenario 5 and $m_\alpha = 0.4$ for scenario 6 allows for a proper evaluation of the identification of the recognized object for most time moments.

Comparing Figures 66–68 with Figures 72–74 and 78–80, conclusion can be drawn that the PCR5 for three BBAs and PCR6 for three BBAs rules provide more stable results of combined belief masses (smaller amplitude of changes). Due to the large dispersion of belief mass changes for scenarios 5 and 6, it is not possible to correctly evaluate the identification of the recognized object for all time moments.

The presented results (Figures 61–78) allow a conclusion to be drawn that the applied methods of removing conflicts in information fusion enables the correct conclusions to be drawn about the real identification of the recognized object.

8. Conclusions

The proposed basic belief assignment model for ESM sensors and radars can be used to build identification information-fusion systems. Models conformable to STANAG 1241 have primary practical significance.

Due to the assumption of conflicts between the ESM sensor declarations in this work, Dezert–Smarandache theory is used to determine the basic belief assignment of declarations as a product of the process of fusion of identification information sent by these sensors. Supplementing standard reports on the detected signals with random identification declarations allows the use of methods of identification information fusion in the information-fusion center. The test results confirm the full usefulness of conflict redistribution rules in reports from ESM sensors developed as a part of Dezert–Smarandache theory, with the best results presented for the PCR5 and PCR6 rule.

The basic belief assignment model for ESM sensors and for combined primary and secondary radars [17] can be applied to build models of different identification data-fusion systems. All the models compatible with STANAG 1241 have primary practical significance as it contains definitions corresponding to intersections of basic identification declarations. Therefore, the paper uses Dezert–Smarandache theory for the calculation of the basic belief assignment.

The conducted research showed that the best results were obtained for the PCR6 rule when reports from three sources (from two sensors and the fusion-system database) were processed simultaneously. This corresponds to the synchronous processing of reports and involves delayed processing of a report from one of the sources. The research confirmed a slight advantage of the PCR6 rule over the PCR5 rule. This was mainly the case when the sensors sent information with a high degree of conflict.

Funding: This work was financed by the Military University of Technology under Research Project UGB 866.

Data Availability Statement: Presented data accessible for authorised staff accordingly to local regulation.

Conflicts of Interest: The author declares no conflict of interest.

References

1. NATO. *NATO Standard Identity Description Structure for Tactical Use*, 5th ed.; Standardization Agreement STANAG No. 1241; North Atlantic Treaty Organization: Brussels, Belgium, 2005.
2. NATO. *NATO Joint Standard Identification Description*, 6th ed.; Standardization Agreement STANAG No. 1241 AO; Ratification Draft 1; North Atlantic Treaty Organization: Brussels, Belgium, 2009.
3. Smets, P. Belief functions: The disjunctive rule of combination and the generalized Bayesian theorem. *Int. J. Approx. Reason.* **1993**, *9*, 1–35. [CrossRef]
4. Smets, P.; Kennes, R. The transferable belief model. *Artif. Intell.* **1994**, *66*, 191–234. [CrossRef]

5. Smarandache, F.; Dezert, J. *Applications and Advances of DSmT for Information Fusion (Collected Works)*; American Research Press (ARP), Rehoboth: Waltham, MA, USA, 2006; Volume 2.
6. Smarandache, F.; Dezert, J. *Applications and Advances of DSmT for Information Fusion (Collected Works)*; American Research Press (ARP), Rehoboth: Waltham, MA, USA, 2004; Volume 1.
7. Pietkiewicz, T.; Kawalec, A. A method of determining the basic belief assignment for combined primary and secondary surveillance radars based on Dezert-Smarandache theory. In Proceedings of the 17th International Radar Symposium (IRS), Kraków, Poland, 10–12 May 2016; pp. 1–6. [CrossRef]
8. Yager, R.R. On the maximum entropy negation of a probability distribution. *IEEE Trans. Fuzzy Syst.* **2014**, *23*, 1899–1902. [CrossRef]
9. Yin, L.; Deng, X.; Deng, Y. The negation of a basic probability assignment. *IEEE Trans. Fuzzy Syst.* **2018**, *27*, 135–143. [CrossRef]
10. Tang, Y.; Chen, Y.; Zhou, D. Measuring Uncertainty in the Negation Evidence for Multi-Source Information Fusion. *Entropy* **2022**, *24*, 1596. [CrossRef] [PubMed]
11. Deng, Y. Uncertainty measure in evidence theory. *Sci. China Inf. Sci.* **2020**, *63*, 210201. [CrossRef]
12. Tang, Y.; Tan, S.; Zhou, D. An Improved Failure Mode and Effects Analysis Method Using Belief Jensen–Shannon Divergence and Entropy Measure in the Evidence Theory. *Arab. J. Sci. Eng.* **2023**, *48*, 7163–7176. [CrossRef]
13. Xiao, F. Gejs: A Generalized Evidential Divergence Measure for Multisource Information Fusion. *IEEE Trans. Syst. Man Cybern. Syst.* **2023**, *53*, 2246–2258. [CrossRef]
14. Xiao, F.; Cao, Z.; Lin, C.-T. A Complex Weighted Discounting Multisource Information Fusion With its Application in Pattern Classification. *IEEE Trans. Knowl. Data Eng.* **2023**, *35*, 8. [CrossRef]
15. Xiao, F.; Pedrycz, W. Negation of the Quantum Mass Function for Multisource Quantum Information Fusion With its Application to Pattern Classification. *IEEE Trans. Pattern Anal. Mach. Intell.* **2023**, *45*, 2054–2070. [CrossRef] [PubMed]
16. Xiao, F. Quantum X-entropy in generalized quantum evidence theory. *Inf. Sci.* **2023**, *643*, 119177. [CrossRef]
17. Pietkiewicz, T.; Kawalec, A.; Wajszczyk, B. Analysis of Fusion Primary Radar, Secondary Surveillance Radar (IFF) and ESM Sensor Attribute Information under Dezert-Smarandache Theory. In Proceedings of the 18th International Radar Symposium (IRS), Prague, Czech Republic, 28–30 June 2017; pp. 1–10. [CrossRef]
18. Dezert, T.; Dezert, J.; Smarandache, F. Improvement of Proportional Conflict Redistribution Rules of Combination of Basic Belief Assignments. *J. Adv. Inf. Fusion* **2021**, *16*, 48–73.
19. NATO. *Tactical Data Exchange—Link 16*, 3rd ed.; Standardization Agreement STANAG No. 5516; North Atlantic Treaty Organization: Brussels, Belgium, 2006.
20. NATO. *Identification Data Combining Process*, 2nd ed.; Standardization Agreement STANAG No. 4162; North Atlantic Treaty Organization: Brussels, Belgium, 2009.
21. Valin, P.; Bossé, E. Using A Priori Databases for Identity Estimation through Evidential Reasoning in Realistic Scenarios. In Proceedings of the RTO IST Symposium on Military Data and Information Fusion, RTO-MP-IST-040, Prague, Czech Republic, 20–22 October 2003.
22. Dezert, J.; Smarandache, F. Importance of Sources Using Repeated Fusion with the Proportional Conflict Redistribution Rules #5 and #6. 2010. hal-00471839. Available online: https://hal.science/hal-00471839 (accessed on 15 June 2023).
23. Stevens, M.C. *Secondary Surveillance Radar*; Artech House: London, UK, 1988.
24. Matuszewski, J.; Dikta, A. Emitter location errors in electronic recognition system. In Proceedings of the XI Conference on Reconnaissance and Electronic Warfare Systems, The International Society for Optical Engineering, Oltarzew, Poland, 21–23 November 2016; Volume 10418, pp. C1–C8. [CrossRef]
25. Djiknavorian, P.; Grenier, D.; Valin, P. DSm theory for fusing highly conflicting ESM reports. In Proceedings of the 12th International Conference on Information Fusion, Seattle, WA, USA, 6–9 July 2009; pp. 1211–1217.
26. Matuszewski, J.; Pietrow, D. Specific Radar Recognition Based on Characteristics of Emitted Radio Waveforms Using Convolutional Neural Networks. *Sensors* **2021**, *21*, 8237. [CrossRef] [PubMed]
27. Djiknavorian, P.; Grenier, D.; Valin, P. Analysis of information fusion combining rules under the DSm theory using ESM input. In Proceedings of the 10th International Conference on Information Fusion, FUSION 2007, Québec, QC, Canada, 9–12 July 2007.

remote sensing

MDPI

Article

Fiber Optic Acoustic Sensing to Understand and Affect the Rhythm of the Cities: Proof-of-Concept to Create Data-Driven Urban Mobility Models

Luz García [1,2,*], Sonia Mota [1,2], Manuel Titos [1,2], Carlos Martínez [1,2], Jose Carlos Segura [1,2] and Carmen Benítez [1,2]

1 Department of Signal Theory, Telematics and Communications, University of Granada, 18071 Granada, Spain; smota@ugr.es (S.M.); mmtitos@ugr.es (M.T.); carlosmc47@correo.ugr.es (C.M.); segura@ugr.es (J.C.S.); carmen@ugr.es (C.B.)

2 Research Center on Information and Communications Technology (CITIC), University of Granada, 18071 Granada, Spain

* Correspondence: luzgm@ugr.es

Abstract: In the framework of massive sensing and smart sustainable cities, this work presents an urban distributed acoustic sensing testbed in the vicinity of the School of Technology and Telecommunication Engineering of the University of Granada, Spain. After positioning the sensing technology and the state of the art of similar existing approaches, the results of the monitoring experiment are described. Details of the sensing scenario, basic types of events automatically distinguishable, initial noise removal actions and frequency and signal complexity analysis are provided. The experiment, used as a proof-of-concept, shows the enormous potential of the sensing technology to generate data-driven urban mobility models. In order to support this fact, examples of preliminary density of traffic analysis and average speed calculation for buses, cars and pedestrians in the testbed's neighborhood are exposed, together with the accidental presence of a local earthquake. Challenges, benefits and future research directions of this sensing technology are pointed out.

Keywords: distributed acoustic sensing; urban mobility patterns; optical fiber; smart cities; massive sensing

1. Introduction

The UN's Sustainable Development Goals Report for 2022 [1] includes the analysis of Goal 11 devoted to sustainable cities and communities stating that 99% of world's urban population breathe polluted air and, depending on the region of the world, few city dwellers have convenient access to public transportation. In addition, often public spaces in congested urban areas play a vital role in social and economic life, but are not widely accessible. The first step to improve actual conditions in cities is learning realistic models of their present mobility patterns usable to monitor urban settlements, implement smart traffic management tools, and create sustainable smart mobility plans.

The paradigms of smart cities [2–4] and multimodal remote sensing [5,6] provide very useful tools to obtain data transformable into knowledge, to face the challenges stated. Massive amounts of data with very diverse formats and origins are analyzed using automatic signal processing and Big Data approaches combined to understand what happens and provide directions of change and improvement. Regarding the analysis of urban traffic, approaches from the massive data retrieval and pattern extraction based on artificial intelligence tools [7,8], traffic prediction models [9,10], to digital-twin based strategies [11,12] are oriented to modify urban traffic once analyzed.

There is a wide range of sensing technologies that contribute to the monitoring of urban traffic like, e.g., unmanned aerial vehicles [13], crowd-sensing of users' mobile phones [14], traffic cameras [15], vehicles GPS [16], or satellite images [17]. The Internet of Vehicles (IoV) approach [18] provides vehicles with smart devices such as wireless sensors, onboard computers, GPS antennas, radar, etc., to collect and process large amounts of data while enabling information interaction between vehicles.

In this multi-modal urban sensing scenario, the usage of communication optical fibers as sensors to monitor mobility patterns has gained great interest. Distributed acoustic sensing [19,20] is an emergent sensing technology based on the Rayleigh scattering phenomenon occurring in an optical fiber when an interrogation light-wave faces its inhomogeneities. Depending on the fiber's refraction index, part of the incoming light-wave is backscattered towards the interrogator and can be analyzed. If a local perturbation occurs along the fiber (e.g., vibrations or changes in the fiber's strain or temperature produced by moving stimuli), its refraction index will change locally providing a proportional change in the properties of the backscattered light-wave coming from the spatial point where the perturbation occurred. This capability of demodulating the magnitude and location of the stimuli affecting the fiber, converts fibers into arrays of sensors adopting the concept of *distributed sensing* versus traditional point sensors. Vibrations and strain or temperature perturbations in the bandwidth of *acoustic* signals (up to the MHz regime) occurring along the fiber are registered.

1.1. Distributed Acoustic Sensing and Urban Traffic Monitoring Overview

Perturbations along the fiber modulate the backscattered light-wave that travels back to the interrogator. Once received, the stimuli demodulation can be performed in the time domain, receiving the name of Optical Time-Domain Reflectometry (OTDR). Conventional OTDR has been widely used to monitor static processes like fiber attenuation for fault detection in telecommunication cables. However, it is not suitable to detect local dynamic changes in the fiber refraction index, as expected in the distributed acoustic sensing. For such a purpose, several approximations based on the analysis of the phase of the backscattered light-wave have been proposed [21]. *Coherent phase-OTDR* [22,23] is based on the complete phase recovery of the interferometry signal provided by optical mixing of the backscattered and reference lights. It provides accurate dynamic measurements of strain at the cost of high system complexity (requisite of laser coherence) and contestable long-term stability. *Phase-sensitive OTDR* [24] is a simpler direct detection approach based only on intensity variations of the interferometry signal, opposite to the phase recovery needed in the coherent detection formulation. As a drawback, intensity variations of the interferometry signal do not show linear dependence with the perturbation applied. Perturbations are detected, but their quantification can only be achieved through a frequency sweep of consecutive probe pulses representing an increase of the measurement time and complexity. *Chirped-pulse phase-sensitive OTDR* (CP-ΦOTDR), mathematically formalized and demonstrated in 2016 [21,25], preserves the direct detection advantages of *Phase-sensitive OTDR*, avoiding the time-consuming frequency sweep needed. Consecutive interrogations are substituted by a single probe pulse with a linear chirp. If the chirp-induced spectral content is much larger than the pulse transform-limited bandwidth, the linear relationship between the time-domain signal and its spectrum allows for the mapping of perturbation-induced spectral shifts in the trace into local temporal trace delays. Then, the empirical mapping of trace delays and ongoing changes of its group refractive index [26] serves to quantify dynamic local perturbations along the fiber expected in distributed acoustic sensing.

The sensing possibilities of the distributed acoustic sensing (DAS) technology are used in a wide range of application fields like active seismology and vertical seismic profiles generation [27], gas or petroleum deposits detection [28], ambient noise interferometries of the Earth's surface [29], passive seismic and volcano-seismic monitoring [30–32], security and perimeters surveillance [33], or big infrastructures health monitoring [34], among others.

In the scope of urban traffic monitoring, the usage of DAS has experienced an important growth in the last years. Its longer monitoring range compared to the spatial sparseness of point sensors due to their higher costs of installation and maintenance, its higher sampling rate compared to GPS or mobile phones, and its independence of weather conditions together with its preservation of anonymity, have made it an attractive option. Table 1 shows the most recent representative approaches of DAS for monitoring moving vehicles and pedestrians. Detection, counting, measuring speed and other traffic flow parameters are common objectives of all works. Signal processing is a key challenge for several reasons: the backscattered ray-trace has low SNR and many events are spatially and temporally overlapped, there are many sources of noise present in the sensing scenarios, and the sensing capacity is very much dependent on the characteristic of the materials solidary to the fiber among others. Frequency analysis and denoising strategies are common approaches. Supervised/unsupervised machine learning approximations are being proposed in the last few years. A new approach will be introduced in our algorithm in order to improve the performance of the system.

Table 1. Traffic monitoring through DAS approaches.

Reference	Objective	Signal Processing	Sensing Scenario
patent, 2016, [35]	vehicles detection, traffic flow, speed measurements	[-]	[-]
journal, 2018, [36]	vehicle detection and counting, speed estimation	wavelet-threshold denoising and dual threshold detection.	200 m. road in the NanShan Iron mine (China) during seismic trial
congress, 2019, [37]	average speed, flow rate, queue detection, congestion detection, journey times, traffic count	[-]	[-]
journal, 2020, [38]	signatures of floats, bands, motorcycles	detrending, filtering, noise removal, frequency analysis	2.5 km of fiber underneath the Rose Parade route, Pasadena(USA)
congress, 2020, [39]	detect pedestrian footstep	convolutional neural network	5km Pennsylvania State University campus
journal, 2020, [40]	vehicle detection and classification, vehicle count, speed measurement	wavelet denoising, dual-threshold detection, feature extraction, vehicle classification with SVM	320 m. campus road of Beijing Jiaotong University (China)
journal, 2020, [41]	vehicle detection, counting and characterization	frequency analysis, template matching	4 km. Telecom. cable running through Palo Alto, CA, leased from Stanford University IT Services (USA)
journal, 2020, [42]	human locomotion detection (walking, running, different shoes)	frequency analysis, shallow and deep Neural Networks	15-m-long hallway.
journal, 2021, [43]	vehicle counting, traffic volume, average speed	detrending, filtering, noise removal, frequency analysis	37 km. Caltech-Pasadena City DAS array (USA).
conference, 2021, [44]	estimation of individual simultaneous vehicles velocity in multiple lane roads	frequency domain MUSIC beamforming	commercial telecom. cable parallel to a main road in Toulon(France).
journal, 2022, [45]	speed and volume estimate of traffic flow	frequency analysis, F-K filtering for noise removal	50 km. of telecom. cable inside the city of Hangzhou (China).
journal, 2022, [46]	counting and velocity estimation for individual vehicles in challenging scenarios without spatial/temporal separation	self-supervised deconvolution autoencoder	14 km. commercial telecomm. along a main road connecting Alba-la-Romaine, Saint-Thomé, and Valvignères (France).

1.2. Contributions of This Work

In the general technology framework presented, this work describes a distributed acoustic sensing experiment deployed in the vicinity of the School of Techonology and Telecommunication Engineering of the University of Granada, Spain. For several months the mobility activity around the building has been recorded to explore the capacities of DAS to extract urban mobility patterns. The contributions we present and the rest of the work are organized as follows:

i. An implementation of the DAS technology in an urban environment with a wide variety of dynamic mobility patterns is presented. Section 2.1 describes the testbed used.
ii. The signal processing needed, the different types of mobile elements sensed and feature extraction possibilities are exposed in Sections 2.2–2.4, respectively.
iii. Example applications derived from the processing of information obtained are shown in Section 3 followed by a reflection about this sensing approach and its possibilities and applications in Section 4.

2. Materials and Methods

The experiment described in this section lasted from the September 2022 until the 20 January 2023. Its main objective has been exploring the technology and obtaining preliminary strategy conclusions applicable to further sensing campaigns.

2.1. Testbed Description and Calibration Process

A dark fiber double-loop was buried for the specific sensing objective around the School of Technology and Telecommunication Engineering of the University of Granada, Spain (ETSIIT). A High-Fidelity Distributed Acoustic Sensor based on the CP-ΦOTDR technology ([21]) manufactured by the Spanish company Aragón Photonics^{TM} has been used. The sensor has a 1 n strain sensitivity, 6 m minimum spatial resolution (gauge length) and up to 70 km reach. The setup provides strain-type data on near a kilometer of fiber, with 10 m spatial sampling and 250 Hz temporal sampling.

Figure 1 shows the triangle-shaped outer fiber loop comprising 2 streets of 140 and 170 m of length (red and blue double arrows), a concrete wall of 140 m of length (yellow double arrow), and an internal loop of 220 m (green oval arrow) surrounding a garden and two prefabricated lecture rooms. The fiber is always buried except for concrete wall section on which the fiber is uncovered, solidary to the wall for research purposes. Sampling points (P1-P32 and M1-M20) are depicted as a result of a calibration process carried our before monitoring activities. The sensor registers strain variations in the fiber resulting from stimuli like pedestrians, public or private buses, cars, bicycles, etc. These data are pre-processed for noise removal (see Section 2.2). Monitored strain registers can be directly processed or converted into 2D energy maps, commonly known as energy waterfalls, used as input to potential automatic labeling or classification systems.

Figure 2 shows the energy waterfall corresponding to the sensing circuit depicted in Figure 1 for 25 min. The X-axis represents time, while the Y-axis represents the spatial points of the extended double loop of the fiber. The color scale represents the strain's energy on each spatial point along time. The colored double arrows on the sides of the waterfall indicate the portions of the school perimeter corresponding to each part of the Y-axis in the waterfall. It is notable that the internal loop sensing the *inside garden* suffers a kind of mechanical superconductivity. High energy appears simultaneously on many spatial points, connected to the existence of a mobile event in other region of the waterfall. This might be due to the existence of a deep concrete platform on top of which the garden and prefabricated lecture rooms were located. This simultaneous energy transmission becomes an specially challenging overlapping noise for sensing points M1-M20 . Energy footprints related to events occurring in the inside garden are overlapped with useless mechanical conduction footprints related to activity in other areas. The criterion used to distinguish real mobility activity in the area from mechanical energy superconductivity is that while the first one will present a certain small slope (space will be gone through in a certain time), the former one will occur simultaneously in spatial positions separated from each other (that is, footprints would have somewhat *infinite slope*). Automatic event detection approaches applied to the registers for counting applications (see Section 3) will face this difficulty with the help of template matching strategies that favor real plausible slope values.

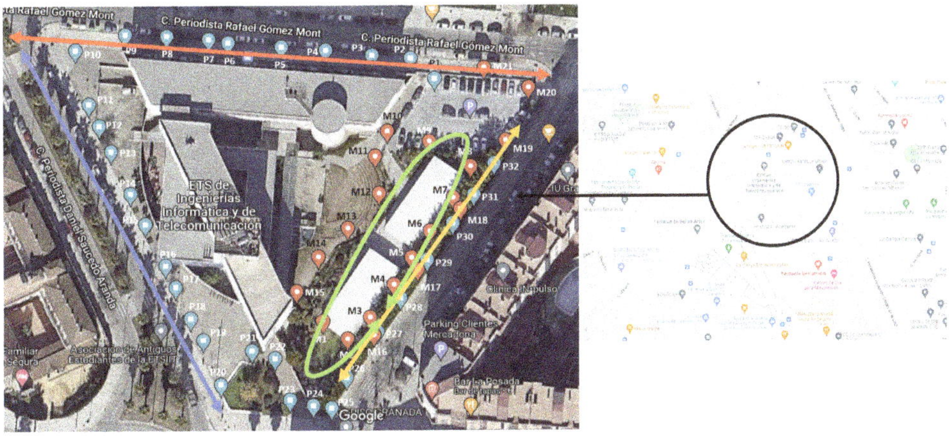

Figure 1. Google Map^{TM} view of sensing testbed installed in the Telecommunication and Computer Science Engineering School of the University of Granada, Spain. Sensing points calibrated in the fiber with spacial resolution of 10 m are depicted. Red markpoints correspond to the internal fiber ring, while blue markpoints correspond to the external fiber ring. Four sensing areas are differentiated: Periodista Rafael Gómez Mont street (red arrows), Periodista Daniel Saucedo Aranda street (blue arrows), internal gardens of the School (green ellipsoid) and concrete wall in a side of the School perimeter (yellow arrows). Sensing points P1 and P10 are the respective entrances/exits of a surface and underground parking.

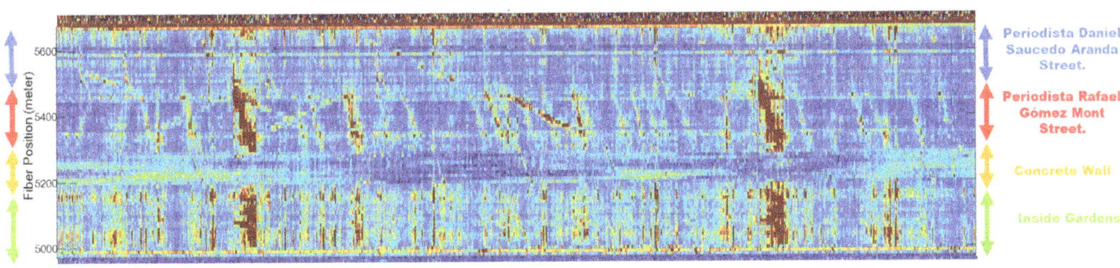

Figure 2. Example of energy waterfall of 25 min for the fiber deployment described in Figure 1. All sensing points are depicted in the Y-axis, with the side color arrows indicating the spatial area corresponding to each segment of the Y-axis.

2.2. Signal Processing

DAS technology has several sources of noise due to optical noises and ground to fiber transfer effects, dependent on the fiber and characteristics and coupling [47]. In addition, backscattered traces are low power signals. For these reasons, denoising approaches are important to achieve quality SNRs. Added to these challenges, the occurrence of time and space overlapped stimuli is other source of noise that can mask the mobile events searched for. Our work presents a preliminary common denoising strategy devoted to the acquisition of a baseline database of mobility patterns usable in further applications. Approaches based on machine learning like [48] are considered for future implementations. Figure 3 shows the four denoising steps followed proposed by the sensor manufacturer

Aragon Photonics: signal thresholding is carried out to compensate for spurious strain peaks based on the study of cumulative values. Then, signal variations are compared to those of a reference portion of the fiber without stimuli. For this purpose, the output of two consecutive median and mean filters applied to the reference strain is subtracted, obtaining the strain variation $\Delta\varepsilon$ signal used in the analysis. Finally, using an iterative process on both time and space dimension, temporal and spatial discontinuities are smoothed, again making use of a median filter.

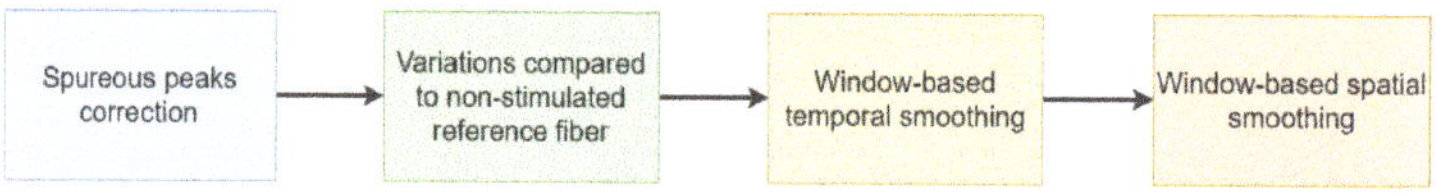

Figure 3. Steps for baseline noise reduction in strain registers.

Once the denoising step has been completed, a frequency analysis is performed. When a moving event approaches a given sensing point, there are two simultaneous effects taking place [43,46]. First there is a low-frequency (<3 Hz.) quasi-static deformation of the subsurface due its weight pressing down on the road/ground. Such deformation is transferred to the fiber leading to a strain of measurable amplitude, traveling at the speed of the mobile event and more easily localized in time. Secondly, the interaction between the vehicle tires/pedestrian and the road/ground generates high frequency (>3 Hz.) surface waves that travel away from the source point at seismic speeds usable in interferometry analysis. We have performed the reported two bands analysis during the experiments. Its results will be shown in Sections 2.3 and 2.4.

2.3. *Types of Events Registered*

The School of Technology and Telecommunication Engineering is located in the north-west of the city of Granada, relatively close a communication hub connecting the inner city to a several of highways around it. It is inserted in the middle of a neighborhood with buildings of homes. Public urban bus n° 9 goes through street Periodista Rafael Gómez Montero (see Figure 1). Mobility patterns of workers and students relating to the School have been continuously registered during the experiment together with those of the people living in the area or traveling through it. There are footprints of different types of vehicles interacting among them or with pedestrians often also monitored entering or exiting bus n° 9. Under a first approach, we have distinguished three basic types of events: buses, cars and pedestrians, with the objective of performing automatic detection and counting and creating a master database with labeled examples. Such a baseline database will permit further machine learning probabilistic approaches to find data classifiable as similar or different types of events, mixtures of them, out-of-distribution events, etc. Figure 4 shows three example representations of the footprint registered for a bus (Figure 4a), a car (Figure 4b) and a pedestrian (Figure 4c) moving parallel to the fiber in the testbed deployed. Their waterfalls, corresponding strain variation matrices along time and space and the spatially-averaged frequency spectrograms for the same footprints are depicted. Figures show that buses have higher energy due their higher weight producing higher strain variations. A basic speed calculation based on the slope of the footprint (space divided by time) shows that, as expected, the bus and the car have higher speeds than the pedestrian. Spatial-average power spectral densities suggest different frequency contents for the three types events that are further analyzed.

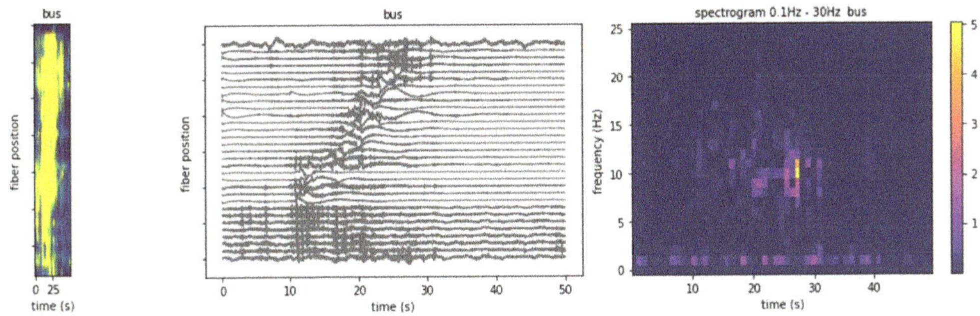

(a)

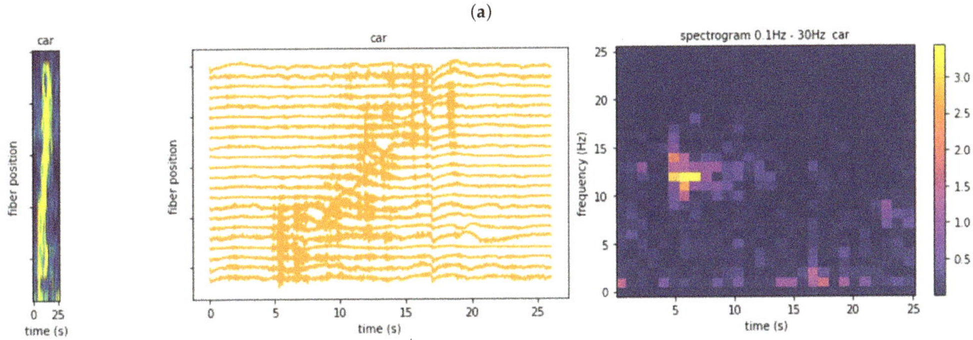

(b)

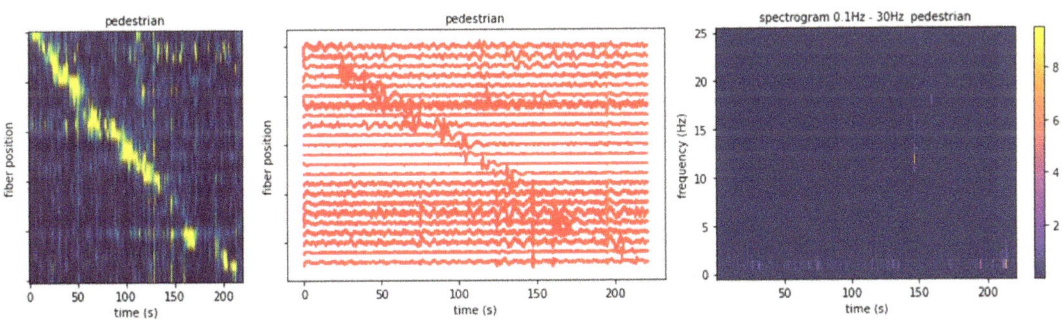

(c)

Figure 4. Example visualizations of the canonic events detected in the monitoring testbed (bus, car and pedestrian). (**a**) Energy waterfall, strain variation and spatial-average power spectral density for a canonic *bus* example. (**b**) Energy waterfall, strain variation and spatial-average power spectral density for a canonic *car* example. (**c**) Energy waterfall, strain variation and spatial-average power spectral density for a canonic *pedestrian* example.

Figure 5 depicts the distribution of the frequencies with maximum energy for a small database of buses, cars and pedestrians registered in the testbed. The analysis is performed for the whole band of frequencies involved in the activity (from 0.1 Hz to 30 Hz) in the left column subfigure, for the quasi-static band of activity (band 0.1–5 Hz) in the central subfigure, and the high-frequency band (5 to 20 Hz) originated by the surface waves. Buses show a higher content of frequencies around 10 Hz that are not that present in cars nor pedestrians which generate very little surface waves. The low frequency band (center subfigure) often used because its simpler analysis and time location, might not be the optimal band when distinguishing different types of events. Analyzing the whole band provides more discriminative differences between events at the price of introducing some noise.

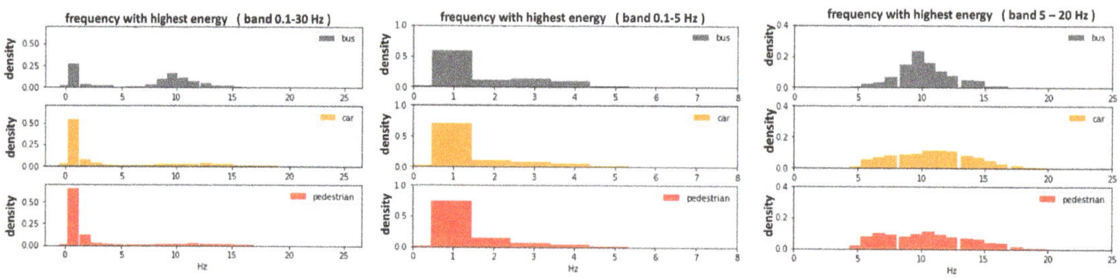

Figure 5. Histograms of the frequency with highest energy for the three baseline events analyzed. Three frequency bands are studied: complete band from 0.1 Hz to 30 Hz (**left**), quasi-static band from 0.1 to 5 Hz (**center**) and high frequency band 5Hz to 20 Hz (**right**).

2.4. Characterization of the Events

There are several approximations to study the strain-variation time series registered. A possible analysis might include feature extraction, combination and measurement of their discriminative potential, and their contribution to the interpretability of the data. Otherwise, in the framework of Information Theory, many approaches focus on the complexity of the time series searching for information contents from a mathematical viewpoint without semantic analysis. In this framework, complexity is a magnitude widely used to quantify the intricacy of a time series allowing choice of the forecasting methods to be applied [49]. The higher the complexity, the more information provided by the time series. That is, complexity is low in regular time series and grows in chaotic ones. There are several methods to measure complexity, Shannon Entropy [50] being a very commonly used one. In recent literature, several other measures have been developed to quantify the changes in complexity for biological signals [51] like electroencephalograms (EEG) [52], electrocardiograms (ECG) [53,54] or magnetoencephalograms (MEG) [55]. Biological time series of a healthy person are more regular than those of a diseased person that become more complex. The same approximation is used in the fault diagnosis in machinery [56] or in financial time series analysis [57].

In the context of our proposal and due the nature of the signals, events generating strain variations in the fiber's backscattered light (cars, buses or pedestrians passing by) will produce changes in the complexity measures. Based on this hypothesis, approximate entropy [58] (see Figure 6) and Hjorth parameters [59] (see Figures 7 and 8) are analyzed by searching for their potential for mobile events discrimination and characterization. The well-known Hjorth parameters of activity, mobility and complexity, transversely used in all mentioned disciplines, are added to amplify the statistical information in the analysis.

Approximate entropy is calculated in the time domain. It measures the matches of a pattern along the signal, calculating then the logarithmic frequency of repeatable patterns. Time series containing many repetitive patterns have relatively small approximate entropy values (the time series is more regular), while more chaotic or complex processes show higher values. Hjorth parameters, although calculated in the time domain, also provide meaning in the frequency domain. Activity gives a measure of the squared standard deviation of the amplitude of the signal, being high if higher frequencies are present; mobility is obtained as the square root of variance of the first derivative of the signal divided by its variance. Complexity, defined as the ratio between the mobility of the first derivative and the mobility of the signal, indicates how the shape of a signal is similar to a pure sine wave providing an estimation of its bandwidth. Adapting window sizes to frequency bands and possible range of events duration, complexity measures have been analyzed for two frequency bands (0.1–2 Hz and 5–20 Hz) following the hypothesis of the different activity and events discriminability pointed out in Section 2.2. The result of the analysis is shown for the same strain variation segment in Figures 6–8. Several events detected have been indicated with different color arrows, with gray, orange and red indicating the corresponding presence of a bus, a car or a pedestrian.

Results show the interesting potential of approximate entropy and Hjorth activity to highlight the presence of a mobile event, removing noise in the strain variation matrices to perform more accurate event detection. Exact event timing, important for applications like event's velocity calculation, can be improved through these parameters. Hjorth's mobility and complexity show a certain presence especially in the band 0.1–2 Hz that is under analysis for a better usage.

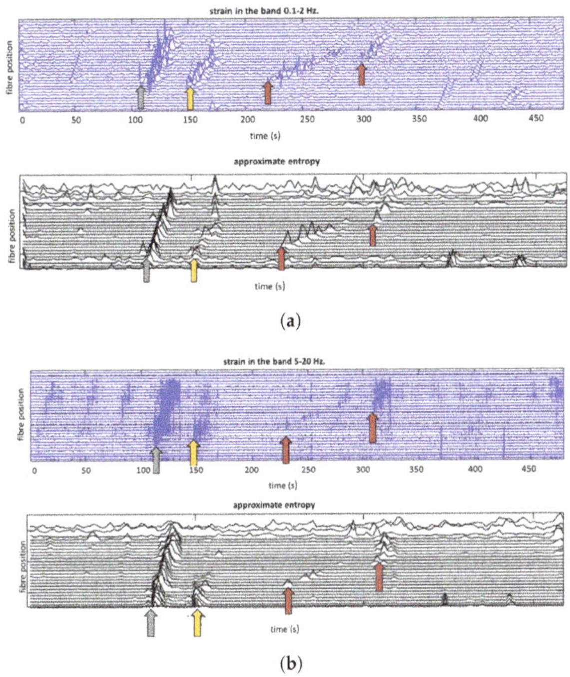

Figure 6. Events detected in a segment of DAS register pointed at with red, yellow and gray arrows indicating the presence of pedestrian, car or bus, respectively. (**a**) shows strain variations and approximate entropy in the band 0.1–2 Hz. (**b**) shows strain variations and approximate entropy in the band from 5–20 Hz.

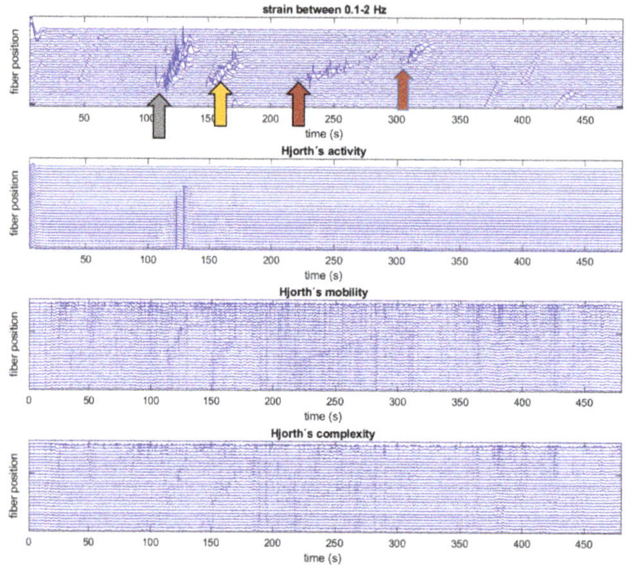

Figure 7. Events detected in a segment of DAS register pointed at with red, yellow and gray arrows indicating the presence of pedestrian, car or bus, respectively. Strain-variation file segment processed in the band 0.1–2 Hz. Corresponding Hjorth parameters.

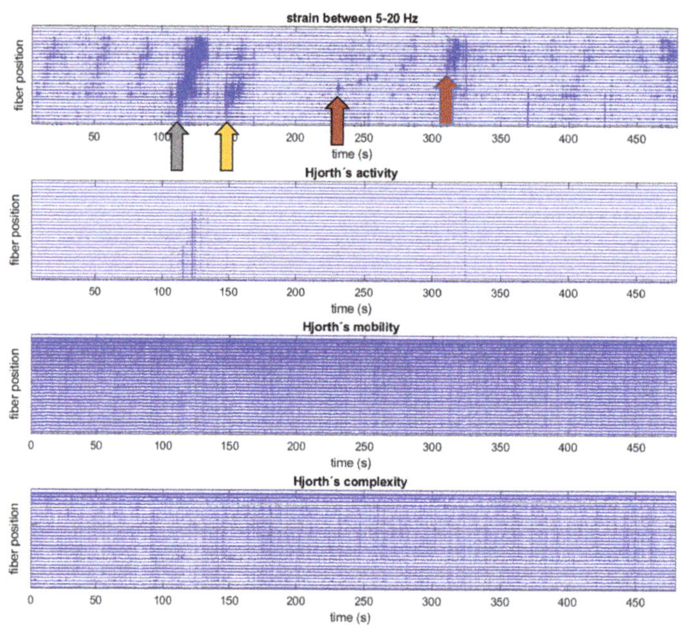

Figure 8. Events detected in a segment of DAS register pointed at with red, yellow and gray arrows indicating the presence of pedestrian, car or bus, respectively. Strain-variation file segment processed in the band 5–20 Hz. Corresponding Hjorth parameters.

3. Results

This section points out potential applications of the DAS monitoring to extract relevant information for data-driven mobility models. Its objective is to show the flavors of what can be accomplished with a deeper analysis of the data obtained. Data monitored in the period from December 2022 to January 2023 have been analyzed and used as example.

3.1. Example of Mobility Changes on New Year's Eve

Continuous monitoring was carried during the evening and night of the 31st of December on New Years eve. Figure A1 in Appendix A shows four example waterfalls of one hour of duration at different times (31 December at 4:00 pm, 9:00 pm and 11:00 pm, and 1 January at 00:00 am). It can be seen that different traffic densities are observed at different times of the day. The last two subfigures show anthropologically interesting information about human behaviors on New Year's Eve. Urban traffic is especially low from the 31 December at 11:00 pm until approximately 1 January at 00:30. Then, many cars start moving during the whole night. This information is highly compatible with the Spanish tradition of welcoming the new year inside homes with family or friends (eating 12 grapes together at exactly 1 January at 00:00) and going out to celebrate afterwards. Figure 9 provides the automatic counting of buses, cars and pedestrians during the mentioned 24 h. The counting has been performed using an image processing multiple template-matching approach over the waterfall images [60]. It is remarkable that the number of cars in the one hour gap starting at 00:58 am is higher than any other time of the day.

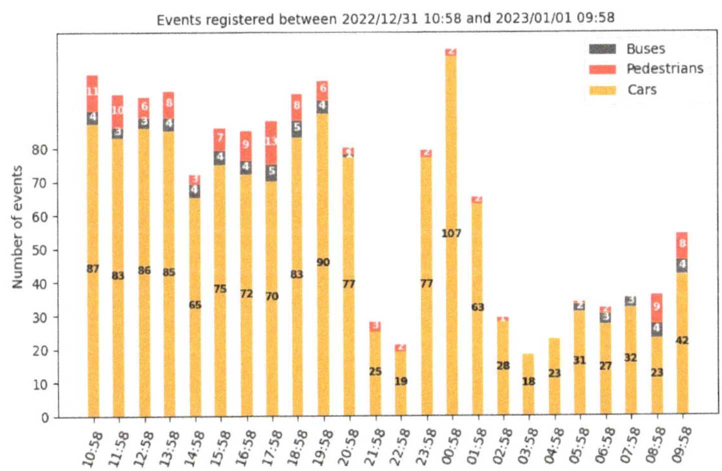

Figure 9. Automatic counting of the number of cars, pedestrians and buses carried out during the monitoring example.

3.2. Example of Mobility during a Work Day

Figure 10 provides the same automatic counting of mobile events performed during a work day (Figure 10, left). Differences compared to the patterns found in New Year's Eve (Section 3.1) are very clear. Traffic peaks are detected from 9:03 to 10:03 am and in the afternoon/evening when the density of pedestrians is also higher. It is remarkable the lower amount of buses and cars in the interval 14:00–15:00. The right subfigure depicts the time interval between buses registered during the same day. Discarding the sporadic presence of private buses that travel through Rafael Gomez Montero street, the figure mainly measures the frequency of public bus n° 9 that commutes this neighborhood to the center of Granada. The approximately constant rhythm of the bus is notable, with slightly higher intervals between buses in the hours with higher traffic density. A deeper analysis could be carried out, correlating these results with the traffic jam hours in other parts of the city. Figure 11 shows a preliminary speed analysis for the three types of events during the monitoring period. Average speeds with their standard deviations are plotted together with the number of events averaged. Speeds were calculated based on the waterfall event detection approach. Further improvements for more exact calculations based on characterization parameters described in Section 2.4 are under analysis.

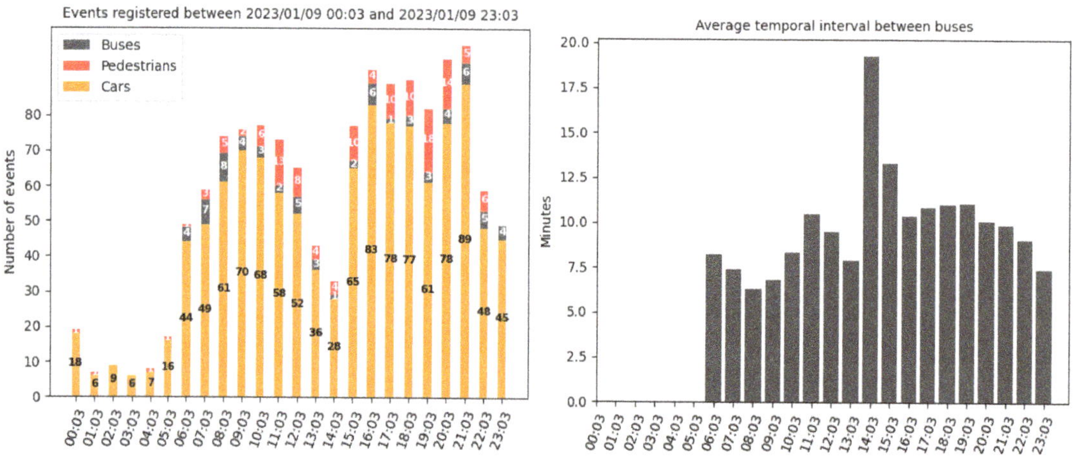

Figure 10. Example hours of traffic during the monitoring example carried out January 9th 2023 from 09/01/2023 00:03:08 to 09/01/2023 23:03. Automatic counting of buses, pedestians and cars (**left** subfigure). Average time interval between buses in minutes (**right** subfigure).

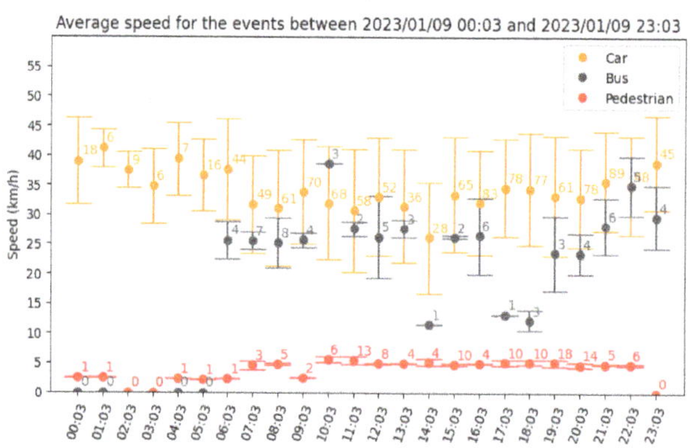

Figure 11. Average speed for the 3 types of events detected during the workday monitoring period.

3.3. *Monitoring Access to the Schools's Surface Parking*

The left-side subfigure in Figure 12 shows the amplified detail of the School of Engineering surface parking depicted in Figure 1. The right-side subfigure shows the strain variation registered at the fiber sensing positions P3, P2, P1, M20, M21, M19, M9 and M8, monitoring the parking and its entrance. The global activation of all sensing points at approximately 460 seconds is due to the presence of an urban bus passing by. Its high weight produces mechanical vibrations monitored by all the sensors under analysis. Entering the parking can only be occur following one of the two routes painted on blue in the left side of Figure 12, and vehicles leaving the parking may only follow the directions marked in red. Strain variations due to the presence of entering or exiting vehicles will be activated at fiber positions M10, M9, and M8 if the vehicle enters the parking. If the vehicle leaves the parking towards the left, fiber positions P1, P2 and P3 will be sequentially activated. That is what can be seen in the left-side subfigure during the seconds 100, 200 and 300, marked with red arrows.

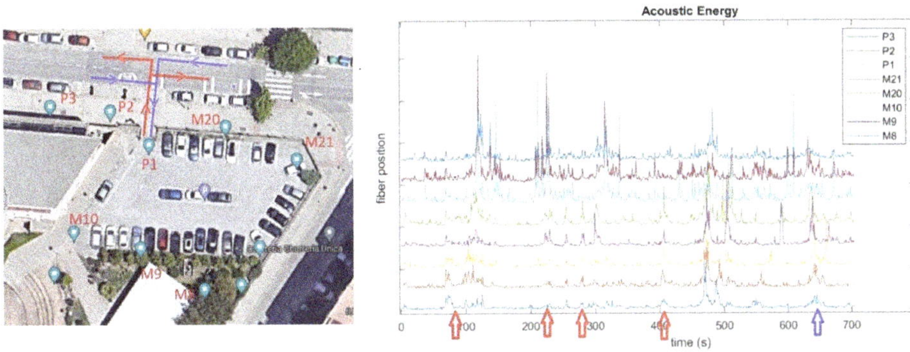

Figure 12. On the **left**, map and fiber position of outdoor parking; and on the **right**, acoustic energy detected at the fiber positions in the map along 600 s: vehicles entering (red arrows) and leaving (blue arrow) the park lot.

Another vehicle leaves the parking around second 400 (see red arrow marked), being fiber positions P1, P2 and P3 inactive. It therefore can be concluded that the vehicle moves towards the right.

Finally, a vehicle entering the parking can be detected at second 650 (marked by a blue arrow). Positions P3, P2, and P1 are sequentially activated, and then positions inside the car park, following the sequence M10, M9, and M8.

3.4. Urban Seismicity Monitoring

During the New Year's Eve monitoring experiment, Figure A2 shows the energy footprint of a local earthquake with an epicenter in the region of Almería (with a distance of around 100 km to the testbed) registered the 31 December 2023 at 08:05:54 am local time, with depth = 0 km and magnitude 4 Mw [61]. Figure 13 shows how simultaneous and energetic strain variations are present in all spatial points with different magnitudes depending on the transmission properties of each ground portion. The concrete wall located approximately in the middle of the waterfall (see Figure 2) cannot register the earthquake being the wall somehow *unlinked* from the Earth's movement. Figure 14 depicts the spatial-average frequency spectrogram during the earthquake, showing the well-known P-wave first arrival with higher frequency contents generated by a fracture source mechanism, followed by an S-wave with lower frequency contents extended longer in time with energy exponential decay [62].

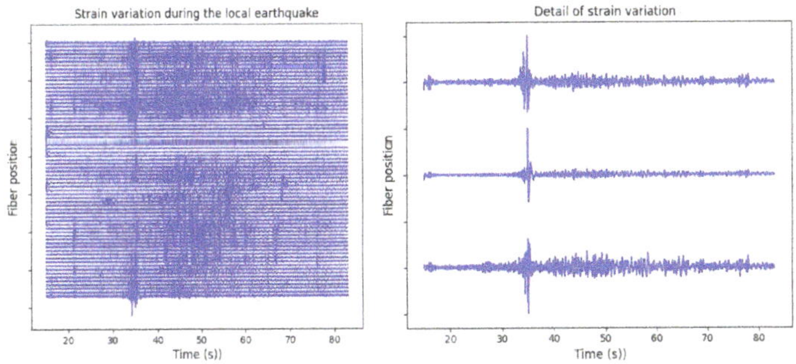

Figure 13. Strain variation of the earthquake registered the 31 December 2023.

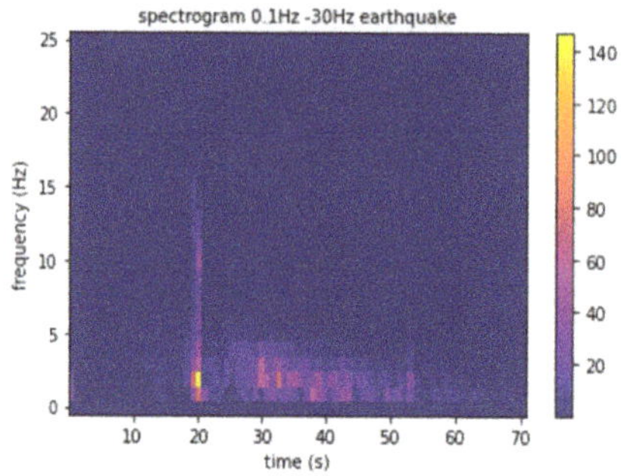

Figure 14. Spatial-average power spectral density for the earthquake observed.

Time-domain analysis of seismic P and S waves using a classic multi-component point geophone would provide separate vertical and horizontal components related to P and S phases, permitting polarization and shear waves analysis. Due to the single-component nature of the DAS array and its measure of strain-rate rather than particle motion or acceleration, it produces a single measurement of the changes in the fiber's group refractive index originated by the projection of the three components along the fiber. Given its interest for the geophysical community, several approaches are under analysis at the moment to overcome this limitation like the usage of helically wound fibers to measure strains in three directions [63], usage of azimuthally varying 2D arrays for horizontal components sensing [64] and machine learning complementary analysis [65].

4. Discussion

The work presents an experimental testbed for distributed acoustic sensing in urban environments, devoted to the analysis of the mobility patterns in the surroundings of the School of Technology and Telecommunication Engineering of the University of Granada. Strain variations registered by the sensor are processed for noise reduction and filtered in convenient frequency bands, identifying three basic types of events (cars, buses and pedestrians) to initiate a preliminary automatic counting process. Hjorth parameters and approximate entropy are explored as possible processing approaches to improve automatic events detection and classification based on template matching. Several example applications of the technology are shown. Time dependent density of traffic, intervals of public bus arrivals, speed of pedestrian vehicles split into classes (to start with high/low weight vehicles) are monitorable without interruption anywhere in the city having an optical fiber installed. In addition, urban seismicity is also recordable with the subsequent interest for urban locations with risk of seismic hazards. The benefits of having data-driven mobility pattern models are many. Green urban planning strategies, sustainable development plans, smart traffic managing applications or emergency evacuation plans, among others, can be designed based on the knowledge provided by them.

Compared to other sensing technologies, the anonymity of the data, independence of weather conditions, no need of maintenance or power supply for point sensors, or long range and high spatial sampling frequency are remarkable advantages. The challenges of distributed acoustic sensing are several, opening an interesting research framework for future works. Strain variations have often low SNR and are dependent on the specific and changing ground and fiber properties. Robust calibration and advanced noise removing approaches are needed. The automatic detection and classification of events that are often overlapped and merged offer the possibility to explore automatic unsupervised and supervised approaches based on state-of-the-art machine learning strategies.

Author Contributions: Conceptualization, L.G., C.B., S.M. and J.C.S.; methodology, L.G., C.B., S.M. and M.T.; software, S.M., C.M., M.T.; formal analysis, L.G., C.B., S.M. and J.C.S.; investigation, L.G., C.B., S.M., M.T.; resources, L.G., J.C.S. and C.B.; writing—original draft preparation, L.G., B.C. and M.S.; writing—review and editing, L.G., S.M., T.M., C.M., S.J.C. and B.C.; funding acquisition, L.G. and C.B. All authors have read and agreed to the published version of the manuscript.

Funding: This research was funded by Grant B-TIC-542-UGR20 funded by "Consejería de Universidad, Investigación e Innovacción de la Junta de Andalucía" and by "ERDF A way of making Europe".

Data Availability Statement: Given the descriptive nature of this work, no data have been generated.

Acknowledgments: We want to thank Aragon Photonics for his technical support. We also want to thank the support and helpful collaboration of the School of Technology and Telecommunication Engineering of the University of Granada, Spain, during the whole installation and experiment.

Conflicts of Interest: The authors declare no conflicts of interest.

Appendix A

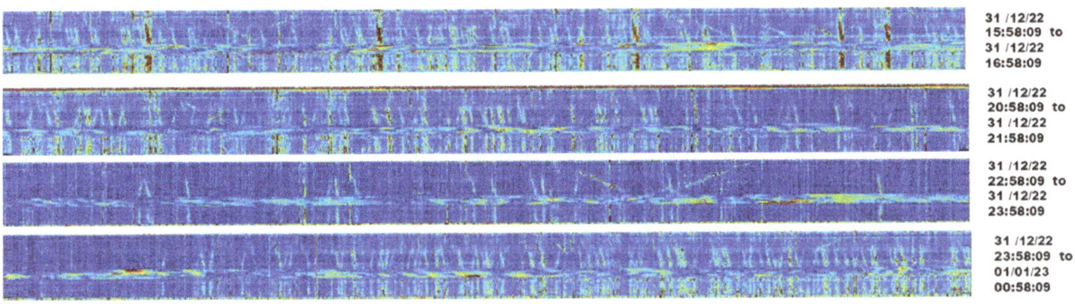

Figure A1. Example hours of traffic during the monitoring example carried out 21 December 2022 from 31 December 2022 10:58:08 to 1 January 2023 9:58:13.

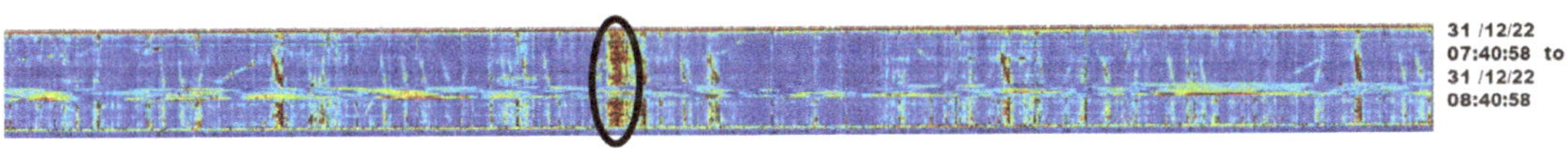

Figure A2. Example of local earthquake with magnitude 4 Mk with epicenter in Almería, registered by the sensor the 31 December 2022.

References

1. United Nations. *The Sustainable Development Goals Report 2022*; United Nations Publications: New York, NY, USA, 2022.
2. Blasi, S.; Ganzaroli, A.; De Noni, I. Smartening sustainable development in cities: Strengthening the linkage between smart cities and SDGs. *Sustain. Cities Soc.* **2022**, *80*, 103793. [CrossRef]
3. Biyik, C.; Abaresho, A.; Paz, A.; Ruiz, R.A.; Battarra, R.; Rogers, C.D.F.; Lizarraga, C. Smart Mobility Adoption: A Review of the Literature. *J. Open Innov. Technol. Mark. Complex.* **2021**, *7*, 146. [CrossRef]
4. Savithramma, R.M.; Ashwini, B.P.; Sumathi, R. Smart Mobility Implementation in Smart Cities: A Comprehensive Review on State-of-art Technologies. In Proceedings of the 4th IEEE International Conference on Smart Systems and Inventive Technology (ICSSIT), Tirunelveli, India, 20–22 January 2022.
5. Runyu, F.; Jun, L.; Weijing, S.; Wei, H.; Jning, Y.; Lizhe, W. Urban informal settlements classification via a transformer-based spatial-temporal fusion network using multimodal remote sensing andtime-series human activity data. *Int. J. Appl. Earth Obs. Geoinf.* **2022**, *111*, 102831.
6. Ahyun, L.; Kang-Woo, L.; Kyong-Ho, K.; Sung-Woong, S. A Geospatial Platform to Manage Large-Scale Individual Mobility for an Urban Digital Twin Platform. *Remote Sens.* **2022**, *14*, 723.
7. Loder, A.; Ambühl, L.; Menendez, M.; Axhausen, K.W. Understanding traffic capacity of urban networks. *Sci. Rep.* **2019**, *9*, 16283. [CrossRef] [PubMed]

8. Serok, N.; Havlin, S.; Blumenfeld Lieberthal, E. Identification, cost evaluation, and prioritization of urban traffic congestions and their origin. *Sci. Rep.* **2022**, *12*, 13026. [CrossRef]
9. Huang, A.J.; Agarwal, S. Physics-Informed Deep Learning for Traffic State Estimation: Illustrations With LWR and CTM Models. *IEEE Open J. Intell. Transp. Syst.* **2022**, *3*, 503–518. [CrossRef]
10. Medina-Salgado, B.; Sánchez-DelaCruz, E.; Pozos-Parra, P.; Sierra, J.E. Urban traffic flow prediction techniques: A review. *Sustain. Comput. Inform. Syst.* **2022**, *35*, 100739. [CrossRef]
11. Jafari, S.; Shahbazi, Z. Designing the Controller-Based Urban Traffic Evaluation and Prediction Using Model Predictive Approach. *Appl. Sci.* **2022**, *12*, 1992. [CrossRef]
12. Wu, J.; Wang, X.; Dang, Y.; Zhihan, L. Digital twins and artificial intelligence in transportation infrastructure: Classification, application, and future research directions. *Comput. Electr. Eng.* **2022**, *101*, 107983. [CrossRef]
13. Butila, E.V.; Boboc, R.G. Urban Traffic Monitoring and Analysis Using Unmanned Aerial Vehicles (UAVs): A Systematic Literature Review. *Remote Sens.* **2022**, *14*, 620. [CrossRef]
14. Liu, Z.; Jiang, S.; Zhou, P.; Li, M. A Participatory Urban Traffic Monitoring System: The Power of Bus Riders. *IEEE Trans. Intell. Transp. Syst.* **2017**, *18*, 2851–2864. [CrossRef]
15. Fredianelli, L.; Carpita, S.; Bernardini, M.; Del Pizzo, L.G.; Brocchi, F.; Bianco, F.; Licitra, G. Traffic Flow Detection Using Camera Images and Machine Learning Methods in ITS for Noise Map and Action Plan Optimization. *Sensors* **2022**, *22*, 1929. [CrossRef] [PubMed]
16. Duan, Z.; Yang, Y.; Zhang, K.; Ni,Y.; Bajgain,S. Improved Deep Hybrid Networks for Urban Traffic Flow Prediction Using Trajectory Data. *IEEE Access* **2018**, *6*, 31820–31827. [CrossRef]
17. Chen, Y.; Qin, R.; Zhang, G.; Albanwan, H. Spatial Temporal Analysis of Traffic Patterns during the COVID-19 Epidemic by Vehicle Detection Using Planet Remote-Sensing Satellite Images. *Remote Sens.* **2021**, *13*, 208. [CrossRef]
18. Ji,B.; Zhang, X.; Mumtaz, S.; Han, C.; Li, C.; Wen, H.; Wang, D. Survey on the Internet of Vehicles: Network Architectures and Applications. *IEEE Commun. Stand. Mag.* **2020**, *4*, 34–41. [CrossRef]
19. Bao, X; Chen, L. Recent progress in Distributed Optic Sensors. *Sensors* **2012**, *12*, 8602–8639. [CrossRef]
20. Lu, P.; Lalam, N.; Badar, M.; Liu, B.; Chorpening, B.; Buric, M.; Ohodnicki, P.R. Distributed optical fibre sensing: Review and perspective. *Appl. Phys. Rev.* **2019**, *6*, 041302. [CrossRef]
21. Pastor-Graekks, J.; Martins, H.F.; Garcia-Ruiz, A.; Martin-Lopez, S.; Gonzalez-Herraez, M.R. Single-shot distributed temperature and strain tracking using direct detection phase-sensitive OTDR with chirped pulses. *Opt. Express* **2016**, *24*, 13122.
22. Tu, G.; Zhang, X.; Zhang, Y.; Zhu, F.; Xia, L.; Nakarmi, B. The development of an Phi-OTDR system for quantitative vibration measurement. *IEEE Photonics Technol. Lett.* **2015**, *27*, 1349–1352. [CrossRef]
23. Wang, Z.; Zhang, L.; Wang, S.; Xue, N.; Peng, F.; Fan, M.; Sun, W.; Qian, X.; Rao, J.; Rao, Y. Coherent Φ-OTDR based on I/Q demodulation and homodyne detection. *Opt. Express* **2016**, *24*, 853–858. [CrossRef] [PubMed]
24. Juarez, J.C.; Taylor, H.F. Polarization discrimination in a phase-sensitive optical time-domain reflectometer intrusion-sensor system. *Opt. Lett.* **2005**, *30*, 3284–3286. [CrossRef] [PubMed]
25. Fernández-Ruiz, M.R.; Costa, L.; Martins, F.H. Distributed Acoustic Sensing Using Chirped-Pulse Phase-Sensitive OTDR Technology. *Sensors* **2019**, *19*, 4368. [CrossRef] [PubMed]
26. Koyamada, Y.; Imahama, M.; Kubota, K.; Hogari, K. Fiber-optic distributed strain and temperature sensing with very high measurand resolution over long range using coherent OTDR. *J. Light. Technol.* **2009**, *27*, 1142–1146. [CrossRef]
27. Martuganova, E.; Stiller, M.; Norden, B.; Henninges, J.; Krawczyk, C.M. 3D deep geothermal reservoir imaging with wireline distributed acoustic sensing in two boreholes. *Solid Earth* **2002**, *13*, 1291–1307. [CrossRef]
28. Young, C.; Shragge, J.; Schultz, W.; Haines, S.; Oren, C.; Simmons, J.; Collett, T.S. Advanced Distributed Acoustic Sensing Vertical Seismic Profile Imaging of an Alaska North Slope Gas Hydrate Field. *Energy Fuels* **2022**, *36*, 3481–3495. [CrossRef]
29. Dou, S.; Lindsey, N.; Wagner, A.M.; Daley, T.M.; Freifeld, B.; Robertson, M.; Peterson, J.; Ulrich, C.; Martin, E.R.; Ajo-Franklin, J.B. Distributed Acoustic Sensing for Seismic Monitoring of The Near Surface: A Traffic-Noise Interferometry Case Study. *Sci. Rep.* **2017**, *7*, 11620. [CrossRef]
30. Zhan, Z. Distributed acoustic sensing turns fiber-optic cables into sensitive seismic antennas. *Seismol. Res. Lett.* **2020**, *91*, 1–15. [CrossRef]
31. Fernández-Ruiz, M.R.; Martins, H.F.; Williams, E.F.; Becerril, C.; Magalhães, R.; Costa, L.; Martin-Lopez, S.; Jia, Z.; Zhan, Z.; González-Herráez, M. Seismic Monitoring with Distributed Acoustic Sensing from the Near-Surface to the Deep Oceans. *J. Light. Technol.* **2022**, *40*, 1453–1463. [CrossRef]
32. Jousset, P.; Currenti, G.; Schwarz, B.; Chalari, A.; Tilmann, F.; Reinsch, T.; Zuccarello, L.; Privitera, E.; Krawczyk, C.M. Fibre optic distributed acoustic sensing of volcanic events. *Nat. Commun.* **2022**, *13*, 1753. [CrossRef]
33. Tejedor, J.; Macias-Guarasa, J.;Martins, H.F.; Pastor-Graells, J.; Corredera, P.; Martin-Lopez, S. Machine learning methods for pipeline surveillance systems based on distributed acoustic sensing: A review. *Appl. Sci.* **2017**, *7*, 841. [CrossRef]
34. Bado, M.F.; Tonelli, D.; Poli, F.; Zonta, D.; Casas, J.R. Digital Twin for Civil Engineering Systems: An Exploratory Review for Distributed Sensing Updating. *Sensors* **2022**, *22*, 3168. [CrossRef] [PubMed]

35. Martin, R.; Bruce, G. Monitoring Traffic Flow. International Patent PCT/GB2016/053330, 26 October 2016.
36. Liu, H.; Ma, J.; Yan, W.; Liu, W.; Zhang, X.; Li, C. Traffic flow detection using distributed fiber optic acoustic sensing. *IEEE Access* **2018**, *6*, 68968–68980. [CrossRef]
37. Hall, A.J.; Minto, C. Using fibre optic cables to deliver intelligent traffic management in smart cities. In Proceedings of the International Conference on Smart Insfrastructure and Construction, Cambridge, UK, 8–10 July 2019.
38. Wang, X.; Williams, E.F.; Karrenbach, M.; González Herráez, M.; Martins , H.F.; Zhan, Z. Rose Parade Seismology: Signatures of Floats and Bands on Optical Fiber. *Seismol. Res. Lett.* **2020**, *91*, 2395–2398. [CrossRef]
39. Jakkampudi, S.; Shen, J.; Weichen, L.; Dev, A.; Zhu, T.; Martin, E. Footstep detection in urban seismic data with a convolutional network. *Lead. Edge* **2020**, *39*, 654–660. [CrossRef]
40. Liu, H.; Ma, J.; Xu, T.; Yan, W.; Ma, L.; Zhang, X. Vehicle Detection and Classification Using Distributed Fiber Optic Acoustic Sensing. *IEEE Trans. Veh. Technol.* **2020**, *69*, 1363–1374. [CrossRef]
41. Lindsey, N.J.; Yuan, S.; Lellouch, A.; Gualtieri, L.; Lecocq, T.; Biondi, B. City-Scale Dark Fiber DAS Measurements of Infrastructure Use During the COVID-19 Pandemic. *Geophys Res. Lett.* **2020**, *47*, e2020GL089931. [CrossRef]
42. Peng, Z.; Wen, H.; Jian, J. Gribok, A.; Wang, M.; Huang, S.; Liu, H.; Mao, Z.H.; Chen, K.P. Identifications and classifications of human locomotion using Rayleigh-enhanced distributed fiber acoustic sensors with deep neural networks. *Sci. Rep.* **2020**, *10*, 21014. [CrossRef]
43. Wang, X.; Zhan, Z.; Williams, E.F.; Herráez, M.G.; Martins, H.F.; Karrenbach, M. Ground vibrations recorded by fiber-optic cables reveal traffic response to COVID-19 lockdown measures in Pasadena, California. *Commun. Earth Environ.* **2021**, *2*, 160. [CrossRef]
44. Ende, M.v.; Ferrari, A.; Sladen, A.; Richard, C. Next-Generation Traffic Monitoring with Distributed Acoustic Sensing Arrays and Optimum Array Processing. In Proceedings of the 2021 55th Asilomar Conference on Signals, Systems, and Computers, Pacific Grove, CA, USA, 31 October–3 November 2021; pp. 1104–1108.
45. Wang, H.; Chen, Y.; Min, R.; Chen, Y. Urban DAS Data Processing and Its Preliminary Application to City Traffic Monitoring. *Sensors* **2022**, *22*, 9976. [CrossRef]
46. Van den Ende, M.; Ferrari, A.; Sladen, A.; Richard, C. Deep Deconvolution for Traffic Analysis with Distributed Acoustic Sensing Data. *IEEE Trans. Intell. Transp. Syst.* **2022**, *24*, 2947–2962. [CrossRef]
47. Lindsey, N.J.; Rademacher, H.; Ajo-Franklin, J.B. On the broadband instrument response of fiber-optic DAS arrays. *J. Geophys. Res. Solid Earth* **2020**, *125*, e2019JB018145. [CrossRef]
48. Van den Ende, M.; Lior, I.; Ampuero, J.P.; Sladen, A. A Self-Supervised Deep Learning Approach for Blind Denoising and Waveform Coherence Enhancement in Distributed Acoustic Sensing Data. *IEEE Trans. Neural Netw. Learn. Syst.* **2021**, *early access*. [CrossRef] [PubMed]
49. Ponce-Flores, M.; Frausto-Solís, J.; Santamaría-Bonfil, G.; Pérez-Ortega, J.; González-Barbosa, J.J. Time Series Complexities and Their Relationship to Forecasting Performance. *Entropy* **2020**, *22*, 89. [CrossRef] [PubMed]
50. Shannon, C.E. A Mathematical Theory of Communication. *Bell Syst. Tech. J.* **1948**, *27*, 379–423. [CrossRef]
51. Gao, J.; Hu, J.; Tung, W. Entropy measures for biological signal analyses. *Nonlinear Dyn.* **2012**, *68*, 431–444. [CrossRef]
52. Fernández, A.; Gómez, G.; Hornero, R.; López-Ibor, J.J. Complexity and schizophrenia. *Prog. Neuro Psychopharmacol. Biol. Psychiatry* **2013**, *45*, 267–276. ISSN 0278-5846. [CrossRef]
53. Chen, C.; Jin, Y.; Lo, I.L.; Zhao, H.; Sun, B.; Zhao, Q.; Zheng, J.; Zhang, X.D. Complexity Change in Cardiovascular Disease. *Int. J. Biol. Sci.* **2017**, *13*, 1320–1328. [CrossRef]
54. Asgharzadeh-Bonab, A.; Chehel, M.; Mehri, A. Spectral entropy and deep convolutional neural network for ECG beat classification. *Biocybern. Biomed. Eng.* **2020**, *40*, 691–700. ISSN 0208-5216. [CrossRef]
55. Rizal, A.; Hidayat, R.; Nugroho, H.A. Entropy measurement as features extraction in automatic lung sound classification. In Proceedings of the 2017 International Conference on Control, Electronics, Renewable Energy and Communications (ICCREC), Yogyakarta, Indonesia, 26–28 September 2017; pp. 93–97. [CrossRef]
56. Li, Y.; Wang, X.; Liu, Z.; Liang, X.; Liang, X.; Si, S. The Entropy Algorithm and Its Variants in the Fault Diagnosis of Rotating Machinery: A Review. *IEEE Access* **2018**, *6*, 66723–66741. [CrossRef]
57. Olbrys, J.; Majewska, E. Approximate entropy and sample entropy algorithms in financial time series analyses. *Procedia Comput. Sci.* **2022**, *207*, 255–264. ISSN 1877-0509. [CrossRef]
58. Pincus, A. Approximate entropy as a measure of system complexity. *Proc. Natl. Acad. Sci. USA* **1991**, *88*, 2297–2301. [CrossRef] [PubMed]
59. Hjorth, B. EEG analysis based on time domain properties. *Electroencephalogr. Clin. Neurophysiol.* **1970**, *9*, 306–310. ISSN 0013-4694. 0)90143-4. [CrossRef] [PubMed]
60. Briechle, K.; Hanebeck, U. Template matching using fast normalized cross correlation. *Proc. SPIE Int. Soc. Opt. Eng.* **2001**, *4387*, 1–8.
61. IGN. Earthquake information. Available online: https://www.ign.es/web/ign/portal/ultimos-terremotos/-/ultimos-terremotos/getDetails?evid=es2022zpswu (accessed on 25 June 2023).
62. García, L.; Alguacil, G.; Titos, M.; Cocina, O.; De la Torre, A.Ç.; Benítez, C. Automatic S-Phase Picking for Volcano-Tectonic Earthquakes Using Spectral Dissimilarity Analysis. *IEEE Geosci. Remote. Sens. Lett.* **2020**, *17*, 874–878. [CrossRef]
63. Baird, A.F. Modelling the response of helically wound DAS cables to microseismic arrivals. In Proceedings of the First EAGE Workshop on Fiber Optic Sensing, Amsterdam, The Netherlands, 9–11 March 2020; European Association of Geoscientists & Engineers: Amsterdam, The Netherlands, 2020; pp. 1–5.

64. Hudson, T.S.; Baird, A.F.; Kendall, J.M.; Kufner, S.K.; Brisbourne, A.M.; Smith, A.M.; Butcher, A.; Chalari, A.; Clarke, A. Distributed Acoustic Sensing (DAS) for natural microseismicity studies: A case study from Antarctica. *J. Geophys. Res. Solid Earth* **2021**, *126*, e2020JB021493. [CrossRef]
65. Jreij, S.F.; Trainor-Guitton, W.J.; Morphew, M.; Chen Ning, I.L. The Value of Information From Horizontal Distributed Acoustic Sensing Compared to Multicomponent Geophones Via Machine Learning. *J. Energy Resour. Technol.* **2021**, *143*, 010902. [CrossRef]

remote sensing

MDPI

Article

Estimation of Handheld Ground-Penetrating Radar Antenna Position with Pendulum-Model-Based Extended Kalman Filter

Piotr Kaniewski * and Tomasz Kraszewski

Faculty of Electronics, Military University of Technology, ul. Gen. S. Kaliskiego 2, 00-908 Warsaw, Poland
* Correspondence: piotr.kaniewski@wat.edu.pl

Abstract: Landmines and explosive remnants of war are a significant threat in tens of countries and other territories, causing the deaths or injuries of thousands of people every year, even long after military conflicts. Effective technical means of remote detecting, localizing, imaging, and identifying mines and other buried explosives are still sought and have a great potential utility. This paper considers a positioning system used as a supporting tool for a handheld ground penetrating radar. Accurate knowledge of the radar antenna position during terrain scanning is necessary to properly localize and visualize the shape of buried objects, which helps in their remote classification and makes demining safer. The positioning system proposed in this paper uses ultrawideband radios to measure the distances between stationary beacons and mobile units. The measurements are processed with an extended Kalman filter based on an innovative dynamics model, derived from the model of a pendulum motion. The results of simulations included in the paper prove that using the proposed pendulum dynamics model ensures a better accuracy than the accuracy obtainable with other typically used dynamics models. It is also demonstrated that our positioning system can estimate the radar antenna position with the accuracy of single centimeters which is required for appropriate imaging of buried objects with the ground penetrating radars.

Keywords: ground-penetrating radar; GPR; position estimation; extended Kalman filter; EKF; ultrawideband radio modules; UWB; landmines detection; imaging

1. Introduction

The presence of landmines and explosive remnants of war (ERW), such as artillery shells, grenades, rockets, bombs, and cluster munition remnants, poses a significant worldwide threat in the areas of current and past military conflicts. It results in deaths and injuries of mostly civilian victims even many years after the wars.

According to the yearly reports of the Landmine Monitor [1,2], providing a global overview of the landmine situation, tens of millions of landmines are still buried underground in at least 60 countries and other territories. Only a single year 2021 brought 7073 casualties of mines/ERW (2492 killed and 4561 injured) in 54 different countries, and 80% of the victims were civilians [1–3].

Considering the significance of the problem, efficient methods of mine clearance are still tough. Currently, various metal detectors (MD) are often used for this purpose, and contemporary MDs offer excellent parameters, enabling the detection of even very small and deeply buried metal objects [4–9]. Paradoxically, this high sensitivity can be also their drawback leading to many false detections which lengthen the time necessary for demining. Moreover, MDs do not offer any way to initially identify or classify the detected objects and every detection must be carefully examined. What is even more problematic and dangerous, not all contemporary landmines and ERWs contain metal elements, which limits the usefulness of MDs in mine clearance operations.

Apart from the MDs, radar technology has been successfully employed for scanning near-underground surfaces in search of mines, improvised explosive devices (IED), and other explosive remnants of war [6–8,10–14]. Ground penetrating radar (GPR) is a general term used with respect to techniques using radio waves, typically in a frequency range from several MHz to several GHz, to acquire information about objects buried underground or hidden under/behind any other concealing obstacles, surfaces, etc. [3,15–21]. These techniques enable non-invasive and non-destructive remote detecting, locating, imaging, and identifying geological structures, cavities, buried objects, and underground man-made infrastructures, which do not have to contain metal parts. The mentioned features make GPRs a very useful tool for demining. They can be used for this purpose alone [12,17,22–24] or integrated with MDs [12,25–30].

In military applications, GPRs can be installed on large armored manned vehicles with enhanced immunity to nearby explosions [31–35]. For increased safety of the crew, the radar antennas are usually attached to the end of long arms in front of the vehicle. A good alternative is mounting GPRs on remotely controlled unmanned wheeled vehicles [36] or tracked vehicles [30,35,37,38], which eliminates the risk for the crew, and reduces the costs of the purchase and the exploitation of such systems. The GPRs on vehicle platforms, however, have limited utility in difficult terrain: mountainous areas, forests, dumps, urban surfaces covered with debris, or interiors of buildings, where landmines and other explosives can be typically found. A good solution applicable in such areas is a handheld version of the ground penetrating radar (HH-GPR) [39–42]. The problem of estimating the antenna position of such type of radar is addressed in this paper.

The GPR operation requires emitting electromagnetic energy in the direction of the ground. The transmitted radio waves penetrate near-surface layers of the soil and encounter on their way various objects and layers of different permittivity ε and conductivity σ, which results in reflecting and scattering back a portion of the transmitted energy. The echo signals are received, collected, and processed to detect and create images of buried objects.

Most contemporary GPRs are pulse radars [15,16,22,24,43,44], transmitting repeatable, very short, high-amplitude pulses and receiving strongly attenuated echo signals reflected or scattered back from layers' boundaries and buried objects [3,45,46] as shown in Figure 1.

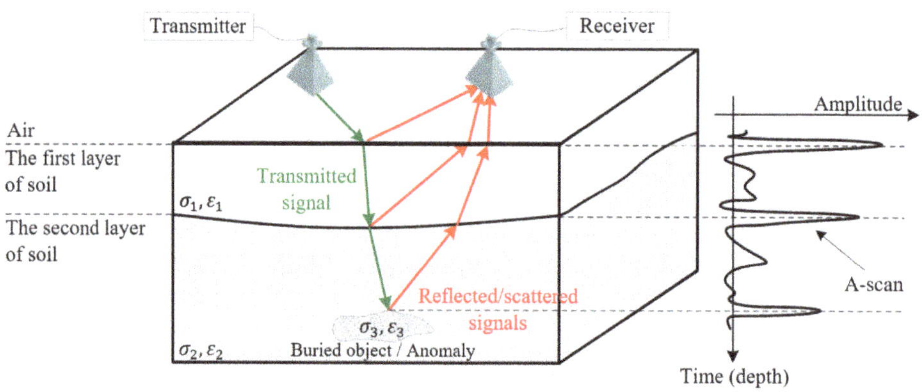

Figure 1. General idea of GPR operation.

Time delays of subsequent peaks in the received echo signals are proportional to the depth of the detected objects or layers of different permittivity. Collecting and joint processing multiple echo signals, so-called echograms or radargrams, for a GPR moving along a predefined scanning path enables locating and imaging those objects and layers [3,15,16].

Three types of visual presentations of GPR radargrams are used in practice [3,14,16,21]. A single echogram was obtained for only one GPR antenna position with coordinates (i, j) is a one-dimensional signal representation, called an A-scan (Figure 2a). Time delays of the signal peaks in the A-scan are usually converted into respective depths and the Z-axis is scaled in the distance units [21,23,39,46,47].

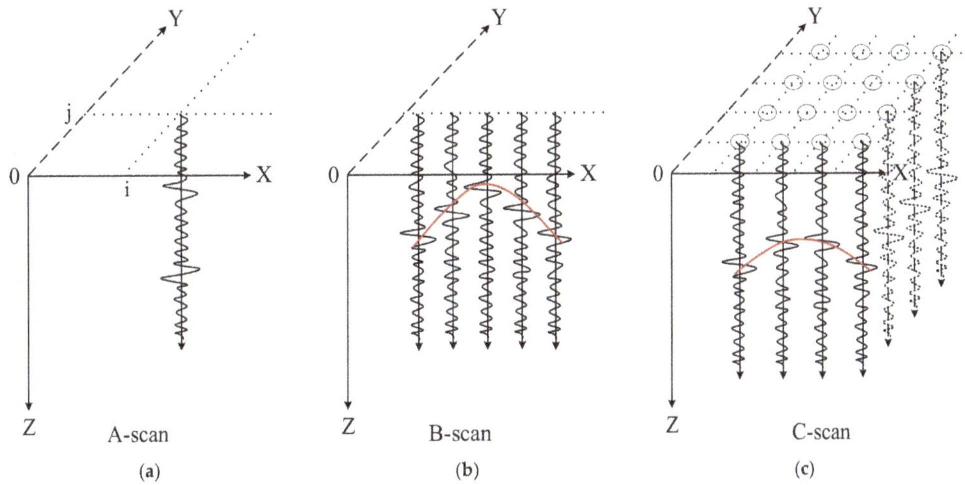

Figure 2. Types of GPR radargrams' visual presentations: (**a**) A-scan; (**b**) B-scan; (**c**) C-scan.

An analysis of GPR data is typically based on a two-dimensional signal representation, called a B-scan, which is a dataset created from many A-scans acquired for various antenna locations along a usually linear scanning path, as shown in Figure 2b. It represents a radar image of a vertical surface intersecting the scanned terrain volume below the scanning path. Due to a relatively large GPR antenna beamwidth, the same buried objects are illuminated many times from different antenna locations and consequently from different distances. Therefore, the echo signals form hyperbolic structures visible in the B-scans [14,23,39,46,47]. An example of such a structure for a single-point object is shown as a red hyperbole in Figure 2b.

Collecting A-scans for multiple antenna locations in the nodes of a grid span onto the OXY surface, one can create another type of GPR signal visual presentation, called a C-scan (Figure 2c). This is a three-dimensional signal representation, which is very useful in visualizing, identifying, and classifying buried objects.

The C-scans are often presented as a set of two-dimensional greyscale or color images, created as horizontal sections through the C-scan volume on various depths [3,21,48]. An example of such a single image is shown in Figure 3.

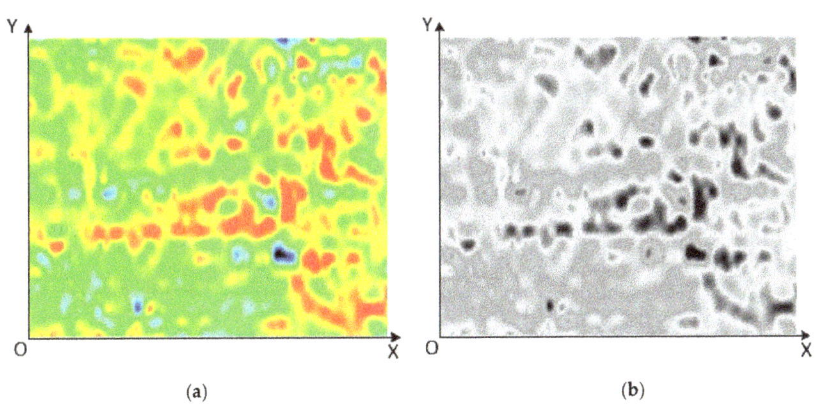

Figure 3. Examples of a horizontal section through a C-scan: (**a**) color image; (**b**) grayscale image.

The presented scans were made using a pulse radar produced by IDS GeoRadar company, containing a DAD K2 control unit and an antenna with a central frequency equal to 900 MHz. This radar made 850 soundings per second, the duration of the probing pulse was about 1 ns, and the obtained spatial resolution was about 5 cm.

Knowledge of accurate positions of a GPR antenna moving along a scanning path is necessary to properly assemble all the acquired radargrams and create high-quality GPR B-scans or C-scans. Several scientific papers [44,49] and patents [50] suggest that the GPR antenna positioning accuracy should be better than one eight of a radar signal wavelength [51]. As typical GPRs work at a frequency range between 400 MHz and 4 GHz (wavelengths from 7.5 to 75 cm) [49,51], the antenna positioning accuracy should be of the order of single centimeters which requires using very high-accuracy navigation systems.

Typically, the navigation devices or systems used for GPR antenna positioning are Global Navigation Satellite Systems (GNSS) receivers [50,52], often with real-time kinematic (RTK) corrections, inertial navigation systems [42,52], wheel odometers [50], visual navigation systems [53], laser scanners [49] or integrated systems combining several of the mentioned devices [42,50,52].

As most of the listed above devices or systems are not adequate for HH-GPRs, due to their large size, weight, specific installation requirements, vulnerability to jamming or signal shadowing, and too low accuracy, the authors of this paper proposed a system based on several ultrawideband (UWB) radio modules. This concept was first described in an authors' conference paper [54], where physical models of a mobile unit and UWB beacons were presented. The mentioned paper also contained a description of an autocalibration procedure, used for self-locating the UWB beacons for quickly establishing a frame of reference before the scanning process, and presented an initial assessment of the system's accuracy which in the scanning zone reaches desired level of 2–3 cm.

In another authors' conference paper [41], it was claimed and demonstrated that the accuracy of the UWB positioning system can be further improved with a properly chosen estimation algorithm. In that paper, using an extended Kalman filter (EKF) based on a GPR antenna motion model, derived from the mathematical pendulum motion model, was proposed. The mentioned paper, however, contained a proof of concept rather than a complete and applicable positioning solution, as the proposed pendulum-based dynamics model used in the EKF was oversimplified to present the main idea only. It assumed that the attachment point of the "pendulum", which is the position of a GPR operator's arm, is initially known and that the angle of orientation of the main axis of the scanning section always equals zero degrees. These assumptions can hardly be met in practice. Moreover, the mentioned conference paper contained only a sketch of the system's model and very limited results of its simulative testing.

This paper can be considered a significantly extended version of the above-mentioned conference paper. It presents an elaborated, practically applicable version of the GPR antenna positioning system using UWB radio modules and includes a complete description of its extended mathematical model and detailed results of its simulative testing. The main novelty of this paper includes:

1. Elaboration and detailed presentation of an advanced and practically applicable dynamics and observation model of the UWB-based GPR antenna positioning system, with relinquished simplifying assumptions of the model presented in [41];
2. Elaboration and detailed presentation of the estimation algorithm used in the proposed GPR antenna positioning system;
3. Presentation of new and detailed results of simulative tests of the positioning system for various realistic system configurations.

This paper is organized as follows. A general concept of the ground penetrating radar, types of GPR data visualizations, accuracy requirements for GPR antenna positioning, technologies used for GPR positioning, previous authors' works in this field, and a discussion of the novelty of this paper are presented in Section 1. The system's description, its mathematical model, and the estimation algorithm elaborated by the authors are presented in Section 2. The methodology and the results of simulative testing of the GPR antenna positioning system are presented in Sections 3 and 4 contain a discussion.

2. Materials and Methods

2.1. Scanning Profiles

As has already been mentioned, creating B-scans requires moving a GPR antenna over the ground, ideally along a linear scanning path (profile) with constant velocity, to collect linearly arranged and uniformly separated radargrams. Creating C-scans requires repeating such scanning (profiling) for many equidistant lines in one direction, as shown in Figure 4a, or bi-directionally, as shown in Figure 4b, where the antenna position is marked as a letter A [7,15,16]. In multichannel GPRs, with several equidistant antennas, the profiling can be realized quicker, unidirectionally (like in Figure 4a) for several scanning paths at a time.

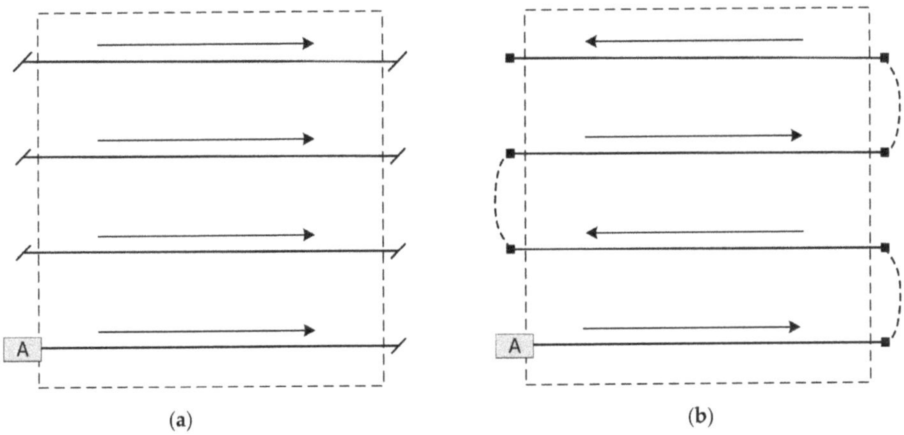

Figure 4. Ideal GPR scanning profiles: (**a**) unidirectional; (**b**) bidirectional.

Although in favorable conditions the profiling shown in Figure 4 can be at least approximately realized with GPRs installed on vehicles (carefully driven or remotely controlled, in non-demanding terrain and with the use of an accurate supporting navigation system), this can hardly be achieved with HH-GPRs. The elements of the scanning path, in this case, are shown in Figure 5, where the letters A and S represent the positions of the antenna and the sapper.

2.2. UWB Positioning System

The structure of the HH-GPR antenna positioning system proposed in this paper is shown in Figure 6. It is composed of four stationary modules $M_1 \div M_4$ serving as radio beacons and two mobile modules M_A and M_S. The M_A module is installed over the GPR antenna and the M_S module over the sapper's shoulder. All the modules contain UWB transceivers. Distance measurements realized by these transceivers are collected and processed using estimation algorithms described in the further part of the paper.

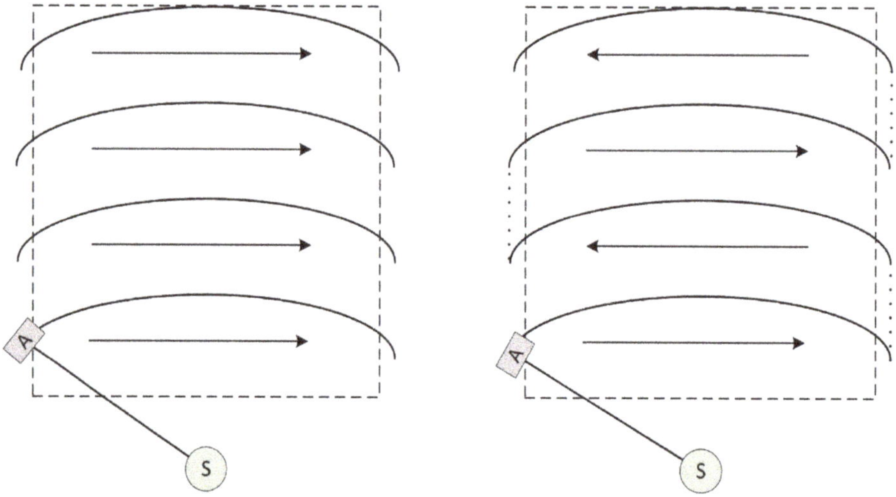

(**a**) (**b**)

Figure 5. Scanning profiles typical for HH-GPR: (**a**) unidirectional; (**b**) bidirectional.

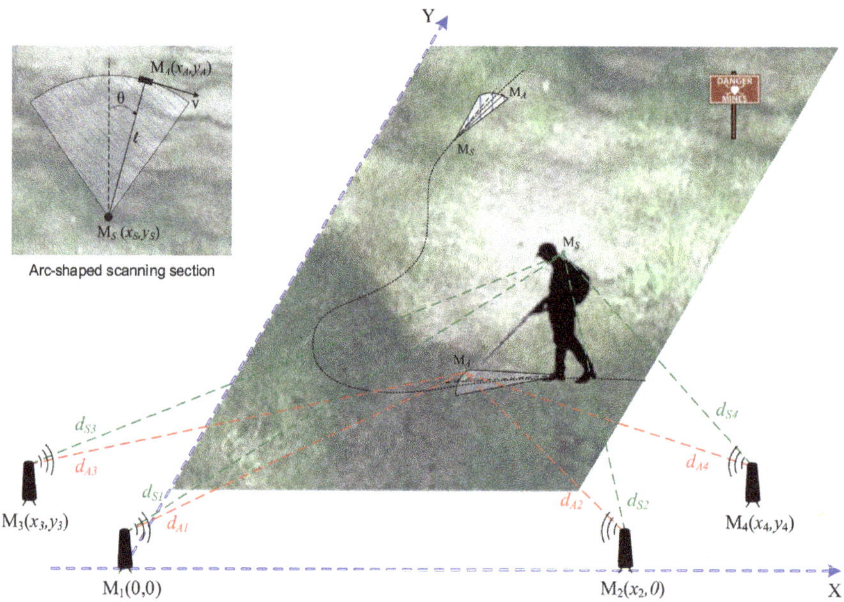

Figure 6. UWB positioning system for HH-GPR antenna.

The following variables are used in Figure 6:

d_{Aj}—distance between a j-th beacon and the antenna module M_A,
d_{Sj}—distance between a j-th beacon and the sapper module M_S,
x_j, y_j—coordinates of a j-th beacon position,
x_A, y_A—coordinates of the M_A module position,
x_S, y_S—coordinates of the M_S module position,
l—length of the HH-GPR handle (horizontal distance between M_S and M_A),

θ—angle between the horizontal projection of the GPR antenna handle and the central axis of the scanning section.

We assumed that the UWB radios used in our system are PulsON P440 modules from TDSR [55]. They use the two-way time-of-flight (TW-TOF) method for ranging and offer an operating range between 300 and 1100 m and a ranging accuracy of about 2 cm in line of sight (LOS) conditions. Such parameters give the potential to build a positioning system with the desired centimeter-level accuracy, required in the considered application of the HH-GPR antenna positioning.

The placement of beacons outside a potentially hazardous area, as shown in Figure 6, is only one of the possible options, suggested for quick and easy deployment of the system in terrain. Other beacons' locations are also possible, and their relative positions with respect to the mobile units M_A and M_S influence the accuracy of the UWB positioning system, which will be discussed in detail in the Results section of the paper.

2.3. Mathematical Model

As can be seen in Figure 5, the scanning profiles are composed of fragments that resemble arcs rather than straight sections. Moreover, the velocity of the HH-GPR antenna is more changeable than in GPRs installed on vehicle platforms, as typically an operator (sapper) performs a swinging motion, initially accelerating and finally decelerating the antenna. Therefore, the collected radargrams are not linearly arranged nor uniformly separated. Nevertheless, the acquired A-scans can be used to create two- or three-dimensional GPR visualizations of buried objects provided that the antenna positions are known for all the collected radargrams [3,15,16].

A single arc belonging to the scanning profile is shown in Figure 7. If we consider the changeable angular velocity of the antenna motion (initially accelerating and finally decelerating), such a trajectory resembles the motion of a mathematical pendulum [56], and can be described by the following formula:

$$\frac{d^2\theta}{dt^2} + \frac{a}{l}\sin\theta = 0, \tag{1}$$

where:

θ—angle between the horizontal projection of the GPR antenna handle and the central axis of the scanning section,

a—acceleration forcing the HH-GPR antenna (M_A module) motion,

l—length of the HH-GPR handle (horizontal distance between M_S and M_A).

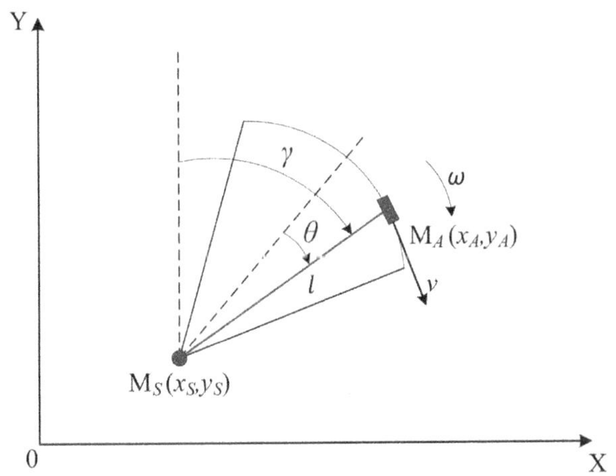

Figure 7. Part of HH-GPR antenna trajectory (a single arc of the scanning profile).

The acceleration a is analogous to the gravity acceleration g in the mathematical pendulum motion model. Contrary to g, which can be considered a constant, the acceleration a is more changeable and to large extent depends on the operator's strength, fatigue, style of HH-GPR operation, etc., thus we treat it as an additional variable to be estimated and we model it as a Wiener stochastic process [57–59]. Considering the geometrical relationships shown in Figure 7, the equations describing the antenna and the sapper's arm motion can be formulated as follows:

$$\begin{cases} \dot{x}_A = \omega l \cos\gamma = \omega(y_A - y_S) \\ \dot{y}_A = -\omega l \sin\gamma = -\omega(x_A - x_S) \\ \dot{x}_S = u_{x_S} \\ \dot{y}_S = u_{y_S} \\ \dot{\theta} = \omega \\ \dot{\omega} = \frac{d^2\theta}{dt^2} = -\frac{a}{l}\sin\theta \\ \dot{a} = u_a \end{cases}, \tag{2}$$

where:

x_A, y_A—coordinates of the HH-GPR antenna (M_A module) position,
x_S, y_S —coordinates of the sapper's arm (M_S module) position,
u_{x_S}, u_{y_S}—Gaussian white noises representing random components of the sapper's arm (M_S module) motion,
l—length of the HH-GPR handle (horizontal distance between M_S and M_A),
θ—angle between the horizontal projection of the GPR antenna handle and the central axis of the scanning section,
γ—angle between the horizontal projection of the GPR antenna handle and the OY axis of the frame of reference,
ω—angular velocity of the HH-GPR antenna (M_A module) motion,
a—acceleration forcing the HH-GPR antenna (M_A module) motion,
u_a—Gaussian white noise representing random changes of a.

Rewriting Equation (2) to fit it into the standard form of a nonlinear continuous dynamics model [60–63]:

$$\dot{\mathbf{x}}(t) = \mathbf{f}[\mathbf{x}(t)] + \mathbf{G}(t)\mathbf{u}(t), \tag{3}$$

one obtains the following detailed version of this model, which has been further used in our estimation algorithm:

$$\underbrace{\begin{bmatrix} \dot{x}_A \\ \dot{y}_A \\ \dot{x}_S \\ \dot{y}_S \\ \dot{\theta} \\ \dot{\omega} \\ \dot{a} \end{bmatrix}}_{\dot{\mathbf{x}}(t)} == \underbrace{\begin{bmatrix} \omega(y_A - y_S) \\ -\omega(x_A - x_S) \\ 0 \\ 0 \\ \omega \\ -\frac{a}{l}\sin\theta \\ 0 \end{bmatrix}}_{\mathbf{f}[x(t)]} + \underbrace{\begin{bmatrix} 0 & 0 & 0 \\ 0 & 0 & 0 \\ 1 & 0 & 0 \\ 0 & 1 & 0 \\ 0 & 0 & 0 \\ 0 & 0 & 0 \\ 0 & 0 & 1 \end{bmatrix}}_{\mathbf{G}(t)} \underbrace{\begin{bmatrix} u_{x_S} \\ u_{y_S} \\ u_a \end{bmatrix}}_{\mathbf{u}(t)}. \tag{4}$$

The nonlinear observation model in the following standard form [60–62]:

$$\mathbf{z}(k) = \mathbf{h}[\mathbf{x}(k)] + \mathbf{v}(k), \tag{5}$$

has been formulated assuming that at every step k the UWB positioning system realizes four distance measurements between a j-th beacon and the antenna module M_A:

$$d_{Aj}(k) = \sqrt{(x_A(k) - x_j)^2 + (y_A(k) - y_j)^2} + v_{Aj}(k), \tag{6}$$

and four distance measurements between a j-th beacon and the sapper module M_S:

$$d_{Sj}(k) = \sqrt{\left(x_S(k) - x_j\right)^2 + \left(y_S(k) - y_j\right)^2 + h^2} + v_{Sj}(k), \tag{7}$$

where:

d_{Aj}—distance between a j-th beacon and the antenna module M_A,
d_{Sj}—distance between a j-th beacon and the sapper module M_S,
x_j, y_j—coordinates of a j-th beacon position,
x_A, y_A—coordinates of the M_A module position,
x_S, y_S—coordinates of the M_S module position,
h—sapper's arm height,
v_{Aj}, v_{Sj}—distance measuring errors for M_A and M_S modules.

A detailed version of the observation model, which has been further used in our estimation algorithm, is as follows:

$$\underbrace{\begin{bmatrix} d_{A1}(k) \\ d_{A2}(k) \\ d_{A3}(k) \\ d_{A4}(k) \\ d_{S1}(k) \\ d_{S2}(k) \\ d_{S3}(k) \\ d_{S4}(k) \end{bmatrix}}_{\mathbf{z}(k)} = \underbrace{\begin{bmatrix} \sqrt{\left(x_A(k) - x_1\right)^2 + \left(y_A(k) - y_1\right)^2} \\ \sqrt{\left(x_A(k) - x_2\right)^2 + \left(y_A(k) - y_2\right)^2} \\ \sqrt{\left(x_A(k) - x_3\right)^2 + \left(y_A(k) - y_3\right)^2} \\ \sqrt{\left(x_A(k) - x_4\right)^2 + \left(y_A(k) - y_4\right)^2} \\ \sqrt{\left(x_S(k) - x_1\right)^2 + \left(y_S(k) - y_1\right)^2 + h^2} \\ \sqrt{\left(x_S(k) - x_2\right)^2 + \left(y_S(k) - y_2\right)^2 + h^2} \\ \sqrt{\left(x_S(k) - x_3\right)^2 + \left(y_S(k) - y_3\right)^2 + h^2} \\ \sqrt{\left(x_S(k) - x_4\right)^2 + \left(y_S(k) - y_4\right)^2 + h^2} \end{bmatrix}}_{\mathbf{h}[x(k)]} + \underbrace{\begin{bmatrix} v_{A1}(k) \\ v_{A2}(k) \\ v_{A3}(k) \\ v_{A4}(k) \\ v_{S1}(k) \\ v_{S2}(k) \\ v_{S3}(k) \\ v_{S4}(k) \end{bmatrix}}_{\mathbf{v}(k)}. \tag{8}$$

As the antenna module M_A is kept close to the soil during scanning, and the differences between slant distances d_{Aj} and their horizontal projections are very small, we assumed that their altitude over the ground can be omitted in the observation model. On the other hand, the M_S module is placed over the ground on the sapper's arm, and its altitude h is non-negligible. In our model, we assumed that it is constant, as its changes in the order of centimeters during the system's operation can be neglected for typical distances from the UWB beacons, which are in the order of tens of meters. In a real system the altitude h can be a settable constant, adjusted before using the system, based on the sapper's height.

2.4. Estimation Algorithm

An extended Kalman filter for HH-GPR antenna position estimation was designed based on the previously described dynamics and observation models and its flowchart is shown in Figure 8.

After initialization of the EKF at step $k = 0$ or after closing each subsequent filter's loop at steps $k > 0$, the filter alternately performs prediction and correction steps. The prediction step requires previous calculations of the fundamental matrix $\mathbf{F}$, the transition matrix $\mathbf{\Phi}$, and the covariance matrix of disturbances $\mathbf{Q}$ at every step k. The method of calculating the $\mathbf{F}$ matrix (more precisely it is $\mathbf{F}_{k-1}$ but to shorten the notation the index $k-1$ will be omitted in further equations) is explained in Appendix A.

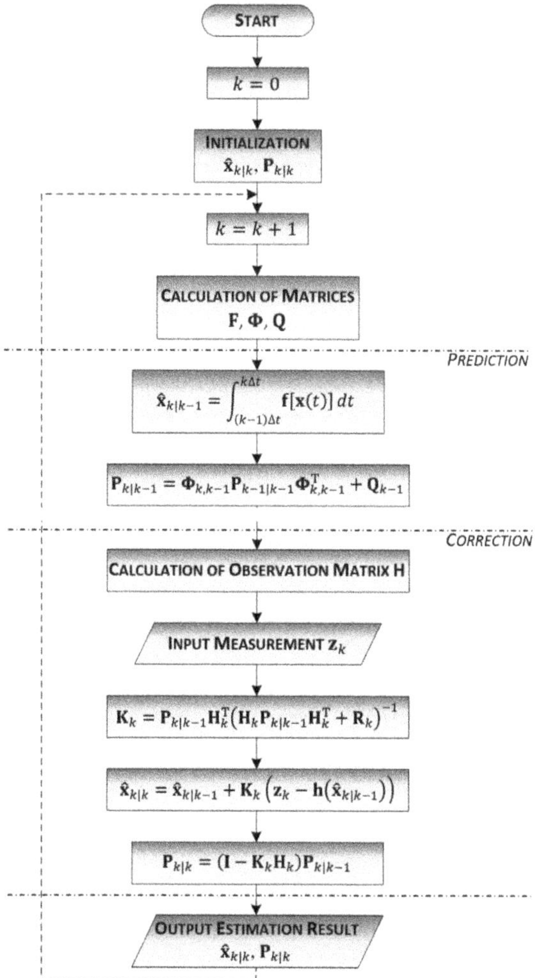

Figure 8. Flowchart of the Extended Kalman Filter used for HH-GPR antenna position estimation.

Using the calculated $\mathbf{F}$ matrix and the $\mathbf{G}$ matrix from the equation (4), $\mathbf{\Phi}_{k,k-1}$ and $\mathbf{Q}_{k-1}$ matrices are obtained as follows [60–62]:

$$\mathbf{\Phi}_{k,k-1} = \mathrm{e}^{\mathbf{F}\Delta t} \approx \mathbf{I} + \mathbf{F}\Delta t + \frac{(\mathbf{F}\Delta t)^2}{2!} \tag{9}$$

$$\begin{aligned}\mathbf{Q}_{k-1} \approx\ & \mathbf{Q}_{c1}\Delta t + \left(\mathbf{F}\mathbf{Q}_{c1} + \mathbf{Q}_{c1}\mathbf{F}^{\mathrm{T}}\right)\tfrac{(\Delta t)^2}{2} + \left[\mathbf{F}^2\mathbf{Q}_{c1} + 2\mathbf{F}\mathbf{Q}_{c1}\mathbf{F}^{\mathrm{T}} + \mathbf{Q}_{c1}\left(\mathbf{F}^{\mathrm{T}}\right)^2\right]\tfrac{(\Delta t)^3}{6} \\ & + \left[\mathbf{F}^3\mathbf{Q}_{c1} + 3\mathbf{F}^2\mathbf{Q}_{c1}\mathbf{F}^{\mathrm{T}} + 3\mathbf{F}\mathbf{Q}_{c1}\left(\mathbf{F}^{\mathrm{T}}\right)^2 + \mathbf{Q}_{c1}\left(\mathbf{F}^{\mathrm{T}}\right)^3\right]\tfrac{(\Delta t)^4}{24},\end{aligned} \tag{10}$$

where:

$$\mathbf{Q}_{c1} = \mathbf{G}\mathbf{Q}_c\mathbf{G}^{\mathrm{T}}, \tag{11}$$

and Δt is a period between two successive time steps $k-1$ and k.

The $\mathbf{Q}_c$ matrix from Equation (11) represents the covariance matrix of continuous disturbances which is a 3-by-3 diagonal matrix containing power spectral densities S_{x_S}, S_{y_S} and S_u of the noises u_{x_S}, u_{y_S} and u_a composing the disturbances vector $\mathbf{u}(t)$ in Equation (4):

$$\mathbf{Q}_c = diag([S_{x_S}, S_{y_S}, S_a]). \tag{12}$$

The predicted state vector $\hat{\mathbf{x}}_{k|k-1}$ is calculated in accordance with the following general equation [64,65]:

$$\hat{\mathbf{x}}_{k|k-1} = \int_{(k-1)\Delta t}^{k\Delta t} \mathbf{f}[\mathbf{x}(t)]dt, \tag{13}$$

but in practical calculations we use Heun's numerical integration method [65–68] which leads to the following formulae:

$$\hat{\mathbf{x}}_{k|k-1} = \hat{\mathbf{x}}_{k-1|k-1} + \frac{1}{2}\left[\mathbf{f}\left(\hat{\mathbf{x}}_{k-1|k-1}\right) + \mathbf{f}\left(\hat{\mathbf{x}}_{k-1|k-1} + \mathbf{f}\left(\hat{\mathbf{x}}_{k-1|k-1}\right)\right)\right], \tag{14}$$

where $\hat{\mathbf{x}}_{k-1|k-1}$ is the final state vector estimate from the previous step $k-1$.

Apart from the predicted state vector $\hat{\mathbf{x}}_{k|k-1}$, the covariance matrix of prediction errors $\mathbf{P}_{k|k-1}$ is calculated based on the covariance matrix of filtration errors $\mathbf{P}_{k-1|k-1}$ from the previous step $k-1$ as follows [60–62]:

$$\mathbf{P}_{k|k-1} = \mathbf{\Phi}_{k,k-1}\mathbf{P}_{k-1|k-1}\mathbf{\Phi}_{k,k-1}^{\mathrm{T}} + \mathbf{Q}_{k-1}, \tag{15}$$

where we use the mentioned matrices $\mathbf{\Phi}$ and $\mathbf{Q}$.

The correction step requires previous calculations of the observation matrix $\mathbf{H}$ at every step k, and the method of its calculation is explained in Appendix B. This step involves a calculation of the Kalman gains matrix $\mathbf{K}_k$, a correction of the predicted state vector $\hat{\mathbf{x}}_{k|k-1}$ based on the current measurement vector z_k, which produces the final estimate $\hat{\mathbf{x}}_{k|k}$ at the step k, as well as calculation the covariance matrix of filtration errors $\mathbf{P}_{k|k}$, and these operations are realized as follows [60–62]:

$$\mathbf{K}_k = \mathbf{P}_{k|k-1}\mathbf{H}_k^{\mathrm{T}}\left(\mathbf{H}_k\mathbf{P}_{k|k-1}\mathbf{H}_k^{\mathrm{T}} + \mathbf{R}_k\right)^{-1}, \tag{16}$$

$$\hat{\mathbf{x}}_{k|k} = \hat{\mathbf{x}}_{k|k-1} + \mathbf{K}_k\left(z_k - \mathbf{h}\left(\hat{\mathbf{x}}_{k|k-1}\right)\right), \tag{17}$$

$$\mathbf{P}_{k|k} = (\mathbf{I} - \mathbf{K}_k\mathbf{H}_k)\mathbf{P}_{k|k-1}. \tag{18}$$

The $\mathbf{R}_k$ matrix in Equation (16) is the covariance matrix of measurement errors [60,61] which is formed as an 8-by-8 diagonal matrix, containing the variances of all eight distance measurements performed between pairs of UWB modules in the positioning system:

$$\mathbf{R}_k = diag\left(\left[\sigma_{A1}^2, \sigma_{A2}^2, \sigma_{A3}^2, \sigma_{A4}^2, \sigma_{S1}^2, \sigma_{S2}^2, \sigma_{S3}^2, \sigma_{S4}^2\right]\right), \tag{19}$$

where σ_{Aj}^2 and σ_{Sj}^2 represent variances of distance measurement between a j-th beacon and the antenna module M_A or the sapper module M_S.

2.5. Alternative Positioning Algorithms

Apart from the proposed pendulum-model-based EKF, simpler algorithms can also be used to estimate the HH-GPR antenna position. One possible solution is a non-linear least squares (NLS) algorithm [57,69,70] which processes a vector $\mathbf{z}(k)$ of distance measurements collected at each step k without using the previously estimated state vector and without filtration. Such an algorithm does not use any dynamics model either. The NLS requires an initialization by assigning at least coarse values to the antenna coordinates x_A and y_A and subsequently, it improves their estimates iteratively. This algorithm is simple but due to lack of filtration, its accuracy is not high.

Better estimation results can be obtained by using EKF filters based on nearly-constant-velocity (CV) or nearly-constant-acceleration (CA) dynamics models [57,71–74], which are typically applied in navigation and radiolocation. The CV model (20) assumes a rectilinear uniform motion, whereas the CA model (21) assumes a uniformly accelerated motion, and, in both cases, small disturbances of these ideal movements are modeled by the vector $\mathbf{u}(t)$:

$$\underbrace{\begin{bmatrix} \dot{x}_A \\ \dot{v}_x \\ \dot{y}_A \\ \dot{v}_y \end{bmatrix}}_{\dot{\mathbf{x}}(t)} == \underbrace{\begin{bmatrix} 0 & 1 & 0 & 0 \\ 0 & 0 & 0 & 0 \\ 0 & 0 & 0 & 1 \\ 0 & 0 & 0 & 0 \end{bmatrix}}_{\mathbf{F}[x(t)]} \underbrace{\begin{bmatrix} x_A \\ v_x \\ y_A \\ v_y \end{bmatrix}}_{\mathbf{x}(t)} + \underbrace{\begin{bmatrix} 0 & 0 \\ 1 & 0 \\ 0 & 0 \\ 0 & 1 \end{bmatrix}}_{\mathbf{G}(t)} \underbrace{\begin{bmatrix} u_{v_x} \\ u_{v_y} \end{bmatrix}}_{\mathbf{u}(t)}, \tag{20}$$

$$\underbrace{\begin{bmatrix} \dot{x}_A \\ \dot{v}_x \\ \dot{a}_x \\ \dot{y}_A \\ \dot{v}_y \\ \dot{a}_y \end{bmatrix}}_{\dot{\mathbf{x}}(t)} == \underbrace{\begin{bmatrix} 0 & 1 & 0 & 0 & 0 & 0 \\ 0 & 0 & 1 & 0 & 0 & 0 \\ 0 & 0 & 0 & 0 & 0 & 0 \\ 0 & 0 & 0 & 0 & 1 & 0 \\ 0 & 0 & 0 & 0 & 0 & 1 \\ 0 & 0 & 0 & 0 & 0 & 0 \end{bmatrix}}_{\mathbf{F}[x(t)]} \underbrace{\begin{bmatrix} x_A \\ v_x \\ a_x \\ y_A \\ \dot{v} \\ a_y \end{bmatrix}}_{\mathbf{x}(t)} + \underbrace{\begin{bmatrix} 0 & 0 \\ 0 & 0 \\ 1 & 0 \\ 0 & 0 \\ 0 & 0 \\ 0 & 1 \end{bmatrix}}_{\mathbf{G}(t)} \underbrace{\begin{bmatrix} u_{a_x} \\ u_{a_y} \end{bmatrix}}_{\mathbf{u}(t)}, \tag{21}$$

where:

x_A, y_A—coordinates of antenna position,
v_x, v_y—components of antenna velocity,
a_x, a_y—components of antenna acceleration,
u_{v_x}, u_{v_y}—Gaussian white noises representing random disturbances of CV motion,
u_{a_x}, u_{a_y}—Gaussian white noises representing random disturbances of CA motion.

Clearly, the CV and CA models do not fit ideally the actual HH-GPR antenna motion but nevertheless, they can be used for prediction in EKFs. Such filters are not optimal, but they are simpler than the EKF presented in Figure 8 because both dynamics models are linear, and the prediction of the state vector is realized like in a linear Kalman filter [60–62]:

$$\hat{\mathbf{x}}_{k|k-1} = \mathbf{\Phi}_{k,k-1}\hat{\mathbf{x}}_{k|k}. \tag{22}$$

Moreover, the transition matrix $\mathbf{\Phi}$ and the covariance matrix of disturbances $\mathbf{Q}$ can be calculated in advance before the filter implementation using the simple formula [62] and they remain constant during the filters' operation. Thus, such EKFs do not require in-run calculation of the Jacobian matrix $\mathbf{F}$ and the matrices $\mathbf{\Phi}$ and $\mathbf{Q}$.

All the mentioned algorithms, NLS and EKFs based on the CV and CA models, have been implemented by the authors and tested to compare them with the previously described pendulum-model-based EKF. Further in the paper, the following acronyms will be used for these algorithms: NLS, EKF-CV, EKF-CA, and EKF-PND.

Although the EKF-CV and the EKF-CA are not optimal, their accuracy can be maximized by choosing appropriate power spectral densities of disturbances S_{v_x}, S_{v_y} of u_{v_x}, u_{v_y} noises in the CV model or S_{a_x}, S_{a_y} of u_{a_x}, u_{a_y} noises in the CA model. Their choice affects the values of the $\mathbf{Q}$ matrices and consequently influences the information quality [75], notably estimation errors of the filters. The process of choosing filters' parameters and minimizing their estimation errors is called "tuning Kalman filter" [76] and it was realized in the case of the EKF-CV and EKF-CA designed in our research.

3. Results

The described HH-GPR antenna positioning system and the proposed pendulum-model-based EKF were implemented and simulatively tested in MATLAB® version R2022b. The simulations included an assessment of the dependence of the system's accuracy on the positions of UWB stationary modules $M_1 \div M_4$ deployed around the area of interest, where the demining process is going to be performed. The results of these analyses are presented in Section 3.1. In further experiments, the accuracy of the EKF-PND filter was analyzed for chosen scanning sections. This accuracy was also compared with the accuracy of the NLS, EKF-CV, and EKF-CA algorithms. The results of these tests and accuracy comparisons are presented in Section 3.2.

3.1. Influence of UWB Beacons' Locations on System's Accuracy

Possible locations of the UWB stationary modules $M_1 \div M_4$ are to large extent dependent on the terrain characteristics and the obstacles present around the scanning area. The sapper usually cannot place it freely when he deploys the system's elements in a previously unsearched and potentially hazardous terrain. When approaching a minefield, he usually knows which part of the terrain is free of explosives and where the search should start. Thus, the most typical and safe locations for placing the UWB beacons lay in front and on the sides of the minefield, as shown in Figure 6. Such a system's geometry is certainly not optimal from the accuracy point of view, but even under the mentioned limitations, the actual placement of $M_1 \div M_4$ may significantly influence the positioning accuracy in various areas of the minefield.

To verify the mentioned dependence of the positioning accuracy on the locations of the UWB stationary modules, three system configurations with different locations of the $M_1 \div M_4$ modules were considered. The assumed positions of the modules are given in Table 1 and are graphically presented in Figure 9.

Table 1. Locations of the UWB stationary modules for different system configurations.

Configuration Number	Coordinates of the UWB Modules			
	M_1	M_2	M_3	M_4
C1	[0,0]	[100,0]	[−50,30]	[150,30]
C2	[30,0]	[70,0]	[10,30]	[90,30]
C3	[40,0]	[60,0]	[30,30]	[70,30]

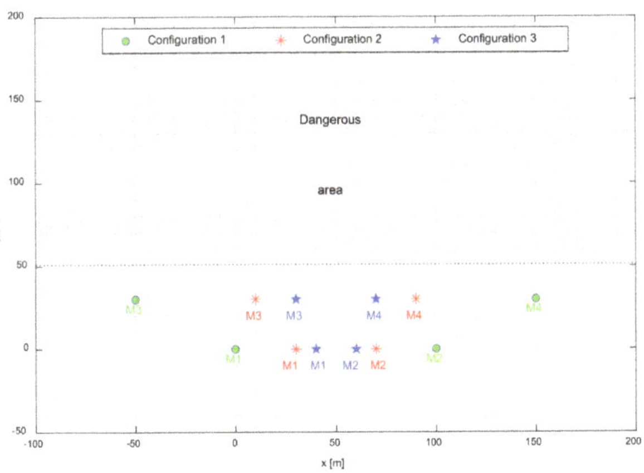

Figure 9. Locations of the UWB stationary modules for different system configurations.

We assumed that an area of 500 m × 500 m lying in front of the UWB beacons is divided by a grid with cells of 10 m × 10 m each. For every node of this grid, a set of ten thousand UWB measurements was generated in MATLAB®, and its position was estimated using a simple iterative NLS algorithm, without any Kalman filtration. Based on the parameters of P440 modules, declared by their producer [55], we assumed that UWB measurement errors have zero-mean Gaussian distribution with a standard deviation of 2 cm. Next, RMS errors (RMSE) for each node were calculated and the obtained results for the three system's configurations are shown as colormaps in Figure 10. As the RMSE is very large in the vicinity of the UWB modules, the colormap is presented for the area where the y coordinate is larger or equal to 50 m. In practice, it means that the actual placement of the UWB beacons should be in the foreground of the minefield, far enough ahead of its border, to ensure that the positioning accuracy in the planned search zone is high.

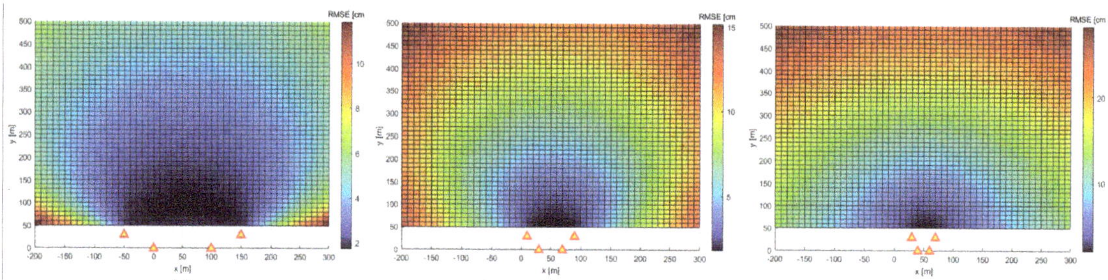

(a) (b) (c)

Figure 10. RMSE for different system configurations: (**a**) C1; (**b**) C2; (**c**) C3.

As can be seen, the smallest positioning errors are achievable in front of the place, where the UWB stationary modules $M_1 \div M_4$ are located. The high-accuracy zone is wider and deeper for a more extended baseline of the positioning system.

Mean and maximal RMSE values for the whole area of 450 $m \times$ 500 m and for smaller areas, limited to the nearest 100 $m \times$ 100 m and 50 m x 50 m respectively, in front of the UWB stationary modules $M_1 \div M_4$ are given in Table 2.

Table 2. Mean and maximal RMSE values for areas of various sizes.

Configuration Number	The Area of Interest					
	450 m x 500 m		100 m x 100 m		50 m x 50 m	
	RMSE Values [cm]					
	Mean	Max	Mean	Max	Mean	Max
C1	4.0786	11.8357	2.0348	2.4603	1.9048	2.1324
C2	8.9495	15.1545	3.6597	4.9420	3.1985	4.0216
C3	15.5555	28.2914	6.6283	9.1209	5.6234	6.9220

As can be seen, the proposed positioning system can provide a centimeter level of accuracy in areas large enough for practical demining tasks and for reasonable and practically realizable systems' configurations. The high-accuracy zone could certainly be extended if the UWB stationary modules were more distributed around the area to be scanned, however for safety reasons it cannot always be achieved.

3.2. Positioning Accuracy

The accuracy of the EKF-PND filter was assessed and compared with the accuracy of EKF-CV and EKF-CA filters as well as with the accuracy of an NLS algorithm in MATLAB® for the C1 configuration of the system. The dynamics and observation models given by Equations (4) and (8) were used to generate the antenna trajectories and the parameters chosen during the simulations are given in Table 3. The choice of these parameters was done in such a way that the shape and duration of the resulting antenna trajectory resemble typical HH-GPR antenna trajectories.

Table 3. Parameters used in dynamics and observation models during simulations.

Parameter Name	Symbol	Value	Unit
Length of the HH-GPR handle	l	1.6	m
Starting angle of the scanning section	θ	−34.2	°
Nominal acceleration	a	0.25	$\frac{m}{s^2}$
Standard deviations of all the distance measuring errors	σ	0.02	m
x-coordinate of the initial sapper position	x_S	80	m
y-coordinate of the initial sapper position	y_S	50	m
Sapper's arm height	h	1.6	m
Period between two successive time steps	t	0.1	s
Power spectral density of disturbances u_{x_S}	S_{x_S}	$4{\cdot}10^{-3}$	$\frac{m^2}{s}$
Power spectral density of disturbances u_{y_S}	S_{y_S}	$4{\cdot}10^{-3}$	$\frac{m^2}{s}$
Power spectral density of disturbances u_a	S_a	$3{\cdot}10^{-3}$	$\frac{m^2}{s^5}$

The same standard deviations of all the distance measuring errors $\sigma_{Aj} = \sigma_{Sj} = \sigma$ and power spectral densities S_{x_S}, S_{y_S}, S_a given in Table 3 were used in the EKF-PND for setting the values of the **R** and **Q** matrices. The EKF-CV and EKF-CA also use $\sigma_{Aj} = \sigma_{Sj} = \sigma$ as given in Table 3, but as their dynamics models are different, their **Q** matrices required finding power spectral densities of different noises S_{v_x}, S_{v_y} or S_{a_x}, S_{a_y}. This was done during the mentioned tuning process and the obtained values are as follows: $S_{v_x} = S_{v_y} = 4.2{\cdot}10^{-3}\ \frac{m^2}{s^3}$ and $S_{a_x} = S_{a_y} = 6.1{\cdot}10^{-3}\ \frac{m^2}{s^5}$.

Firstly, the results of the antenna position estimation with the EKF-PND and the NLS algorithm were compared for various orientations of scanning sections and chosen results of these tests are presented in Figure 11. These experiments confirmed that the EKF-PND filter works properly and achieves a similar accuracy for various orientations of the central axis of the scanning section.

Next, a closer inspection of the estimation results was done for all the implemented algorithms for a chosen orientation of the central axis of the scanning section equal to 45°.

A comparison of HH-GPR antenna positions estimated with NLS, EKF-CV, EKF-CA, and EKF-PND is presented in Figure 12. As can be seen, all these algorithms are capable of properly estimating the antenna position, however, their accuracy is noticeably different and required further analysis, which will be presented further.

At this step of the simulations, the results of the estimation of other elements of the state vector **x** from the dynamics model given by Equation (4) were analyzed and they are presented in Figures 13–15. The angle θ between the horizontal projection of the antenna handle and the central axis of the scanning section, estimated with the EKF-PND filter, is shown in Figure 13. The results of angular velocity estimation are presented in Figure 14. Figure 15 contains an estimate of the acceleration a forcing the HH-GPR antenna movement. All these figures contain only results for the EKF-PND, as other algorithms do not estimate variables such as θ, ω, and a.

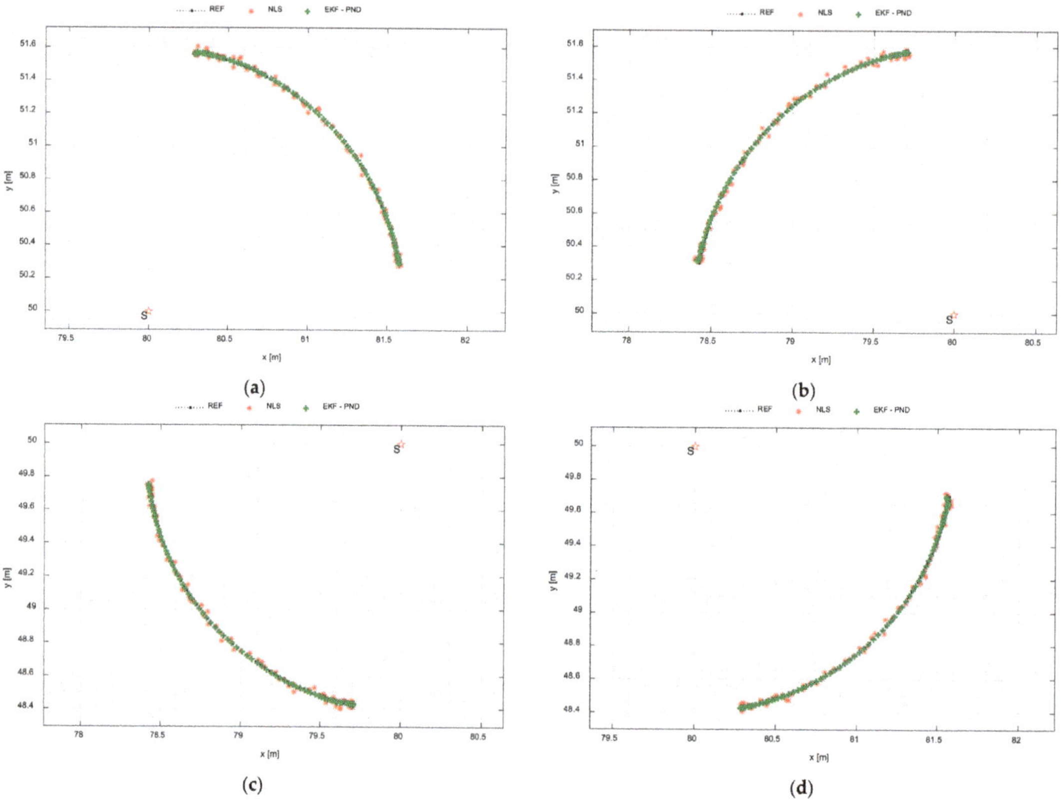

Figure 11. Estimated antenna positions with EKF-PND and NLS, for various orientations of the central axis of the scanning section: (**a**) 45°; (**b**) 135°; (**c**) 225°; (**d**) 315°.

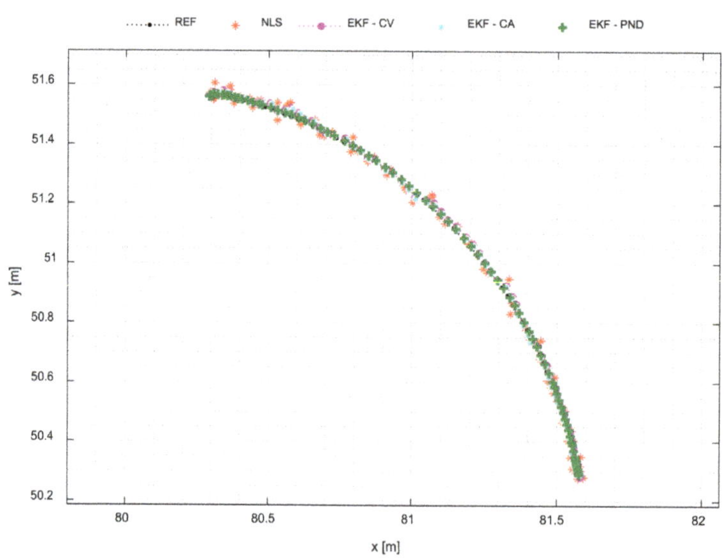

Figure 12. Comparison of antenna positions estimated with NLS, EKF-CV, EKF-CA, and EKF-PND.

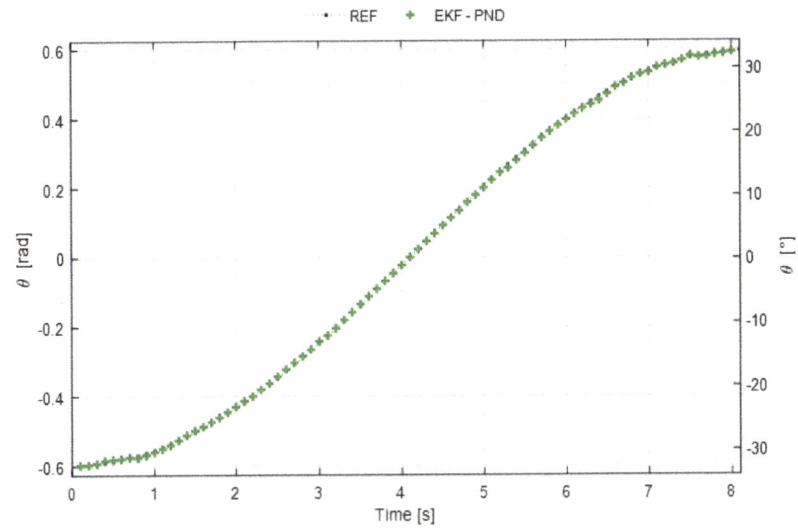

Figure 13. Angle between the horizontal projection of the antenna handle and the central axis of the scanning section estimated with EKF-PND.

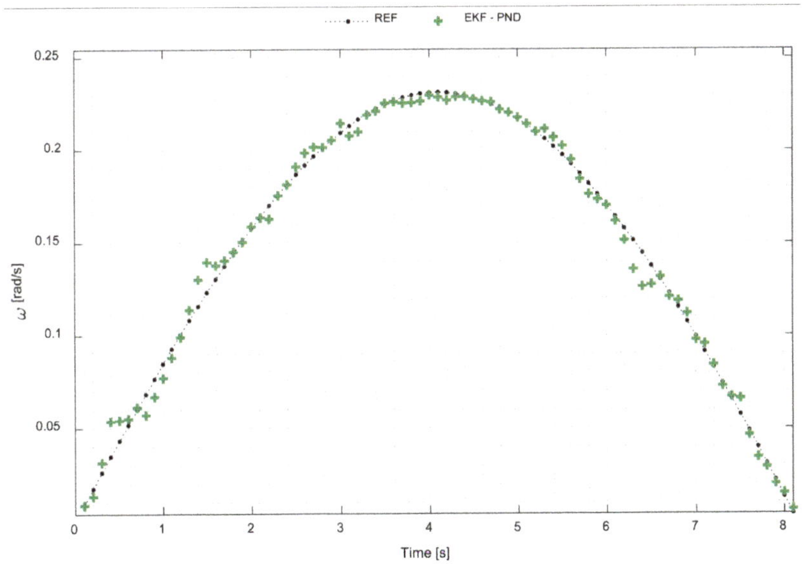

Figure 14. Angular velocity of the antenna motion estimated with EKF-PND.

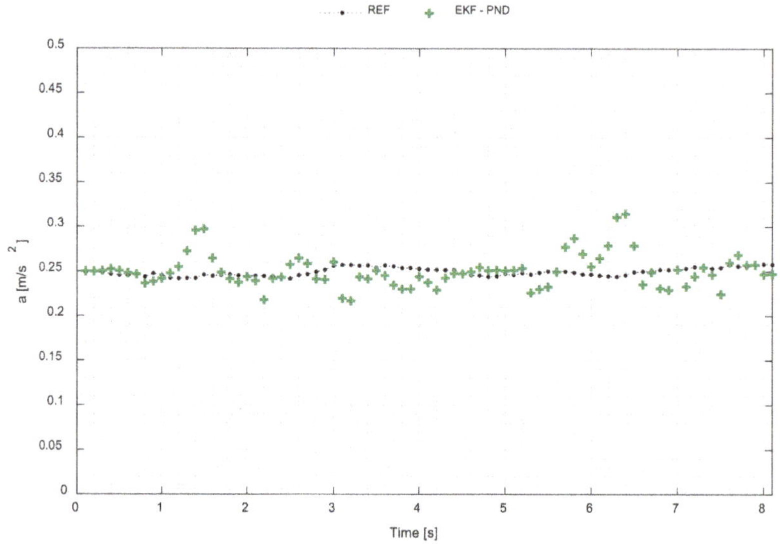

Figure 15. Acceleration estimated with EKF-PND.

To better compare the accuracy of estimation with various algorithms, we conducted a series of ten thousand simulations and calculated average RMS antenna position errors for the whole scanning sections for each realization of the simulations. The obtained RMSE values are shown in Figure 16. Single points in various colors are RMS antenna position errors obtained with NLS, EKF-CV, EKF-CA, and EKF-PND. Although they are changeable in various simulation runs, they form bands on noticeably different levels.

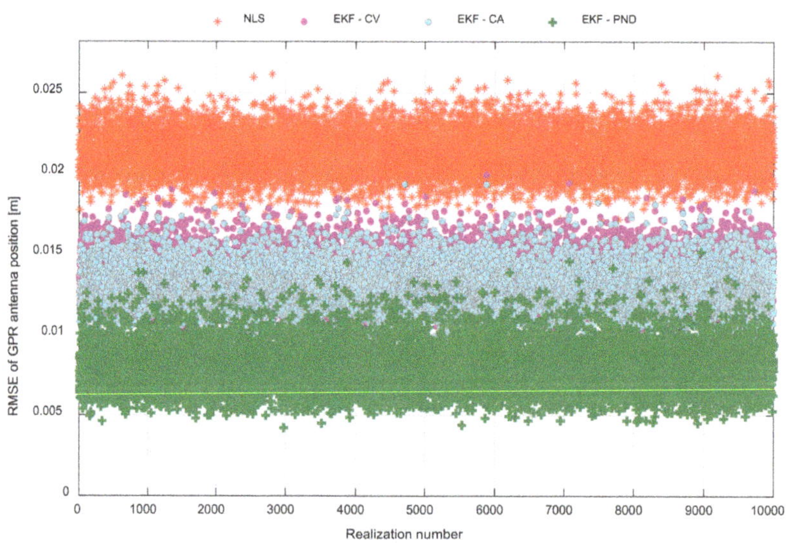

Figure 16. Comparison of RMS antenna position errors for NLS, EKF-CV, EKF-CA, and EKF-PND.

Based on the above results we created a histogram of RMS antenna position errors for NLS, EKF-CV, EKF-CA, and EKF-PND and it is presented in Figure 17. From this and the previous figure, one can conclude that the EKF-PND is more accurate than all other tested algorithms and the EKF-CV and EKF-CA perform similarly, but still better than the NLS algorithm. The EKF-CA is slightly more accurate than the EKF-CV.

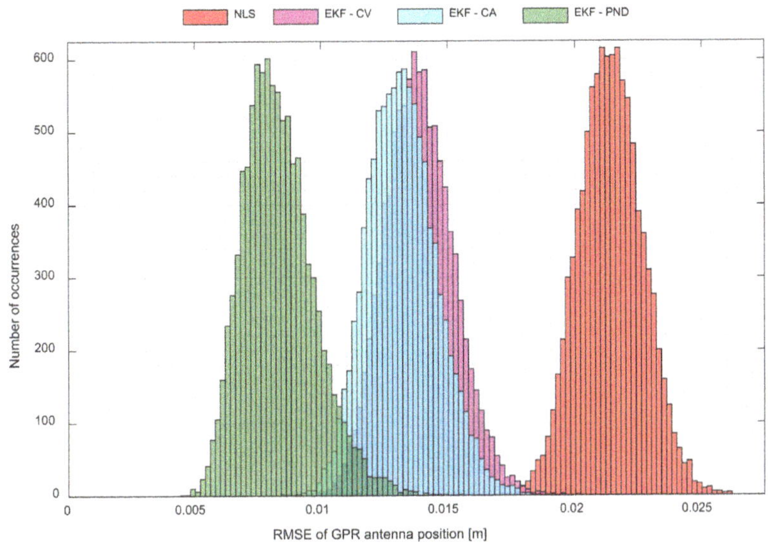

Figure 17. Histogram of RMS antenna position errors for NLS, EKF-CV, EKF-CA, and EKF-PND.

A comparison of numerical values of average RMS antenna position errors for all the realizations and for the NLS, EKF-CV, EKF-CA, and EKF-PND algorithms are given in Table 4. This table also presents percentage improvements of accuracy for EKF-CV and EKF-CA vs. NLS and EKF-PND versus all other algorithms. As can be seen, in the chosen simulation scenario, the EKF-PND provides positioning results about 40% more accurate than other tested EKFs and about 60% better than NLS.

Table 4. Average RMS antenna position errors for NLS, EKF-CV, EKF-CA, and EKF-PND.

	NLS	EKF-CV	EKF-CA	EKF-PND
Mean RMSE [cm]	2.14	1.39	1.32	0.83
Improvement vs. NLS [%]	—	35.2	38.3	61.1
Improvement vs. CV [%]	—	—	—	40.0
Improvement vs. CA [%]	—	—	—	36.9

4. Discussion

In this paper, an accurate positioning system dedicated as a supporting tool for a handheld ground penetrating radar was presented. The system uses ultrawideband radio technology for accurate distance measurements and processes them to estimate the GPR antenna position. Various estimation algorithms were used for this purpose, from NLS, through simple EKFs (EKF-CV and EKF-CA), based on those typically used in radiolocation and navigation CV and CA dynamics models, to the EKF-PND, based on the proposed by the authors' dynamics model derived from the model of a pendulum motion.

The results of simulations included in the paper have demonstrated that the proposed positioning system can provide a desired centimeter level of accuracy in areas large enough for practical demining tasks. They also have shown how the actual placement of UWB beacons influences the system's accuracy. It occurs that the smallest positioning errors are achievable in some distance in front of the area where the beacons are located and that the high-accuracy positioning zone is wider and deeper for a more extended baseline of the system.

Further experiments have confirmed that the EKF-PND filter works properly for various orientations of the central axis of the scanning section and have proved that using the proposed pendulum dynamics model ensures a better accuracy than the accuracy obtainable with other typically used dynamics models CV and CV. The simulations have shown that the EKF-PND provides positioning results about 40% more accurate than other tested EKFs (EKF-CV and EKF-CA) and about 60% better than NLS.

Author Contributions: Conceptualization, P.K. and T.K.; methodology, P.K. and T.K.; software, T.K.; validation, P.K. and T.K.; formal analysis, P.K. and T.K.; investigation, P.K. and T.K.; resources, T.K.; data curation, T.K.; writing—original draft preparation, P.K.; writing—review and editing, P.K. and T.K.; visualization, T.K.; supervision, P.K.; project administration, P.K.; funding acquisition, P.K. All authors have read and agreed to the published version of the manuscript.

Funding: This research was funded by the Military University of Technology under research project UGB 736.

Data Availability Statement: The data presented in this study are available on request from the corresponding author.

Acknowledgments: The authors express their gratitude towards Mateusz Pasternak from the Military University of Technology, Warsaw, Poland, for many fruitful discussions and valuable explanations regarding the ground penetrating radar technology.

Conflicts of Interest: The authors declare no conflict of interest. The funders had no role in the design of the study; in the collection, analyses, or interpretation of data; in the writing of the manuscript; or in the decision to publish the results.

Appendix A

Appendix A explains the method applied by the authors for calculating the fundamental matrix **F**, which is necessary to perform each prediction step in the EKF.

The **F** matrix is a Jacobian of the nonlinear vector-valued function $\mathbf{f}(\mathbf{x})$ from the continuous dynamics model (4). It is obtained by calculating the first-order partial derivatives of the $\mathbf{f}(\mathbf{x})$ function with respect to all the elements of the state vector $\mathbf{x}$ [60–62].

The numerical values of its elements are calculated at each processing step k, based on the state vector $\hat{\mathbf{x}}_{k-1|k-1}$, which is estimated at the previous step $k-1$, and therefore the **F** matrix at this step can be more specifically written as $\mathbf{F}_{k-1}$.

A general formula for calculating the $\mathbf{F}_{k-1}$ matrix is given by (A1) and the equations (A2)–(A10) explain the way of calculating all its individual non-zero elements.

$$\mathbf{F}_{k-1} = \left[\frac{\partial \mathbf{f}(\mathbf{x})}{\partial \mathbf{x}}\right]\Bigg|_{\mathbf{x}=\hat{\mathbf{x}}_{k-1|k-1}} = \begin{bmatrix} \frac{\partial f_1}{\partial x_A} & \frac{\partial f_1}{\partial y_A} & \frac{\partial f_1}{\partial x_S} & \frac{\partial f_1}{\partial y_S} & \frac{\partial f_1}{\partial \theta} & \frac{\partial f_1}{\partial \omega} & \frac{\partial f_1}{\partial a} \\ \frac{\partial f_2}{\partial x_A} & \frac{\partial f_2}{\partial y_A} & \frac{\partial f_2}{\partial x_S} & \frac{\partial f_2}{\partial y_S} & \frac{\partial f_2}{\partial \theta} & \frac{\partial f_2}{\partial \omega} & \frac{\partial f_2}{\partial a} \\ \frac{\partial f_3}{\partial x_A} & \frac{\partial f_3}{\partial y_A} & \frac{\partial f_3}{\partial x_S} & \frac{\partial f_3}{\partial y_S} & \frac{\partial f_3}{\partial \theta} & \frac{\partial f_3}{\partial \omega} & \frac{\partial f_3}{\partial a} \\ \frac{\partial f_4}{\partial x_A} & \frac{\partial f_4}{\partial y_A} & \frac{\partial f_4}{\partial x_S} & \frac{\partial f_4}{\partial y_S} & \frac{\partial f_4}{\partial \theta} & \frac{\partial f_4}{\partial \omega} & \frac{\partial f_4}{\partial a} \\ \frac{\partial f_5}{\partial x_A} & \frac{\partial f_5}{\partial y_A} & \frac{\partial f_5}{\partial x_S} & \frac{\partial f_5}{\partial y_S} & \frac{\partial f_5}{\partial \theta} & \frac{\partial f_5}{\partial \omega} & \frac{\partial f_5}{\partial a} \\ \frac{\partial f_6}{\partial x_A} & \frac{\partial f_6}{\partial y_A} & \frac{\partial f_6}{\partial x_S} & \frac{\partial f_6}{\partial y_S} & \frac{\partial f_6}{\partial \theta} & \frac{\partial f_6}{\partial \omega} & \frac{\partial f_6}{\partial a} \\ \frac{\partial f_7}{\partial x_A} & \frac{\partial f_7}{\partial y_A} & \frac{\partial f_7}{\partial x_S} & \frac{\partial f_7}{\partial y_S} & \frac{\partial f_7}{\partial \theta} & \frac{\partial f_7}{\partial \omega} & \frac{\partial f_7}{\partial a} \end{bmatrix}\Bigg|_{\mathbf{x}=\hat{\mathbf{x}}_{k-1|k-1}} \tag{A1}$$

$$\left(\frac{\partial f_1}{\partial y_A}\right)\Bigg|_{\hat{\mathbf{x}}_{k-1|k-1}} = \frac{\partial}{\partial y_A}(\omega(y_A - y_S))\Bigg|_{\hat{\mathbf{x}}_{k-1|k-1}} = \hat{\omega}_{k-1|k-1}, \tag{A2}$$

$$\left(\frac{\partial f_1}{\partial y_S}\right)\Bigg|_{\hat{\mathbf{x}}_{k-1|k-1}} = \frac{\partial}{\partial y_S}(\omega(y_A - y_S))\Bigg|_{\hat{\mathbf{x}}_{k-1|k-1}} = -\hat{\omega}_{k-1|k-1}, \tag{A3}$$

$$\left(\frac{\partial f_1}{\partial \omega}\right)\bigg|_{\hat{\mathbf{x}}_{k-1|k-1}} = \frac{\partial}{\partial \omega}(\omega(y_A - y_S))\bigg|_{\hat{\mathbf{x}}_{k-1|k-1}} = \hat{y}_{A_{k-1|k-1}} - \hat{y}_{S_{k-1|k-1}}, \tag{A4}$$

$$\left(\frac{\partial f_2}{\partial x_A}\right)\bigg|_{\hat{\mathbf{x}}_{k-1|k-1}} = \frac{\partial}{\partial x_A}(-\omega(x_A - x_S))\bigg|_{\hat{\mathbf{x}}_{k-1|k-1}} = -\hat{\omega}_{k-1|k-1}, \tag{A5}$$

$$\left(\frac{\partial f_2}{\partial x_S}\right)\bigg|_{\hat{\mathbf{x}}_{k-1|k-1}} = \frac{\partial}{\partial x_S}(-\omega(x_A - x_S))\bigg|_{\hat{\mathbf{x}}_{k-1|k-1}} = \hat{\omega}_{k-1|k-1}, \tag{A6}$$

$$\left(\frac{\partial f_2}{\partial \omega}\right)\bigg|_{\hat{\mathbf{x}}_{k-1|k-1}} = \frac{\partial}{\partial \omega}(-\omega(x_A - x_S))\bigg|_{\hat{\mathbf{x}}_{k-1|k-1}} = \hat{x}_{S_{k-1|k-1}} - \hat{x}_{A_{k-1|k-1}}, \tag{A7}$$

$$\left(\frac{\partial f_5}{\partial \omega}\right)\bigg|_{\hat{\mathbf{x}}_{k-1|k-1}} = \frac{\partial \omega}{\partial \omega}\bigg|_{\hat{\mathbf{x}}_{k-1|k-1}} = 1, \tag{A8}$$

$$\left(\frac{\partial f_6}{\partial \theta}\right)\bigg|_{\hat{\mathbf{x}}_{k-1|k-1}} = \frac{\partial}{\partial \theta}\left(-\frac{a}{l}\sin\theta\right)\bigg|_{\hat{\mathbf{x}}_{k-1|k-1}} = -\frac{\hat{a}_{k-1|k-1}\cos\hat{\theta}_{k-1|k-1}}{l}, \tag{A9}$$

$$\left(\frac{\partial f_6}{\partial a}\right)\bigg|_{\hat{\mathbf{x}}_{k-1|k-1}} = \frac{\partial}{\partial a}\left(-\frac{a}{l}\sin\theta\right)\bigg|_{\hat{\mathbf{x}}_{k-1|k-1}} = -\frac{\sin\hat{\theta}_{k-1|k-1}}{l}. \tag{A10}$$

The final shape of the fundamental matrix $\mathbf{F}_{k-1}$ can be obtained by placing all its individual elements given by the Equations (A2)–(A10) at appropriate positions in (A1) and it is given below as Equation (A11).

$$\mathbf{F}_{k-1} = \begin{bmatrix} 0 & \hat{\omega}_{k-1|k-1} & 0 & -\hat{\omega}_{k-1|k-1} & 0 & \hat{y}_{A_{k-1|k-1}} - \hat{y}_{S_{k-1|k-1}} & 0 \\ -\hat{\omega}_{k-1|k-1} & 0 & \hat{\omega}_{k-1|k-1} & 0 & 0 & \hat{x}_{S_{k-1|k-1}} - \hat{x}_{A_{k-1|k-1}} & 0 \\ 0 & 0 & 0 & 0 & 0 & 0 & 0 \\ 0 & 0 & 0 & 0 & 0 & 0 & 0 \\ 0 & 0 & 0 & 0 & 0 & 1 & 0 \\ 0 & 0 & 0 & 0 & -\frac{\hat{a}_{k-1|k-1}\cos\hat{\theta}_{k-1|k-1}}{l} & 0 & -\frac{\sin\hat{\theta}_{k-1|k-1}}{l} \\ 0 & 0 & 0 & 0 & 0 & 0 & 0 \end{bmatrix} \tag{A11}$$

Appendix B

Appendix B explains the method applied by the authors for calculating the observation matrix $\mathbf{H}$, which is necessary to perform each correction step in the EKF.

The $\mathbf{H}$ matrix is a Jacobian of the nonlinear vector-valued function $\mathbf{h}(\mathbf{x})$ from the observation model (8). It is obtained by calculating the first-order partial derivatives of the $\mathbf{h}(\mathbf{x})$ function with respect to all the elements of the state vector $\mathbf{x}$ [60–62].

The numerical values of its elements are calculated at each processing step k, based on the predicted state vector $\hat{\mathbf{x}}_{k|k-1}$. Thus, the $\mathbf{H}$ matrix at the step k can be more specifically written as $\mathbf{H}_k$.

A general formula for calculating the $\mathbf{H}_k$ matrix is given by (A12) and the equations (A13)–(A16) explain the way of calculating all its individual non-zero elements.

$$\mathbf{H}_k = \left[\frac{\partial \mathbf{h}(\mathbf{x})}{\partial \mathbf{x}}\right]\Bigg|_{\mathbf{x}=\hat{\mathbf{x}}_{k|k-1}} = \begin{bmatrix} \frac{\partial d_{A1}}{\partial x_A} & \frac{\partial d_{A1}}{\partial y_A} & \frac{\partial d_{A1}}{\partial x_S} & \frac{\partial d_{A1}}{\partial y_S} & \frac{\partial d_{A1}}{\partial \theta} & \frac{\partial d_{A1}}{\partial \omega} & \frac{\partial d_{A1}}{\partial a} \\ \frac{\partial d_{A2}}{\partial x_A} & \frac{\partial d_{A2}}{\partial y_A} & \frac{\partial d_{A2}}{\partial x_S} & \frac{\partial d_{A2}}{\partial y_S} & \frac{\partial d_{A2}}{\partial \theta} & \frac{\partial d_{A2}}{\partial \omega} & \frac{\partial d_{A2}}{\partial a} \\ \frac{\partial d_{A3}}{\partial x_A} & \frac{\partial d_{A3}}{\partial y_A} & \frac{\partial d_{A3}}{\partial x_S} & \frac{\partial d_{A3}}{\partial y_S} & \frac{\partial d_{A3}}{\partial \theta} & \frac{\partial d_{A3}}{\partial \omega} & \frac{\partial d_{A3}}{\partial a} \\ \frac{\partial d_{A4}}{\partial x_A} & \frac{\partial d_{A4}}{\partial y_A} & \frac{\partial d_{A4}}{\partial x_S} & \frac{\partial d_{A4}}{\partial y_S} & \frac{\partial d_{A4}}{\partial \theta} & \frac{\partial d_{A4}}{\partial \omega} & \frac{\partial d_{A4}}{\partial a} \\ \frac{\partial d_{S1}}{\partial x_A} & \frac{\partial d_{S1}}{\partial y_A} & \frac{\partial d_{S1}}{\partial x_S} & \frac{\partial d_{S1}}{\partial y_S} & \frac{\partial d_{S1}}{\partial \theta} & \frac{\partial d_{S1}}{\partial \omega} & \frac{\partial d_{S1}}{\partial a} \\ \frac{\partial d_{S2}}{\partial x_A} & \frac{\partial d_{S2}}{\partial y_A} & \frac{\partial d_{S2}}{\partial x_S} & \frac{\partial d_{S2}}{\partial y_S} & \frac{\partial d_{S2}}{\partial \theta} & \frac{\partial d_{S2}}{\partial \omega} & \frac{\partial d_{S2}}{\partial a} \\ \frac{\partial d_{S3}}{\partial x_A} & \frac{\partial d_{S3}}{\partial y_A} & \frac{\partial d_{S3}}{\partial x_S} & \frac{\partial d_{S3}}{\partial y_S} & \frac{\partial d_{S3}}{\partial \theta} & \frac{\partial d_{S3}}{\partial \omega} & \frac{\partial d_{S3}}{\partial a} \\ \frac{\partial d_{S4}}{\partial x_A} & \frac{\partial d_{S4}}{\partial y_A} & \frac{\partial d_{S4}}{\partial x_S} & \frac{\partial d_{S4}}{\partial y_S} & \frac{\partial d_{S4}}{\partial \theta} & \frac{\partial d_{S4}}{\partial \omega} & \frac{\partial d_{S4}}{\partial a} \end{bmatrix}\Bigg|_{\mathbf{x}=\hat{\mathbf{x}}_{k|k-1}}, \tag{A12}$$

$$\left(\frac{\partial d_{Aj}}{\partial x_A}\right)\Bigg|_{\hat{\mathbf{x}}_{k|k-1}} = \frac{\partial}{\partial x_A}\left(\sqrt{(x_A(k)-x_j)^2+(y_A(k)-y_j)^2}\right)\Bigg|_{\hat{\mathbf{x}}_{k|k-1}} = \frac{\hat{x}_{A_{k|k-1}}-x_j}{\sqrt{\left(\hat{x}_{A_{k|k-1}}-x_j\right)^2+\left(\hat{y}_{A_{k|k-1}}-y_j\right)^2}}, \tag{A13}$$

$$\left(\frac{\partial d_{Aj}}{\partial y_A}\right)\Bigg|_{\hat{\mathbf{x}}_{k|k-1}} = \frac{\partial}{\partial y_A}\left(\sqrt{(x_A(k)-x_j)^2+(y_A(k)-y_j)^2}\right)\Bigg|_{\hat{\mathbf{x}}_{k|k-1}} = \frac{\hat{y}_{A_{k|k-1}}-y_j}{\sqrt{\left(\hat{x}_{A_{k|k-1}}-x_j\right)^2+\left(\hat{y}_{A_{k|k-1}}-y_j\right)^2}}, \tag{A14}$$

$$\left(\frac{\partial d_{Sj}}{\partial x_S}\right)\Bigg|_{\hat{\mathbf{x}}_{k|k-1}} = \frac{\partial}{\partial x_S}\left(\sqrt{(x_S(k)-x_j)^2+(y_S(k)-y_j)^2+h^2}\right)\Bigg|_{\hat{\mathbf{x}}_{k|k-1}} = \frac{\hat{x}_{S_{k|k-1}}-x_j}{\sqrt{\left(\hat{x}_{S_{k|k-1}}-x_j\right)^2+\left(\hat{y}_{S_{k|k-1}}-y_j\right)^2+h^2}}, \tag{A15}$$

$$\left(\frac{\partial d_{Sj}}{\partial y_S}\right)\Bigg|_{\hat{\mathbf{x}}_{k|k-1}} = \frac{\partial}{\partial y_S}\left(\sqrt{(x_S(k)-x_j)^2+(y_S(k)-y_j)^2+h^2}\right)\Bigg|_{\hat{\mathbf{x}}_{k|k-1}} = \frac{\hat{y}_{S_{k|k-1}}-y_j}{\sqrt{\left(\hat{x}_{S_{k|k-1}}-x_j\right)^2+\left(\hat{y}_{S_{k|k-1}}-y_j\right)^2+h^2}}. \tag{A16}$$

The final shape of the observation matrix $\mathbf{H}_k$, obtained by placing all its individual elements given by the Equations (A13)–(A16) at appropriate positions in (A12) is given below as Equation (A17).

$$\mathbf{H}_k = \begin{bmatrix} \frac{\hat{x}_{A_{k|k-1}}-x_1}{\hat{d}_{A1_{k|k-1}}} & \frac{\hat{y}_{A_{k|k-1}}-y_1}{\hat{d}_{A1_{k-1}}} & 0 & 0 & 0 & 0 & 0 \\ \frac{\hat{x}_{A_{k|k-1}}-x_2}{\hat{d}_{A2_{k|k-1}}} & \frac{\hat{y}_{A_{k|k-1}}-y_2}{\hat{d}_{A2_{k-1}}} & 0 & 0 & 0 & 0 & 0 \\ \frac{\hat{x}_{A_{k|k-1}}-x_3}{\hat{d}_{A3_{k|k-1}}} & \frac{\hat{y}_{A_{k|k-1}}-y_3}{\hat{d}_{A3_{k-1}}} & 0 & 0 & 0 & 0 & 0 \\ \frac{\hat{x}_{A_{k|k-1}}-x_4}{\hat{d}_{A4_{k|k-1}}} & \frac{\hat{y}_{A_{k|k-1}}-y_4}{\hat{d}_{A4_{k-1}}} & 0 & 0 & 0 & 0 & 0 \\ 0 & 0 & \frac{\hat{x}_{S_{k|k-1}}-x_1}{\hat{d}_{S1_{k|k-1}}} & \frac{\hat{y}_{S_{k|k-1}}-y_1}{\hat{d}_{S1_{k|k-1}}} & 0 & 0 & 0 \\ 0 & 0 & \frac{\hat{x}_{S_{k|k-1}}-x_2}{\hat{d}_{S2_{k|k-1}}} & \frac{\hat{y}_{S_{k|k-1}}-y_2}{\hat{d}_{S2_{k|k-1}}} & 0 & 0 & 0 \\ 0 & 0 & \frac{\hat{x}_{S_{k|k-1}}-x_3}{\hat{d}_{S3_{k|k-1}}} & \frac{\hat{y}_{S_{k|k-1}}-y_3}{\hat{d}_{S3_{k|k-1}}} & 0 & 0 & 0 \\ 0 & 0 & \frac{\hat{x}_{S_{k|k-1}}-x_4}{\hat{d}_{S4_{k|k-1}}} & \frac{\hat{y}_{S_{k|k-1}}-y_4}{\hat{d}_{S4_{k|k-1}}} & 0 & 0 & 0 \end{bmatrix} \tag{A17}$$

To keep the notation of Equation (A17) more compact, the following auxiliary variables were introduced in the denominators of respective fractions:

$$\hat{d}_{Aj_{k|k-1}} = \sqrt{\left(\hat{x}_{A_{k|k-1}} - x_j\right)^2 + \left(\hat{y}_{A_{k|k-1}} - y_j\right)^2}, \text{ for } j = 1\ldots4, \tag{A18}$$

$$\hat{d}_{Sj_{k|k-1}} = \sqrt{\left(\hat{x}_{S_{k|k-1}} - x_j\right)^2 + \left(\hat{y}_{S_{k|k-1}} - y_j\right)^2 + h^2}, \text{ for } j = 1\ldots4. \tag{A19}$$

References

1. Landmine Monitor 2017. *International Campaign to Ban Landmines*; Cluster Munition Coalition (ICBL-CMC): Geneva, Switzerland, 2017.
2. Landmine Monitor 2021. *23rd Annual ed., International Campaign to Ban Landmines*; Cluster Munition Coalition (ICBL-CMC): Geneva, Switzerland, 2021.
3. Daniels, D.J. *Ground Penetrating Radar*, 2nd ed.; IEE Radar, Sonar and Navigation Series 15; The Institution of Electrical Engineers: London, UK, 2004. [CrossRef]
4. Kruger, H.; Ewald, H. Handheld metal detector with online visualisation and classification for the humanitarian mine clearance. In Proceedings of the 2008 IEEE SENSORS, Lecce, Italy, 26–29 October 2008. [CrossRef]
5. Bryakin, I.V.; Bochkarev, I.V.; Khramshin, V.R.; Khramshina, E.A. Developing a Combined Method for Detection of Buried Metal Objects. *Machines* **2021**, *9*, 92. [CrossRef]
6. Guelle, D.; Smith, A.; Lewis, A.; Bloodworth, T. *Metal Detector Handbook for Humanitarian Demining*; Office for Official Publications of the European Communities: Luxembourg, 2003.
7. Furuta, K.; Ishikawa, J. *Anti-Personnel Landmine Detection for Humanitarian Demining: The Current Situation and Future Direction for Japanese Research and Development*; Springer-Verlag London Limited: London, UK, 2009. [CrossRef]
8. Barrowes, B.; Prishvin, M.; Jutras, G.; Shubitidze, F. High-Frequency Electromagnetic Induction (HFEMI) Sensor Results from IED Constituent Parts. *Remote Sens.* **2019**, *11*, 2355. [CrossRef]
9. Bruschini, C.; Gros, B.; Guerne, F.; Piece, P.-Y.; Carmona, O. Ground penetrating radar and induction coil sensor imaging for antipersonnel mines detection. In Proceedings of the 6th International Conference on Ground Penetrating Radar (GPR'96), Sendai, Japan, 30 September–3 October 1996.
10. Szynkarczyk, P.; Wrona, J.; Pasternak, M.; Rubiec, A.; Serafin, P. Unmanned Ground Vehicle Equipped with Ground Penetrating Radar for Improvised Explosives Detection. *J. Autom. Mob. Robot. Intell. Syst.* **2022**, *15*, 20–31. [CrossRef]
11. Daniels, D. *Unexploded Ordnance Detection and Mitigation*; NATO Science for Peace and Security Series B: Physics and Biophysics; Byrnes, J., Ed.; Springer Science & Business Media: Dordrecht, The Netherlands, 2008. [CrossRef]
12. Siegel, R. Land mine detection. *IEEE Instrum. Meas. Mag.* **2002**, *5*, 22–28. [CrossRef]
13. Bechtel, T.; Truskavetsky, S.; Capineri, L.; Pochanin, G.; Simic, N.; Viatkin, K.; Sherstyuk, A.; Byndych, T.; Falorni, P.; Bulletti, A.; et al. A survey of electromagnetic characteristics of soils in the Donbass region (Ukraine) for evaluation of the applicability of GPR and MD for landmine detection. In Proceedings of the 16th International Conference on Ground Penetrating Radar (GPR), Hong Kong, China, 13–16 June 2016. [CrossRef]
14. Ebrahim, S.M.; Medhat, N.I.; Mansour, K.K.; Gaber, A. Examination of soil effect upon GPR detectability of landmine with different orientations. *NRIAG J. Astron. Geophys.* **2018**, *7*, 90–98. [CrossRef]
15. Jol, H.M. *Ground Penetrating Radar. Theory and Applications*, 1st ed.; Elsevier Science: Amsterdam, The Netherlands, 2009.
16. Pasternak, M. *Radarowa Penetracja Gruntu*; Wydawnictwa Komunikacji i Łączności WKŁ: Sulejówek, Poland, 2015.
17. Li, W.; Cui, X.; Guo, L.; Chen, J.; Chen, X.; Cao, X. Tree Root Automatic Recognition in Ground Penetrating Radar Profiles Based on Randomized Hough Transform. *Remote Sens.* **2016**, *8*, 430. [CrossRef]
18. Karsznia, K.R.; Onyszko, K.; Borkowska, S. Accuracy Tests and Precision Assessment of Localizing Underground Utilities Using GPR Detection. *Sensors* **2021**, *21*, 6765. [CrossRef]
19. Sun, H.; Pashoutani, S.; Zhu, J. Nondestructive Evaluation of Concrete Bridge Decks with Automated Acoustic Scanning System and Ground Penetrating Radar. *Sensors* **2018**, *18*, 1955. [CrossRef]
20. Dong, Z.; Ye, S.; Gao, Y.; Fang, G.; Zhang, X.; Xue, Z.; Zhang, T. Rapid Detection Methods for Asphalt Pavement Thicknesses and Defects by a Vehicle-Mounted Ground Penetrating Radar (GPR) System. *Sensors* **2016**, *16*, 2067. [CrossRef]
21. Kelly, T.B.; Angel, M.N.; O'Connor, D.E.; Huff, C.C.; Morris, L.E.; Wach, G.D. A novel approach to 3D modelling ground-penetrating radar (GPR) data—A case study of a cemetery and applications for criminal investigation. *Forensic Sci. Int.* **2021**, *325*, 110882. [CrossRef]
22. Harari, Z. Ground-penetrating radar (GPR) for imaging stratigraphic features and groundwater in sand dunes. *J. Appl. Geophys.* **1996**, *36*, 43–52. [CrossRef]
23. Zan, Y.; Li, Z.; Su, G.; Zhang, X. An innovative vehicle-mounted GPR technique for fast and efficient monitoring of tunnel lining structural conditions. *Case Stud. Nondestruct. Test. Eval.* **2016**, *6*, 63–69. [CrossRef]
24. Ahmed, S.A.; El Qassas, R.A.Y.; El Salam, H.F.A. Mapping the possible buried archaeological targets using magnetic and ground penetrating radar data, Fayoum, Egypt. *Egypt. J. Remote Sens. Space Sci.* **2020**, *23*, 321–332. [CrossRef]

25. Marsh, L.A.; van Verre, W.; Davidson, J.L.; Gao, X.; Podd, F.J.W.; Daniels, D.J.; Peyton, A.J. Combining Electromagnetic Spectroscopy and Ground-Penetrating Radar for the Detection of Anti-Personnel Landmines. *Sensors* **2019**, *19*, 3390. [CrossRef]
26. Ivashov, S.I.; Makarenkov, V.; Masterkov, A.V.; Razevig, V.V.; Sablin, V.N.; Sheyko, A.P.; Vasilyev, I.A. Remote control mine-detection system with GPR and metal detector. In Proceedings of the SPIE 4084, 8th International Conference on Ground Penetrating Radar, Gold Coast, Australia, 23–26 May 2000. [CrossRef]
27. Sato, M.; Yokota, Y.; Takahashi, K. ALIS: GPR System for Humanitarian Demining and Its Deployment in Cambodia. *J. Korean Inst. Electromagn. Eng. Sci.* **2012**, *12*, 55–62. [CrossRef]
28. Daniels, D.J.; Curtis, P.; Lockwood, O. Classification of landmines using GPR. In Proceedings of the 2008 IEEE Radar Conference, Rome, Italy, 26–30 May 2008. [CrossRef]
29. De Benedetto, D.; Montemurro, F.; Diacono, M. Mapping an Agricultural Field Experiment by Electromagnetic Induction and Ground Penetrating Radar to Improve Soil Water Content Estimation. *Agronomy* **2019**, *9*, 638. [CrossRef]
30. Frigui, H.; Zhang, L.; Gader, P.D. Context-Dependent Multisensor Fusion and Its Application to Land Mine Detection. *IEEE Trans. Geosci. Remote Sens.* **2010**, *48*, 2528–2543. [CrossRef]
31. AMULET Vehicle-Mounted EODS. Available online: https://www.chelton.com/land/explosive-ordnance-detection-systems/amulet-vehicle-mounted-eods/ (accessed on 29 August 2022).
32. Super Buffalo. Available online: https://www.defensemedianetwork.com/stories/general-dynamics-develops-super-buffalo-to-enhance-counter-ied-operations-part-i-multi-functionality/ (accessed on 29 August 2022).
33. Ground Penetrating Radar System. Available online: https://www.exponent.com/experience/ground-penetrating-radar-system, (accessed on 29 August 2022).
34. Husky Mounted Detection System (HMDS). Available online: https://www.militaryaerospace.com/sensors/article/14175811/groundpenetrating-radar-ied-detection (accessed on 29 August 2022).
35. Chemring Sensors & Electronic System. Available online: https://www.chemring.com/what-we-do/sensors-and-information/ied-detection, (accessed on 4 September 2022).
36. Arvaniti, A.; Orzeł-Tatarczuk, E.; Popkowski, J.; Kaczmarek, P.; Kawalec, A.M.; Pasternak, M.L. Mobilna platforma z ultraszerokopasmowym radarem do wykrywania ukrytych w ziemi niebezpiecznych obiektów. In *Urządzenia i systemy radioelektroniczne. Wybrane problemy 2*; Wojskowa Akademia Techniczna: Warszawa, Poland, 2012.
37. The GPR for UGV project. Available online: https://www.australiandefence.com.au/land/sme-proves-radar-equipped-ugv-concept-for-mine-and-ied-detection (accessed on 29 August 2022).
38. Defence Research & Development Organisation. Available online: https://www.drdo.gov.in/muntra-m (accessed on 29 August 2022).
39. Knox, M.; Torrione, P.; Collins, L.; Morton, K., Jr. Buried threat detection using a handheld ground penetrating radar system. In Proceedings of the SPIE 9454, Detection and Sensing of Mines, Explosive Objects, and Obscured Targets XX, 94540F, Baltimore, MD, USA, 20–23 April 2015. [CrossRef]
40. Ho, K.C.; Collins, L.M.; Huettel, L.G.; Gader, P.D. Discrimination mode processing for EMI and GPR sensors for hand-held land mine detection. *IEEE Trans. Geosci. Remote Sens.* **2004**, *42*, 249–263. [CrossRef]
41. Kaniewski, P.; Kraszewski, T. Novel Algorithm for Position Estimation of Handheld Ground-Penetrating Radar Antenna. In Proceedings of the 21st International Radar Symposium (IRS), Warsaw, Poland, 5–8 October 2020. [CrossRef]
42. Pasternak, M.; Miluski, W.; Czarnecki, W.; Pietrasiński, J. An optoelectronic-inertial system for handheld GPR positioning. In Proceedings of the 15th International Radar Symposium (IRS), Gdansk, Poland, 16–18 June 2014. [CrossRef]
43. Zoubir, A.M.; Chant, I.J.; Brown, C.L.; Barkat, B.; Abeynayake, C. Signal processing techniques for landmine detection using impulse ground penetrating radar. *IEEE Sens. J.* **2002**, *2*, 41–51. [CrossRef]
44. Lee, J.S.; Nguyen, C.; Scullion, T. A novel, compact, low-cost, impulse ground-penetrating radar for nondestructive evaluation of pavements. *IEEE Trans. Instrum. Meas.* **2004**, *53*, 1502–1509. [CrossRef]
45. Suksmono, A.B.; Bharata, E.; Lestari, A.A.; Yarovoy, A.G.; Ligthart, L.P. Compressive Stepped-Frequency Continuous-Wave Ground-Penetrating Radar. *IEEE Geosci. Remote Sens. Lett.* **2010**, *7*, 665–669. [CrossRef]
46. Iftimie, N.; Savin, A.; Steigmann, R.; Dobrescu, G.S. Underground Pipeline Identification into a Non-Destructive Case Study Based on Ground-Penetrating Radar Imaging. *Remote Sens.* **2021**, *13*, 3494. [CrossRef]
47. Bigman, D.P.; Day, D.J. Ground penetrating radar inspection of a large concrete spillway: A case-study using SFCW GPR at a hydroelectric dam. *Case Stud. Constr. Mater.* **2022**, *16*, e00975. [CrossRef]
48. Jing, H.; Vladimirova, T. Novel algorithm for landmine detection using C-scan ground penetrating radar signals. In Proceedings of the Seventh International Conference on Emerging Security Technologies (EST), Canterbury, UK, 6–8 September 2017; pp. 68–73. [CrossRef]
49. Grasmueck, M.; Viggiano, D.A. Integration of Ground-Penetrating Radar and Laser Position Sensors for Real-Time 3-D Data Fusion. *IEEE Trans. Geosci. Remote Sens.* **2007**, *45*, 130–137. [CrossRef]
50. Doerksen, K.; McNaughton, A. Positioning system for ground penetrating radar instruments. U.S. Patent US 2003/0112170 A1, 19 June 2003.
51. Pasternak, M.; Kaczmarek, P. Continuous wave ground penetrating radars: State of the art. In Proceedings of the SPIE, Event: XII Conference on Reconnaissance and Electronic Warfare Systems, Oltarzew, Poland, 19–21 November 2018; SPIE: Bellingham, WA, USA, 2019; pp. 1–7. [CrossRef]

52. Ferrara, V.; Pietrelli, A.; Chicarella, S.; Pajewski, L. GPR/GPS/IMU system as buried objects locator. *Measurement* **2018**, *114*, 534–541. [CrossRef]
53. Barzaghi, R.; Cazzaniga, N.E.; Pagliari, D.; Pinto, L. Vision-Based Georeferencing of GPR in Urban Areas. *Sensors* **2016**, *16*, 132. [CrossRef]
54. Kaniewski, P.; Kraszewski, T.; Pasek, P. UWB-Based Positioning System for Supporting Lightweight Handheld Ground-Penetrating Radar. In Proceedings of the IEEE International Conference on Microwaves, Antennas, Communications and Electronic Systems (COMCAS), Tel Aviv, Israel, 4–6 November 2019; pp. 1–4. [CrossRef]
55. TDSR. *Data Sheet/User Guide P440 UWB Module*; TDSR: Petersburg, TN, USA, 2020.
56. Awrejcewicz, J. Mathematical and Physical Pendulum. In *Classical Mechanics. Advances in Mechanics and Mathematics, 2012th Edition*; Springer: New York, NY, USA, 2012; Volume 29. [CrossRef]
57. Bar-Shalom, Y.; Li, X.R.; Kirubarajan, T. *Estimation with Applications to Tracking and Navigation: Theory, Algorithms and Software*; Wiley-Interscience; John Wiley & Sons, Inc.: Hoboken, NJ, USA, 2001.
58. Balakrishnan, A.V. *Introduction to Random Processes in Engineering*; John Wiley & Sons: Hoboken, NJ, USA, 1995.
59. Szabados, T. An elementary introduction to the Wiener process and stochastic integrals. In *Studia Scientiarum Mathematicarum Hungarica*; Akadémiai Kiadó: Budapest, Hungary, 2010.
60. Brown, R.G.; Hwang, P.Y.C. *Introduction to Random Signals and Applied Kalman Filtering with Matlab Exercises*, 4th ed.; John Wiley & Sons, Inc.: Hoboken, NJ, USA, 2012.
61. Kaniewski, P.T. *Struktury, Modele i Algorytmy w Zintegrowanych Systemach Pozycjonujących i Nawigacyjnych*; Wojskowa Akademia Techniczna: Warszawa, Poland, 2010.
62. Zarchan, P.; Musoff, H. *Fundamentals of Kalman Filtering: A Practical Approach*, 3rd ed.; American Institute of Aeronautics and Astronautics, Inc.: Reston, VA, USA, 2009.
63. Setoodeh, P.; Habibi, S.; Haykin, S. *Nonlinear Filters: Theory and Applications*; John Wiley & Sons: Hoboken, NJ, USA, 2022.
64. Frogerais, P.; Bellanger, J.-J.; Senhadji, L. Various Ways to Compute the Continuous-Discrete Extended Kalman Filter. *IEEE Trans. Autom. Control.* **2012**, *57*, 1000–1004. [CrossRef]
65. Takeno, M.; Katayama, T. A numerical method for continuous-discrete unscented Kalman filter. *Int. J. Innov. Comput. Inf. Control.* **2012**, *8*, 2261–2274.
66. Chapra, S.V.R. *Numerical Methods for Engineers*; McGraw Hill: New York, NY, USA, 2010.
67. Hellevik, L.F. *Numerical Methods for Engineers*; Department of Structural Engineering, Norwegian University of Science and Technology: Trondheim, Norway, 2020.
68. Pańczyk, B.; Łukasik, E.; Sikora, J.; Guziak, T. *Metody Numeryczne w Przykładach*; Politechnika Lubelska: Lublin, Poland, 2012.
69. Cheung, K.W.; So, H.C.; Ma, W.-K.; Chan, Y.T. Least squares algorithms for time-of-arrival-based mobile location. *IEEE Trans. Signal Process.* **2004**, *52*, 1121–1130. [CrossRef]
70. Wu, X.; Tan, S. Error Estimation of Iterative Localization Based on Non-linear Least Square Residuals. In Proceedings of the International Conference on Computer Science and Service System, Nanjing, China, 11–13 August 2012; pp. 1579–1582. [CrossRef]
71. Rong, X.L.; Jilkov, V.P. Survey of maneuvering target tracking. Part I. Dynamic models. *IEEE Trans. Aerosp. Electron. Syst.* **2003**, *39*, 1333–1364. [CrossRef]
72. Kolat, M.; Törő, O.; Bécsi, T. Performance Evaluation of a Maneuver Classification Algorithm Using Different Motion Models in a Multi-Model Framework. *Sensors* **2022**, *22*, 347. [CrossRef]
73. Pant, B.; Alkin, O. Correlated movement mobility model and constant acceleration model for EKF-based tracking applications. In Proceedings of the IEEE 8th International Conference on Wireless and Mobile Computing, Networking and Communications (WiMob), Barcelona, Spain, 8–10 October 2012; pp. 869–874. [CrossRef]
74. Blackman, S.; Popoli, R. Design and Analysis of Modern Tracking Systems. Artech House Inc.: Norwood, MA, USA, 1999.
75. Stawowy, M.; Duer, S.; Paś, J.; Wawrzyński, W. Determining Information Quality in ICT Systems. *Energies* **2021**, *14*, 5549. [CrossRef]
76. Candy, J.V. *Bayesian Signal Processing: Classical, Modern, and Particle Filtering Methods*, 2nd ed.; John Wiley & Sons, Inc.: Hoboken, NJ, USA, 2016.

MDPI AG
Grosspeteranlage 5
4052 Basel
Switzerland
Tel.: +41 61 683 77 34

Remote Sensing Editorial Office
E-mail: remotesensing@mdpi.com
www.mdpi.com/journal/remotesensing

www.ingramcontent.com/pod-product-compliance
Lightning Source LLC
Chambersburg PA
CBHW040743250926
51850CB00007BB/1398
* 9 7 8 3 7 2 5 8 5 0 6 9 3 *